DOUBLEDAY
CELEBRATES
100 YEARS OF
EXCELLENCE

Other Books by Michio Kaku

HYPERSPACE
BEYOND EINSTEIN

MICHIO KAKU

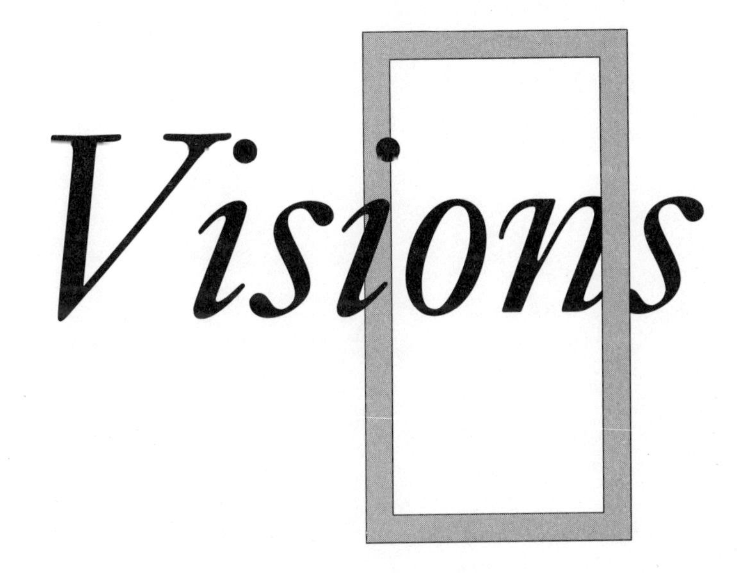

# *Visions*

## How Science Will Revolutionize the 21st Century

ANCHOR BOOKS
DOUBLEDAY
New York London Toronto Sydney Auckland

An Anchor Book
PUBLISHED BY DOUBLEDAY
a division of Bantam Doubleday Dell Publishing Group, Inc.
1540 Broadway, New York, New York 10036



Book design by Paul Randall Mize

Library of Congress Cataloging-in-Publication Data

Kaku, Michio.
Visions: how science will revolutionize the 21st century
Michio Kaku.
p. cm.
Includes index.
1. Science—Methodology. 2. Science—Forecasting. 3. Technological—Forecasting.
4. Quantum theory. 5. Molecular biology. 6. Computers and civilization. 7. Twenty-first
century—Forecasting. I. Title.
Q175.K157 1997
501′.12—dc21 97-18493
CIP

ISBN 0-385-48498-4

Printed in the United States of America
First Anchor Books Edition: October 1997

10 9 8 7 6 5 4 3 2 1

This book is dedicated to my parents

# Contents

Preface ix

**PART ONE VISIONS**

CHAPTER 1
Choreographers of Matter, Life, and Intelligence 3

**PART TWO THE COMPUTER REVOLUTION**

CHAPTER 2
The Invisible Computer 23

CHAPTER 3
The Intelligent Planet 43

CHAPTER 4
Machines That Think 70

CHAPTER 5
Beyond Silicon: Cyborgs and the Ultimate Computer 99

CHAPTER 6
Second Thoughts: Will Humans Become Obsolete? 118

**PART THREE THE BIOMOLECULAR REVOLUTION**

CHAPTER 7
Personal DNA Codes 139

CHAPTER 8
Conquering Cancer—Fixing Our Genes 162

CHAPTER 9
Molecular Medicine and the Mind/Body Link 181

CHAPTER 10
To Live Forever? 200

CHAPTER 11
Playing God: Designer Children and Clones 220

CHAPTER 12
Second Thoughts: The Genetics of a Brave New World? 241

**PART FOUR THE QUANTUM REVOLUTION**

CHAPTER 13
The Quantum Future 265

CHAPTER 14
To Reach for the Stars 295

CHAPTER 15
Toward a Planetary Civilization 322

CHAPTER 16
Masters of Space and Time 338

Notes 356
Recommended Reading 389
Index 394

# Preface

THIS IS A BOOK about the limitless future of science and technology, focusing on the next 100 years and beyond.

A book with the proper scope, depth, and accuracy necessary to summarize the exciting and fast-paced progress of science could not be written without the insights and wisdom of the scientists who are making the future possible.

Of course, no one person can invent the future. There is simply too much accumulated knowledge, there are too many possibilities and too many specializations. In fact, most predictions of the future have floundered because they have reflected the eccentric, often narrow viewpoints of a single individual.

The same is not true of *Visions*. In the course of writing numerous books, articles, and science commentaries, I have had the rare privilege of interviewing over 150 scientists from various disciplines during a ten-year period.

On the basis of these interviews, I have tried to be careful to delineate the time frame over which certain predictions will or will not be realized. Scientists expect some predictions to come about by the year 2020; others will not materialize until much later—from 2050 to the year 2100. As a result, not all predictions are created equal—some are more forward-looking and necessarily more speculative than others. The time frames I've identified in the book, of course, are to be taken only as guidelines, to give readers a sense of when certain trends and technologies can be expected to emerge.

The outline for the book is as follows: In Part I of *Visions*, I discuss the remarkable developments that await us in the computer revolution, which are already beginning to transform business, communications, and our lifestyles, and which I believe will one day give us the power to place intelligence in every part of our planet. In Part II, I turn to the biomolecular revolution, which will ultimately give us the power to alter and synthesize new forms of life, and create new medicines and therapies. Part III focuses on the quantum revolution, perhaps the most profound of the three, which will give us control over matter itself.

I wish to thank the following scientists who have given me their time, advice, and invaluable insights in the course of writing this book:

Walter Gilbert, Nobel Laureate in Chemistry, Harvard University
Murray Gell-Mann, Nobel Laureate in Physics, Santa Fe Institute
Henry Kendall, Nobel Laureate in Physics, MIT
Leon Lederman, Nobel Laureate in Physics, Illinois Institute of Technology
Steven Weinberg, Nobel Laureate in Physics, University of Texas
Joseph Rotbalt, physicist, Nobel Laureate in Peace
Carl Sagan, Director, Laboratory for Planetary Studies, Cornell University
Steven Jay Gould, Professor of Biology, Harvard University
Douglas Hofstadter, Pulitzer Prize winning author, Indiana University
Michael Dertouzos, Director of MIT Laboratory for Computer Sciences
Paul Davies, author and cosmologist, University of Adelaide
Hans Moravec, Robotics Institute, Carnegie-Mellon University
Daniel Crevier, AI expert, CEO of Coreco, Inc.
Jeremy Rifkin, founder of Foundation for Economic Trends
Philip Morrison, Professor of Physics, MIT
Miguel Virasoro, Director, International Center for Theoretical Physics, Trieste, Italy
Mark Weiser, Xerox PARC
Larry Tesler, chief scientist, Apple Computer
Paul Ehrlich, environmentalist, Stanford University
Paul Saffo, Director, Institute for the Future
Francis Collins, Director, National Center for Human Genome Research (NCHG), National Institutes of Health
Michael Blaese, Clinical Gene Therapy Branch (NCHG), National Institutes of Health
Lawrence Brody, Laboratory of Gene Transfer (NCHG), National Institutes of Health
Eric Green, Diagnostic Development Branch (NCHG), National Institutes of Health
Jeffrey Trent, Director, Division of Intramural Research (NCHG), National Institutes of Health
Paul Meltzer, Laboratory of Cancer Genetics (NCHG), National Institutes of Health
Leslie Biesecker, Laboratory of Genetic Disease Research (NCHG), National Institutes of Health
Anthony Wynshaw-Boris, Laboratory of Genetic Disease Research (NCHG), National Institutes of Health
Steven Rosenberg, Head of Surgery, National Institutes of Health

Lieutenant Colonel Robert Bowman, Director, Institute for Space and Security Studies
Paul Hoffman, Editor in Chief, *Discover* magazine
Leonard Hayflick, Professor of Anatomy at the University of California at San Francisco School of Medicine
Edward Witten, physicist, Institute for Advanced Study, Princeton
Cumrun Vafa, physicist, Harvard University
Paul Townsend, physicist, Cambridge University
Alan Guth, cosmologist, MIT
Barry Commoner, environmentalist, Queens College, CUNY
Rodney Brooks, Associate Director, Artificial Intelligence Laboratory, MIT
Robert Irie, Artificial Intelligence Laboratory, MIT
James McLurkin, Artificial Intelligence Laboratory, MIT
Jay Jaroslav, Artificial Intelligence Laboratory, MIT
Peter Dilworth, Artificial Intelligence Laboratory, MIT
Mike Wessler, Artificial Intelligence Laboratory, MIT
Neal Gershenfeld, Principal Investigator, Physics and Media Group, MIT Media Laboratory
Pattie Maes, Principal Investigator, Autonomous Agents Group, MIT Media Laboratory
David Riquier, Associate Director of Communications and Sponsor Relations, MIT Media Laboratory
Bradley Rhodes, MIT Media Laboratory
Donna Shirley, Jet Propulsion Laboratory, Manager of the Mars Exploration Mission
Frank Von Hipple, physicist, Princeton University
John Pike, Federation of American Scientists
Steve Aftergood, Federation of American Scientists
John Horgan, science writer, *Scientific American*
Lester Brown, Director and Founder, World Watch Institute
Christopher Flavin, World Watch Institute
Neil Tyson, Director, Hayden Planetarium, American Museum of Natural History
Brian Sullivan, project designer, Hayden Planetarium
Michael Oppenheimer, chief scientist, Environmental Defense Fund
Rebecca Goldberg, chief scientist, Environmental Defense Fund
Clifford Stoll, computer analyst
John Lewis, Co-director, NASA/University of Arizona Space Engineering Research Center
Richard Muller, Professor of Physics, University of California at Berkeley
Larry Krauss, Chairman of the Physics Department, Case Western Reserve University
David Gelertner, Associate Professor of Computer Science, Yale University
Ted Taylor, atomic bomb designer, Los Alamos
David Nahamoo, Senior Manager, Human Language Technology, IBM

Paul Shuch, Executive Director, SETI League
Arthur Caplan, Director, Center for Bioethics, University of Pennsylvania
Yolanda Moses, President, American Anthropological Association, and President, City College of New York
Meredith Small, Associate Professor of Anthropology, Cornell University
Freeman Dyson, Professor of Physics, Institute for Advanced Study, Princeton
Michael Jacobson, Executive Director, Center for Science in the Public Interest
Robert Alvarez, Department of Energy staff
Steve Cook, NASA spokesman
Karl Grossman, Professor of Journalism, SUNY Old Westbury
Helen Caldicott, pediatrician and peace activist
Jay Gould, former EPA official
Arjun Makhijani, President, Institute for Energy and Environmental Research
Thomas Cochran, senior scientist, Natural Resources Defense Council (NRDC)
Ashok Gupta, Senior Energy Policy Analyst, NRDC
David Schwarzbach, Project Associate for Nuclear Policy, NRDC
Richard Gott, cosmologist, Princeton University
Karl Drlica, Professor of Biology and Microbiology, New York University
Wendy McGoodwyn, Executive Director, Council for Responsible Genetics
Andrew Kimbrell, former Policy Director of the Foundation on Economic Trends
Jerome Glenn, Millennium Project
Jane Rissler, senior staff scientist, Union of Concerned Scientists
Charles Pillar, author of *Gene Wars*
Eric Chivian, International Physicians for the Prevention of Nuclear War
Jack Geiger, co-founder, Physicians for Social Responsibility
Gordon Thompson, Director, Institute for Resource and Security Studies

I also wish to thank those individuals who have given me encouragement and have read large portions of this book, including Karl Drlica, Joel Gersten, Mike and Iris Anshel, Tadmiri Venkatesh, and others. I would especially like to thank my agent, Stuart Krichevsky, who has guided several of my popular books from conception to the bookshelf, and of course my editor at Anchor Books, Roger Scholl, whose sharp, critical eye has vastly improved the presentation of the manuscript and also helped to focus its message with clarity and thoughtfulness.

Michio Kaku
New York, N.Y.

# Part One

# Visions

# 1

# Choreographers of Matter, Life, and Intelligence

"There are three great themes in science in the twentieth century—the atom, the computer, and the gene."
—HAROLD VARMUS, NIH Director

"Prediction is very hard, especially when it's about the future."
—YOGI BERRA

THREE CENTURIES AGO, Isaac Newton wrote: ". . . to myself I seem to have been only like a boy playing on a seashore, and diverting myself in now and then finding a smoother pebble or a prettier shell than ordinary, whilst the great ocean of truth lay all undiscovered before me." When Newton surveyed the vast ocean of truth which lay before him, the laws of nature were shrouded in an impenetrable veil of mystery, awe, and superstition. Science as we know it did not exist.

Life in Newton's time was short, cruel, and brutish. People were illiterate for the most part, never owned a book or entered a classroom, and rarely ventured beyond several miles of their birthplace. During the day, they toiled at backbreaking work in the fields under a merciless sun. At night, there was usually no entertainment or relief to comfort them except the empty sounds of the night. Most people knew firsthand the gnawing pain of hunger and chronic, debilitating disease. Most people would live not much longer than age thirty, and would see many of their ten or so children die in infancy.

But the few wondrous shells and pebbles picked up by Newton and other scientists on the seashore helped to trigger a marvelous chain of events. A profound transformation occurred in human society. With Newton's mechanics came powerful machines, and eventually the steam engine, the motive force which reshaped the world by overturning agrarian society, spawning factories and stimulating commerce, unleashing the industrial revolution, and opening up entire continents with the railroad.

By the nineteenth century, a period of intense scientific discovery was well underway. Remarkable advances in science and medicine helped to lift people out of wretched poverty and ignorance, enrich their lives, empower them with knowledge, open their eyes to new worlds, and eventually unleash complex forces which would topple the feudal dynasties, fiefdoms, and empires of Europe.

By the end of the twentieth century, science had reached the end of an era, unlocking the secrets of the atom, unraveling the molecule of life, and creating the electronic computer. With these three fundamental discoveries, triggered by the quantum revolution, the DNA revolution, and the computer revolution, the basic laws of matter, life, and computation were, in the main, finally solved.

That epic phase of science is now drawing to a close; one era is ending and another is only beginning.

This book is about this new dynamic era of science and technology which is now unfolding before our eyes. It focuses on science in the next 100 years, and beyond. The next era of science promises to be an even deeper, more thoroughgoing, more penetrating one than the last.

Clearly, we are on the threshold of yet another revolution. Human knowledge is doubling every ten years. In the past decade, more scientific knowledge has been created than in all of human history. Computer power is doubling every eighteen months. The Internet is doubling every year. The number of DNA sequences we can analyze is doubling every two years. Almost daily, the headlines herald new advances in computers, telecommunications, biotechnology, and space exploration. In the wake of this technological upheaval, entire industries and lifestyles are being overturned, only to give rise to entirely new ones. But these rapid, bewildering changes are not just quantitative. They mark the birth pangs of a new era.

Today, we are again like children walking on the seashore. But the ocean that Newton knew as a boy has largely disappeared. Before us lies a new ocean, the ocean of endless scientific possibilities and applications, giving us the potential for the first time to manipulate and mold these forces of Nature to our wishes.

For most of human history, we could only watch, like bystanders, the beautiful dance of Nature. But today, we are on the cusp of an epoch-making transition, from being *passive observers of Nature to being active choreographers of Nature.* It is this tenet that forms the central message of *Visions.* The era now unfolding makes this one of the most exciting times to be alive, allowing us to reap the fruits of the last 2,000 years of science. The Age of Discovery in science is coming to a close, opening up an Age of Mastery.

### Emerging Consensus Among Scientists

What will the future look like? Science fiction writers have sometimes made preposterous predictions about the decades ahead, from vacationing on Mars to banishing all diseases. And even in the popular press, all too often an eccentric social critic's individual prejudices are substituted for the consensus within the scientific community. (In 1996, for example, *The New York Times Magazine* devoted an entire issue to life in the next 100 years. Journalists, sociologists, writers, fashion designers, artists, and philosophers all submitted their thoughts. Remarkably, *not a single scientist* was consulted.)

The point here is that predictions about the future made by professional scientists tend to be based much more substantially on the realities of scientific knowledge than those made by social critics, or even those by scientists of the past whose predictions were made before the fundamental scientific laws were completely known.

It is, I think, an important distinction between *Visions,* which concerns an emerging consensus among the scientists themselves, and the predictions in the popular press made almost exclusively by writers, journalists, sociologists, science fiction writers, and others who are *consumers* of technology, rather than by those who have helped to shape and *create* it. (One is reminded of the prediction made by Admiral William Leahy to President Truman in 1945: "That is the biggest fool thing we have ever done. . . . The [atomic] bomb will never go off, and I will speak as an expert in explosives." The admiral, like many "futurists" today, was substituting his own prejudices for the consensus of physicists working on the bomb.)

As a research physicist, I believe that physicists have been particularly successful at predicting the broad outlines of the future. Professionally, I work in one of the most fundamental areas of physics, the quest to complete Einstein's dream of a "theory of everything." As a result, I am constantly reminded of the ways in which quantum physics touches many of the key discoveries that shaped the twentieth century.

In the past, the track record of physicists has been formidable: we have been intimately involved with introducing a host of pivotal inventions (TV, radio, radar, X-rays, the transistor, the computer, the laser, the atomic bomb), decoding the DNA molecule, opening new dimensions in probing the body with PET, MRI, and CAT scans, and even designing the Internet and the World Wide Web. Physicists are by no means seers who can foretell the future (and we certainly haven't been spared our share of silly predictions!). Nonetheless, it is true that some of the shrewd observations and penetrating insights of leading physicists in the history of science have opened up entirely new fields.

There undoubtedly will be some astonishing surprises, twists of fate, and embarrassing gaps in this vision of the future: I will almost inevitably overlook some important inventions and discoveries of the twenty-first century. But by focusing on the interrelations between the three great scientific revolutions, and by consulting with the scientists who are actively bringing about this revolution and examining their discoveries, it is my hope that we can see the direction of science in the future with considerable insight and accuracy.

Over the past ten years, while working on this book, I have had the rare privilege of interviewing over 150 scientists, including a good many Nobel Laureates, in part during the course of preparing a weekly national science radio program and producing science commentaries.

These are the scientists who are tirelessly working in the trenches, who are laying the foundations of the twenty-first century, many of whom are opening up new avenues and vistas for scientific discovery. In these interviews, as well as through my own work and research, I was able to go back over the vast panorama of science laid out before me and draw from a wide variety of expertise and knowledge. These scientists have graciously opened their offices and their laboratories and shared their most intimate scientific ideas with me. In this book, I've tried to return the favor by capturing the raw excitement and vitality of their scientific discoveries, for it is essential to instill the romance and excitement of science in the general public, especially the young, if democracy is to remain a vibrant and resonating force in an increasingly technological and bewildering world.

The fact is that there *is* a rough consensus emerging among those engaged in research about how the future will evolve. Because the laws behind the quantum theory, computers, and molecular biology are now well established, it is possible for scientists to generally predict the paths of scientific progress in the future. *This is the central reason why the predictions made here, I feel, are more accurate than those of the past.*

What is emerging is the following.

## The Three Pillars of Science

Matter. Life. The Mind.

These three elements form the pillars of modern science. Historians will most likely record that the crowning achievement of twentieth-century science was unraveling the basic components underlying these three pillars, culminating in the splitting of the nucleus of the atom, the decoding of the nucleus of the cell, and the development of the electronic computer. With our basic understanding of matter and life largely complete, we are witnessing the close of one of the great chapters in the history of science. (This does not mean that all the laws of these three pillars are completely known, only the most fundamental. For example, although the laws of electronic computers are well known, only some of the basic laws of artificial intelligence and the brain are known.)

The first of these twentieth-century revolutions was the *quantum revolution*, the most fundamental of all. It was the quantum revolution that later helped to spawn the two other great scientific revolutions, the *biomolecular revolution* and the *computer revolution.*

### THE QUANTUM REVOLUTION

Since time immemorial, people have speculated what the world was made of. The Greeks thought that the universe was made of four elements: water, air, earth, and fire. The philosopher Democritus believed that even these could be broken down into smaller units, which he called "atoms." But attempts to explain how atoms could create the vast, wondrous diversity of matter we see in Nature always faltered. Even Newton, who discovered the cosmic laws which guided the motion of planets and moons, was at a loss to explain the bewildering nature of matter.

All this changed in 1925 with the birth of the quantum theory, which has unleashed a thundering tidal wave of scientific discovery that continues to surge unabated to this day. The quantum revolution has now given us an almost complete description of matter, allowing us to describe the seemingly infinite multiplicity of matter we see arrayed around us in terms of a handful of particles, in the same way that a richly decorated tapestry is woven from a few colored strands.

The quantum theory, created by Erwin Schrödinger, Werner Heisenberg, and many others, reduced the mystery of matter to a few postulates. First, that energy is not continuous, as the ancients thought, but occurs in discrete bundles, called "quanta." (The photon, for example, is a quantum or packet of light.) Second, that subatomic particles have both particle and wavelike qualities, obeying a well-defined equation, the celebrated Schrödinger wave equation, which determines the probability that certain events occur. With this equation, we can mathematically predict the properties of a wide variety of substances before creating them in the laboratory. The culmination of the quantum theory is the Standard Model, which can predict the properties of everything from tiny subatomic quarks to giant supernovas in outer space.

In the twentieth century, the quantum theory has given us the ability to understand the matter we see around us. In the next century, the quantum revolution may open the door to the next step: the ability to manipulate and choreograph new forms of matter, almost at will.

### THE COMPUTER REVOLUTION

In the past, computers were mathematical curiosities; they were supremely clumsy, messy contraptions, consisting of a complex mass of gears, levers, and cogs. During World War II, mechanical computers were replaced by vacuum tubes, but they were also monstrous in size, filling up entire rooms with racks of thousands of vacuum tubes.

The turning point came in 1948, when scientists at Bell Laboratories discovered the transistor, which made possible the modern computer. A decade after that, the laser was discovered, which is essential to the Internet and the information highway. Both are quantum mechanical devices.

In the quantum theory, electricity can be understood as the movement of electrons, just as droplets of water can make a river. But one of the surprises of the quantum theory is that there are "bubbles" or "holes" in the current, corresponding to vacancies in electron states, which act as if they are electrons with positive charge. The motion of these currents of both holes and electrons allows transistors to amplify tiny electrical signals, which forms the basis of modern electronics.

Today, tens of millions of transistors can be crammed into an area the size of a fingernail. In the future, our lifestyles will be irrevocably changed when microchips become so plentiful that intelligent systems are dispersed by the millions into all parts of our environment.

In the past, we could only marvel at the precious phenomenon called intelligence; in the future, we will be able to manipulate it according to our wishes.

### THE BIOMOLECULAR REVOLUTION

Historically, many biologists were influenced by the theory of "vitalism"—i.e., that a mysterious "life force" or substance animated living things. This view was challenged when Schrödinger, in his 1944 book *What Is Life?*, dared to claim that life could be explained by a "genetic code" written on the molecules within a cell. It was a bold idea: that the secret of life could be explained by using the quantum theory.

James Watson and Francis Crick, inspired by Schrödinger's book, eventually proved his conjecture by using X-ray crystallography. By analyzing the pattern of X-rays scattered off a DNA molecule, they were able to reconstruct the detailed atomic structure of DNA and identify its double-helical nature. Since the quantum theory also gives us the precise bonding angles and bonding strength between atoms, it enables us to determine the position of practically all the individual molecules in the genetic code of a complex virus like HIV.

The techniques of molecular biology will allow us to read the genetic code of life as we would read a book. Already, the complete DNA code of several living organisms, like viruses, single-cell bacteria, and yeast, have been completely decoded, molecule for molecule.

The complete human genome will be decoded by the year 2005, giving us an "owner's manual" for a human being. This will set the stage for twenty-first century science and medicine. Instead of watching the dance of life, the biomolecular revolution will ultimately give us the nearly god-like ability to manipulate life almost at will.

## From Passive Bystanders to Active Choreographers of Nature

Some commentators, witnessing these historic advances in science over the past century, have claimed that we are seeing the demise of the scientific enterprise. John Horgan, in his book *The End of Science*, writes: "If one believes in science, one must accept the possibility—even the probability—that the great era of scientific discovery is over. . . . Further research may yield no more great revelations or revolutions, but only incremental, diminishing returns."

In one limited sense, Horgan is right. Modern science has no doubt uncovered the fundamental laws underlying most of the disciplines of science: the quantum theory of matter, Einstein's theory of space-time, the Big Bang theory of cosmology, the Darwinian theory of evolution, and the molecular basis of DNA and life. Despite some notable exceptions (e.g., determining the nature of consciousness and proving that superstring theory, my particular field of specialization, is the fabled unified field theory), the "great ideas" of science, for the most part, have probably been found.

Likewise, the era of reductionism—i.e., reducing everything to its smallest components—is coming to a close. Reductionism has been spectacularly successful in the twentieth century, unlocking the secrets of the atom, the DNA molecule, and the logic circuits of the computer. But reductionism has probably, in the main, run its course.

However, this is just the beginning of the romance of science. These scientific milestones certainly mark a significant break with the ancient past, when Nature was interpreted through the prism of animism, mysticism, and spiritualism. But they only open the door to an entirely new era of science.

The next century will witness an even more far-reaching scientific revolution, as we make the transition from unraveling the secrets of Nature to becoming masters of Nature.

Sheldon Glashow, a Nobel Laureate in physics, describes this difference metaphorically when he tells the story of a visitor named Arthur from another planet meeting earthlings for the first time:

"Arthur [is] an intelligent alien from a distant planet who arrives at Washington Square [in New York City] and observes two old codgers playing chess. Curious, Arthur gives himself two tasks: to learn the rules of the game, and to become a grand master." By carefully watching the moves, Arthur is gradually able to reconstruct the rules of the game: how pawns advance, how queens capture knights, and how vulnerable kings are. However, just knowing the rules does not mean that Arthur has become a grand master! As Glashow adds: "Both kinds of endeavors are important—one more 'relevant,' the other more 'fundamental.' Both represent immense challenges to the human intellect."

In some sense, science has finally decoded many of the fundamental "rules of Nature," but this does not mean that we have become grand masters. Likewise, the dance of elementary particles deep inside stars and the rhythms of DNA molecules coiling and uncoiling within our bodies have been largely deciphered, but this does not mean that we have become master choreographers of life.

In fact, the end of the twentieth century, which ended the first great phase in the history of science, has only opened the door to the exciting developments of the next. We are now making the transition from amateur chess players to grand masters, from observers to choreographers of Nature.

## From Reductionism to Synergy

Similarly, this is creating a new approach in the way in which scientists view their own discipline. In the past, the reductionist approach has paid off handsomely, eventually establishing the foundation for modern physics, chemistry, and biology.

At the heart of this success was the discovery of the quantum theory, which helped to spark the other two revolutions.

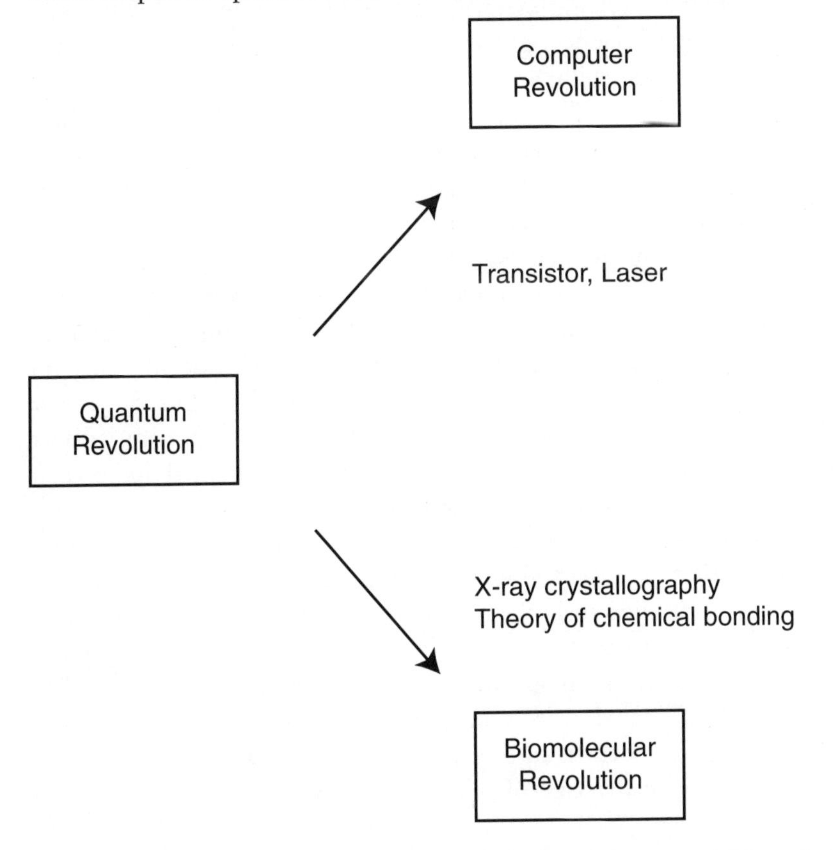


The quantum revolution gave birth to the computer and biomolecular revolutions via the transistor, laser, X-ray crystallography, and the theory of molecular bonds.

But since the quantum theory helped to initiate these other revolutions in the 1950s, they have since matured and grown on their own, largely independent of physics and of each other. The watchword was specialization, as scientists probed deeper and deeper into their subdisciplines, smugly ignoring the developments in other fields. But now the heyday of reductionism has probably passed. Seemingly impenetrable obstacles have been encountered which cannot be solved by the simple reductionist approach. This is heralding a new era, one of *synergy* between the three fundamental revolutions.

This is the second main theme of this book.

The twenty-first century, unlike the previous ones, will be typified by synergy, the cross-fertilization between all three fields, which will mark a sharp turning point in the development of science. The cross-pollination between these three revolutions will be vastly accelerated and will enrich the development of science, giving us unprecedented power to manipulate matter, life, and intelligence.

In fact, it will be difficult to be a research scientist in the future without having some working knowledge of all these three areas. Already, scientists who do not have some understanding of these three revolutions are finding themselves at a distinct competitive disadvantage.

The new relationship between the three revolutions is an intensely dynamic one. Often, when an impasse is reached in one area, usually a totally unexpected development in another field is found to contain the solution. For example, biologists once despaired of ever deciphering the millions of genes which contain the blueprint for life. But the recent torrent of genes being discovered in our laboratories is being driven largely by a development in another field: the exponential increase in computer power, which is mechanizing and automating the gene-sequencing process. Similarly, silicon computer chips will eventually hit a roadblock as they become too clumsy for the computer of the next century. But new advances in DNA research are making possible a new type of computer architecture which actually computes on organic molecules. Thus, discoveries in one field nourish and fertilize discoveries in totally unrelated fields. The whole is more than the sum of its parts.

One of the consequences of this intense synergy between these revolutions is that the steady pace of scientific discovery is accelerating at an ever increasing rate.

## The Wealth of Nations

This acceleration of science and technology into the next century will necessarily have vast repercussions on the wealth of nations and our standard of living. For the past three centuries, wealth was usually accumulated by those nations which were endowed with rich natural resources or which amassed large amounts of capital. The rise of the Great Powers of

Europe in the nineteenth century and the United States in the twentieth century follows this classic textbook principle.

However, as Lester C. Thurow, former dean of MIT's Sloan School of Management, has stressed, in the coming century, there will be a historic movement in wealth away from nations with natural resources and capital. In the same way that shifts in the earth's tectonic plates can generate powerful earthquakes, this seismic shift in wealth will reshape the distribution of power on the planet. Thurow writes: "In the twenty-first century, brainpower and imagination, invention, and the organization of new technologies are the key strategic ingredients." In fact, many nations which are richly endowed with abundant natural resources will find their wealth vastly reduced because, in the marketplace of the future, commodities will be cheap, trade will be global, and markets will be linked electronically. Already, the commodity prices of many natural resources plummeted some 60 percent from the 1970s to the 1990s, and, in Thurow's estimation, will plummet another 60 percent by 2020.

Even capital itself will be reduced to a commodity, racing around the globe electronically. Many nations which are barren of natural resources will flourish in the next century because they placed a premium on those technologies which can give them a competitive edge in the global marketplace. "Today, knowledge and skills now stand alone as the only source of comparative advantage," Thurow asserts.

As a consequence, some nations have drawn up lists of the key technologies which will serve as the engines of wealth and prosperity into the next century. A typical list was compiled in 1990 by Japan's Ministry of International Trade and Industry. That list included:

- microelectronics
- biotechnology
- the new material science industries
- telecommunications
- civilian aircraft manufacturing
- machine tools and robots
- computers (hardware and software)

Without exception, every one of the technologies singled out to lead the twenty-first century are deeply rooted in the quantum, computer, and DNA revolutions.

The point is that these three scientific revolutions are not only the key to scientific breakthroughs in the next century; they are also the dynamic engines of wealth and prosperity. *Nations may rise and fall on their ability to master these three revolutions.* In any activity, there are winners and losers. The winners will likely be those nations which fully grasp the vital importance of these three scientific revolutions. Those who would scoff at the power of these revolutions may find themselves marginalized in the global marketplace of the twenty-first century.

## Time Frames for the Future

In making predictions about the future, it is crucial to understand the time frame being discussed, for, obviously, different technologies will mature at different times. The time frames of the predictions made in *Visions* fall into three categories: those breakthroughs and technologies that will evolve between now and the year 2020, those that will evolve from 2020 to 2050, and those that will emerge from 2050 to the end of the twenty-first century. (These are not absolute time frames; they represent only the general period in which certain technologies and sciences will reach fruition.)

### TO THE YEAR 2020

From now to the year 2020, scientists foresee an explosion in scientific activity such as the world has never seen before. In two key technologies, computer power and DNA sequencing, we will see entire industries rise and fall on the basis of breathtaking scientific advances. Since the 1950s, the power of our computers has advanced by a factor of roughly *ten billion.* In fact, because both computer power and DNA sequencing double roughly every two years, one can compute the rough time frame over which many scientific breakthroughs will take place. This means that predictions about the future of computers and biotechnology can be quantified with reasonable statistical accuracy through the year 2020.

For computers, this staggering growth rate is quantified by Moore's law, which states that computer power doubles roughly every eighteen months. (This was first stated in 1965 by Gordon Moore, co-founder of the Intel Corp. It is not a scientific law, in the sense of Newton's laws, but a rule-of-thumb which has uncannily predicted the evolution of computer power for several decades.) Moore's law, in turn, determines the fate of multibillion-dollar computer corporations, which base their future projections and product lines on the expectation of continued growth. By 2020, microprocessors will likely be as cheap and plentiful as scrap paper, scattered by the millions into the environment, allowing us to place intelligent systems everywhere. This will change everything around us, including the nature of commerce, the wealth of nations, and the way we communicate, work, play, and live. This will give us smart homes, cars, TVs, clothes, jewelry, and money. We will speak to our appliances, and they will speak back. Scientists also expect the Internet will wire up the entire planet and evolve into a membrane consisting of millions of computer networks, creating an "intelligent planet." The Internet will eventually become a "Magic Mirror" that appears in fairy tales, able to speak with the wisdom of the human race.

Because of revolutionary advances in our ability to etch ever-smaller transistors onto silicon wafers, scientists expect this relentless drive to continue to generate newer and more powerful computers up to 2020, when the iron laws of quantum physics eventually take over once again. By then, the size of microchip components will be so small—roughly on the scale of molecules—that quantum effects will necessarily dominate and the fabled Age of Silicon will end.

The growth curve for biotechnology will be equally spectacular in this period. In biomolecular research, what is driving the remarkable ability to decode the secret of life is the introduction of computers and robots to automate the process of DNA sequencing. This process will continue unabated until roughly 2020, until literally thousands of organisms will have their complete DNA code unraveled. By then, it may be possible for anyone on earth to have their personal DNA code stored on a CD. We will then have the Encyclopedia of Life.

This will have profound implications for biology and medicine. Many genetic diseases will be eliminated by injecting people's cells with the correct gene. Because cancer is now being revealed to be a series of genetic mutations, large classes of cancers may be curable at last, without invasive surgery or chemotherapy. Similarly, many of the microorganisms involved in infectious diseases will be conquered in virtual reality by locating the molecular weak spots in their armor and creating agents to attack those weak spots. Our molecular knowledge of cell development will be so advanced that we will be able to grow entire organs in the laboratory, including livers and kidneys.

### FROM 2020 TO 2050

The prediction of explosive growth of computer power and DNA sequencing from now through 2020 is somewhat deceptive, in that both are driven by known technologies. Computer power is driven by packing more and more transistors onto microprocessors, while DNA sequencing is driven by computerization. Obviously, these technologies cannot indefinitely continue to grow exponentially. Sooner or later, a bottleneck will be hit.

By around 2020, both will encounter large obstacles. Because of the limits of silicon chip technology, eventually we will be forced to invent new technologies whose potentials are largely unexplored and untested, from optical computers, molecular computers, and DNA computers to quantum computers. Radically new designs must be developed, based on the quantum theory, which will likely disrupt progress in computer science. Eventually, the reign of the microprocessor will end, and new types of quantum devices will take over.

If these difficulties in computer technology can be overcome, then the period 2020 to 2050 may mark the entrance into the marketplace of an entirely new kind of technology: true robot automatons that have common sense, can understand human language, can recognize and manipulate objects in their environment, and can learn from their mistakes. It is a development that will likely alter our relationship with machines forever.

Similarly, biotechnology will face a new set of problems by 2020. The field will be flooded with millions upon millions of genes whose basic functions are largely unknown. Even before 2020, the focus will shift away from DNA sequencing to understanding the basic functions of these genes, a process which cannot be computerized, and to understand polygenic diseases and traits—i.e., those involving the complex interaction of multiple genes. The shift to polygenic diseases may prove to be the key to solving some of the most pressing chronic diseases facing humanity, including heart disease, arthritis, autoimmune diseases, schizophrenia, and the like. It may also lead to cloning humans and to isolating the fabled "age genes" which control our aging process, allowing us to extend the human life span.

Beyond 2020, we also expect some amazing new technologies germinating in physics laboratories to come to fruition, from new generations of lasers and holographic three-dimensional TV to nuclear fusion. Room-temperature superconductors may find commercial applications and generate a "second industrial revolution." The quantum theory will give us the ability to manufacture machines the size of molecules, thereby opening up an entirely new class of machines with unheard-of properties called nanotechnology. Eventually, we may be able to build ionic rocket engines that may ultimately make interplanetary travel commonplace.

### FROM 2050 TO 2100 AND BEYOND

Last, *Visions* makes predictions about breakthroughs in science and technology from 2050 to the dawn of the twenty-second century. Although any predictions this far into the future are necessarily vague, it is a period that will likely be dominated by several new developments. Robots may gradually attain a degree of "self-awareness" and consciousness of their own. This could greatly increase their utility in society, as they are able to make independent decisions and act as secretaries, butlers, assistants, and aides. Similarly, the DNA revolution will have advanced to the point where biogeneticists are able to create new types of organisms involving the transfer of not just a few but even hundreds of genes, allowing us to increase our food supply and improve our medicines and our health. It may also give us the ability to design new life forms and to orchestrate the physical and perhaps even the mental makeup of our children, which raises a host of ethical questions.

The quantum theory, too, will exert a powerful influence in the next century, especially in the area of energy production. We may also see the beginnings of rockets that can reach the nearby stars and plans to form the first colonies in space.

Beyond 2100, some scientists see a further convergence of all three revolutions, as the quantum theory gives us transistor circuits and entire machines the size of molecules, allowing us to duplicate the neural patterns of the brain on a computer. In this era, some scientists have given serious thought to extending life by growing new organs and bodies, by manipulating our genetic makeup, or even by ultimately merging with our computerized creations.

## Toward a Planetary Civilization

When confronted with dizzying scientific and technological upheaval on this scale, there are some voices that say we are going too far, too fast, that unforeseen social consequences will be unleashed by these scientific revolutions.

I will try to address these legitimate questions and concerns by carefully exploring the sensitive social implications of these powerful revolutions, especially if they aggravate existing fault lines within society.

In addition, we will address an even more far-reaching question: to where are we rushing? If one era of science is ending and another is just beginning, then where is this all leading to?

This is exactly the question asked by astrophysicists who scan the heavens searching for signs of extraterrestrial civilizations which may be far more advanced than ours. There are about 200 billion stars in our galaxy, and trillions of galaxies in outer space. Instead of wasting millions of dollars randomly searching all the stars in the heavens for signs of extraterrestrial life, astrophysicists engaged in this search have tried to focus their efforts by theorizing about the energy characteristics and signatures of civilizations several centuries to millennia more advanced than ours.

Applying the laws of thermodynamics and energy, astrophysicists who scan the heavens have been able to classify hypothetical extraterrestrial civilizations into three types, based on the ways they utilize energy. Russian astronomer Nikolai Kardashev and Princeton physicist Freeman Dyson label them Type I, II, and III civilizations.

Assuming a modest yearly increase in energy consumption, one can extrapolate centuries into the future when certain energy supplies will be exhausted, forcing society to advance to the next level.

A Type I civilization is one that has mastered all forms of terrestrial energy. Such a civilization can modify the weather, mine the oceans, or extract energy from the center of their planet. Their energy needs are so large that they must harness the potential resources of the entire planet. Harnessing and managing resources on this gigantic scale requires a sophisticated degree of cooperation among their individuals with elaborate planetary communication. This necessarily means that they have attained a truly planetary civilization, one that has put to rest most of the factional, religious, sectarian, and nationalistic struggles that typify their origin.

Type II civilizations have mastered stellar energy. Their energy needs are so great that they have exhausted planetary sources and must use their sun itself to drive their machines. Dyson has speculated that, by building a giant sphere around their sun, such a civilization might be able to harness their sun's total energy output. They have also begun the exploration and possible colonization of nearby star systems.

Type III civilizations have exhausted the energy output of a single star. They must reach out to neighboring star systems and clusters, and eventually evolve into a galactic civilization. They obtain their energy by harnessing collections of star systems throughout the galaxy.

(To give a sense of scale, the United Federation of Planets described in *Star Trek* probably qualifies for an emerging Type II status, as they have just attained the ability to ignite stars and have colonized a few nearby star systems.)

This system of classifying civilizations is a reasonable one because it relies on the available supply of energy. Any advanced civilization in space will eventually find three sources of energy at their disposal: their planet, their star, and their galaxy. There is no other choice.

With a modest growth rate of 3 percent per year—the growth rate typically found on earth—one can calculate when our planet might make the transition to a higher status in the galaxy. For example, astrophysicists estimate that, based on energy considerations, a factor of ten billion may separate the energy demands between the various types of civilizations. Although this staggering number at first seems like an insurmountable obstacle, a steady 3 percent growth rate can overcome even this factor. In fact, we can expect to reach Type I status within a century or two. To reach Type II status may require no more than about 800 years. But attaining Type III status may take on the order of 10,000 years or more (depending on the physics of interstellar travel). But even this is nothing but the twinkling of an eye from the perspective of the universe.

Where are we now? you might ask. At present, we are a Type 0 civilization. Essentially, we use dead plants (coal and oil) to energize our machines. On this planetary scale, we are like children, taking our first hesitant and clumsy steps into space. But by the close of the twenty-first century, the sheer power of the three scientific revolutions will force the nations of the earth to cooperate on a scale never seen before in history. By the twenty-second century, we will have laid the groundwork of a Type I civilization, and humanity will have taken the first step toward the stars.

Already the information revolution is creating global links on a scale unparalleled in human history, tearing down petty, parochial interests while creating a global culture. Just as the Gutenberg printing press made people aware of worlds beyond their village or hamlet, the information revolution is building and forging a common planetary culture out of thousands of smaller ones.

What this means is that our headlong journey into science and technology will one day lead us to evolve into a true Type I civilization—a planetary civilization which harnesses truly planetary forces. The march to a planetary civilization will be slow, accomplished in fits and starts, undoubtedly full of unexpected twists and setbacks. In the background always lurks the possibility of a nuclear war, the outbreak of a deadly pandemic, or a collapse of the environment. Barring such a collapse, however, I think it is safe to say that the progress of science has the potential to create forces which will bind the human race into a Type I civilization.

Far from witnessing the end of science, we see that the three scientific revolutions are unleashing powerful forces which may eventually elevate our civilization to Type I status. So when Newton first gazed alone at the vast, uncharted ocean of knowledge, he probably never realized that the chain reaction of events that he and others initiated would one day affect all of modern society, eventually forging a planetary civilization and propelling it on its way to the stars.

# Part Two

# The Computer Revolution

# 2

# The Invisible Computer

"Long-term, the PC and workstation will wither because computing access will be everywhere: in the walls, on wrists, and in 'scrap computers' (like scrap paper) lying about to be grabbed as needed."
—MARK WEISER, Xerox PARC

THE XEROX PARC (Palo Alto Research Center) lies nestled in the quiet, rolling hills overlooking Silicon Valley, surrounded by acres of golden-brown fields lying under a brilliant sky. With a herd of horses grazing quietly nearby, one might never suspect that Xerox PARC sits in the eye of a hurricane that may help to reshape the twenty-first century. Anyone who doubts Xerox PARC's uncanny ability to predict the future of computer technology need only examine its remarkable history of picking winners.

Outside the front entrance, there is no sign or poster that properly introduces visitors to the historic significance of this lab. But by rights, Xerox PARC could claim, "The PC was invented here," not to mention laying the foundation for the laser printer and the program that eventually became the basis for the Macintosh and Windows operating systems.

Even in the fiercely competitive Silicon Valley, Xerox PARC has built up a formidable reputation in an industry moving at breakneck speed. If we are experiencing a tidal wave of new products and high-tech gadgets coming from Silicon Valley, it's because Xerox PARC built the foundation that led to their invention.

If anyone has seen the future, it is Mark Weiser, former head of the

Computer Science Laboratory of Xerox PARC, and his team of engineers. They belong to an elite cadre of highly select computer scientists located in Silicon Valley and Cambridge who have the rare knack of combining raw technical ability with creative, artistic virtuosity. Weiser is short, with thinning hair, with an animated, engaging style and an impish smile. (He also has a mischievous side; he bangs on the drums for a raucous rock and roll band—appropriately called Severe Tire Damage—which is famous for its prankster-like antics on the Internet.) When he is not jamming with his rock and roll band, he is busily constructing the computer architecture of the twenty-first century. The goal of his team is to foresee the next stage in the evolution of the computer.

Because microchips are becoming so powerful and so cheap, Weiser and computer scientists like him believe that microchips will quietly disappear by the thousands into the fabric of our lives, and will be incorporated into the walls, the furniture, our appliances, our home, our car, even our jewelry. A simple necktie may one day contain more computing power than today's supercomputer. Already, prototypes of these devices have been built which silently follow our movements from room to room and building to building, seamlessly carrying out our commands invisibly.

The computer, far from being the demanding taskmaster it can be today, will be a truly liberating force in our lives. As Weiser notes: "Machines that fit the human environment instead of forcing humans to enter theirs will make using a computer as refreshing as taking a walk in the woods." These invisible devices will communicate with each other and tap in automatically to the Internet; gradually, they will become intelligent and will be able to anticipate our wishes and, by accessing the Internet, bring the wisdom of the planet to us.

The full implications of this vision are astounding. By comparison, the PC is just a computing appliance.

The ideas of the people at places like Xerox PARC have attracted enormous attention because the fortunes of a multibillion-dollar industry may one day ride on the silly doodlings and idle daydreams of these engineering wizards. A consensus is growing among America's top computer experts. Computers, instead of becoming the rapacious monsters featured in science fiction movies, will become so small and ubiquitous that they will be invisible, everywhere and nowhere, so powerful that they will disappear from view. Weiser has christened this idea "ubiquitous computing."

## The Disappearing PC

This push toward invisibility may well be a universal law of human behavior. As Weiser says: "Disappearance is a fundamental consequence not of technology but of human psychology. Whenever people learn something sufficiently well, they cease to be aware of it."

If that seems far-fetched, think of the evolution of electricity and the electric motor. In the nineteenth century, electricity and the electric motor were so precious that entire factories were designed to accommodate the presence of lightbulbs and bulky motors. The placement of workers, machine parts, tables, and so on were all designed around the needs of electricity and the motor.

Today, however, electricity is everywhere, hidden in the walls and stored in tiny batteries. Motors are so small and prevalent that scores of them are concealed inside the frame of a car, moving the windows, mirrors, the radio dial, the tape deck, and antennas. Yet while we are driving, we are blissfully unaware that we are surrounded by up to twenty-two motors and twenty-five solenoids.

To use an analogy, the next stage of the computer can be compared to the evolution of writing. Several thousand years ago, writing was a secret art jealously controlled by a small caste of scribes who were trained to write on clay tablets. These tablets were very scarce and were laboriously baked and carefully guarded by the king's soldiers. When paper was first invented, it too was an extremely precious commodity, taking hundreds of hours to produce a simple scroll. Paper was so expensive that only royalty had access to it. Most people only rarely caught fleeting glimpses of paper in their lives.

Today, we are not even aware that we are surrounded by a world brimming with paper and writing. Strolling down the street, we see nothing special in the writing on billboards, gum wrappers, or street signs. Every day, we grab scrap sheets of paper, scribble on them, and then throw them away. Writing has progressed from being a labor-intensive, sacred form of communication jealously guarded by kings and scribes into becoming invisible, disposable, and ubiquitous. (In fact, one of the single largest sources of waste in modern society is paper, almost all of it with writing on it.)

While this vision of powerful but invisible computers hidden in our environment sounds impractical and very expensive, that's an illusion. With the falling cost of microchips relentlessly driving down the costs of computers, computers will be so cheap, Weiser claims, "that we'll think nothing of going to the grocery store to pick up a six-pack of computers, just like we pick up batteries today."

In the computer industry, it takes roughly fifteen years, on the average, from the conception of an idea to its entering the marketplace. The first PC, for example, was built at Xerox PARC in 1972, but it wasn't until the late 1980s that the PC caught the public's fancy. Ubiquitous computing was conceived in 1988; it may take until the year 2003 to begin to see these ideas affect our lives in an appreciable way. And it may be years after that before they reach "critical mass" and ignite the marketplace. But by 2010, one can expect to see ubiquitous computing becoming of age. By 2020, it will dominate our lives.

## Three Phases of Computing

It may be helpful to put the evolution of the computer into a larger historical context. Many computer analysts divide the history of the computer into three or more distinct phases.

The first phase was dominated by the clunky but powerful mainframe computer, pioneered by IBM, Burroughs, Honeywell, and others. Computers were so expensive that an entire division of scientists and engineers was forced to share one giant mainframe. The ratio of computers to people was often one computer for a hundred scientists. "Machines were so scarce and so expensive that man approached the computer the way an ancient Greek approached an oracle," said John Kemeny, former president of Dartmouth University. "There was a certain degree of mysticism in the relationship, even to the extent that only specially selected acolytes were allowed to have direct communication with the computer."

As with clay tablets, an entire "priesthood" developed around servicing and programming each computer; to outsiders, they seemed to jealously guard their power and their access to the mainframe by dreaming up inscrutable incantations and rituals.

The second phase of computing began in the early 1970s, when the engineers at Xerox PARC realized that computer power was exploding even as the size of chips was imploding. They envisioned a ratio of computers to people that would eventually reach one-to-one. To test their ideas, in 1972 they created ALTO, the first PC ever built.

The engineers at Xerox PARC realized that one bottleneck to computing was the fiendishly complicated commands and clumsy manuals, often as thick as the Manhattan telephone book and just as illuminating. Computers were not "user-friendly." They were "user-belligerent." Why not, they mused, create a computer screen which was based entirely on pictures or "icons," where you would simply point a "mouse" at these pictograms to open programs and manipulate them.

In one masterful stroke, computers, instead of being a painful rite of passage, could be operated from scratch even by children. Using a computer could become a pleasant, playful, even enjoyable journey of discovery navigating through unexplored and uncharted menus and playful icons.

Later, the ideas of Xerox PARC were borrowed by Apple Computer, which eventually created the Macintosh computer. Eventually, the Microsoft Corporation adopted Xerox PARC's ideas again in their Windows program, which has since become almost the universal operating system for IBM-based computers sold throughout the world. One wag dubbed this process of raiding the ideas of Xerox PARC "inventing the past." (Ironically enough, Apple tried to sue Microsoft for piracy of its Macintosh operating system, which, in turn, was pirated from Xerox PARC.)

The transition between these phases has never been easy. Even giant multibillion-dollar corporations have been mercilessly crushed like eggshells because they were unable or unwilling to understand and adapt to these phases of computing. Not very long ago, the IBM, Digital, and Wang corporations were the towering giants of the computer business, with lucrative markets in the mainframe, minicomputer, and word processing businesses, respectively. But they mistakenly thought this phase would last forever. Like lumbering dinosaurs, all three thought the personal computer was a passing fad. In the end, all three were shaken to their foundations. Wang is all but bankrupt, and both IBM and Digital were forced to throw out their corporate leadership after devastating and humiliating multibillion-dollar losses.

## The Third Phase and Beyond

The third phase of computers is now known as ubiquitous computing, which refers to a time when computers are all connected to each other and the ratio of computers to people flips the other way, with as many as one hundred computers for every person.

Even today's software giant, Microsoft, is trembling in the face of the tidal wave of the third phase that began with the Internet. As Bill Gates admitted: "It's a little scary that as computer technology has moved ahead there's never been a leader from one era who was also a leader in the next. Microsoft has been a leader in the PC era." Suddenly realizing that Microsoft could be relegated to the dustbin of history by the Internet, Gates wrenched his giant corporation completely around to accommodate the new advances in computer networks, a move he hadn't anticipated in the original edition of his 1995 book, *The Road Ahead.*

By 2020, the era of ubiquitous computing should be in full flower. However, even this phase cannot continue forever. Beyond 2020, it is likely that the reign of silicon will end and entirely new computer architectures will have been created.

Some computer analysts believe that this will lead to a fourth phase, the introduction of artificial intelligence into computing systems. From 2020 to 2050, the world of computers may well be dominated by invisible, networked computers which have the power of artificial intelligence: reason, speech recognition, even common sense.

Some commentators believe that computing devices may even enter a fifth phase beyond 2050, when machines become self-aware and even conscious. The computing world from 2020 to 2100 will be discussed in more detail in later chapters.

The implications of these phases are truly profound, affecting every aspect of our lives. A few of the technological wonders which await us, especially in the next ten years, have already been profiled in the media, as in Gates's book, such as the wallet PC and the smart home. Readers may be familiar with some of these developments, which I will briefly review in this chapter. However, I will go far beyond this, focusing on developments which will take us well past the coming decade to the end of the twenty-first century.

## Moore's Law

To appreciate the remarkable increase in computer power that is propelling us from one phase to the next, it is important to remember that from 1950 to the present, there has been an increase in computer power by a factor of about *ten billion*. At the heart of this explosive growth is Moore's law, which states that computer power doubles every eighteen months. A rapid increase in power on this scale is almost unheard of in the history of technology.

In order to better appreciate the size of this massive increase, it is helpful to realize that it is larger than the transition from chemical explosives to the hydrogen bomb! In fact, if we go back eighty years, computer power has increased by a factor of *one trillion*. These are the astronomical numbers that are inevitably driving us into the third phase of computing. Using Moore's law, we can reasonably predict the future of computer technology for the next twenty-five years. Moore's law is deceptive because our brains function linearly, rather than exponentially. In the short term, we often see very little change year by year, so we erroneously conclude that not much is happening. But over a period of five to ten years the changes can become monumental.

Two of the most powerful forces in the world favor the long-term vision of ubiquitous computers: the laws of economics and the laws of physics.

As the price of microprocessors continues its plunge, many predict that the sheer power of economics will drive the computer industry into the next phase. Ron Bernal, president of MIPS Technologies, predicts that the price of the microchip will drop to 10 cents by the year 2000, 4 cents in 2005, and 2 cents in 2010. Thomas George, general manager of semiconductor products with Motorola, basically agrees, estimating that the microchip will cost 50 cents in 2000, 7 cents in 2005, and 1 cent in 2010. Eventually, microprocessors will be as *cheap as scrap paper*, and just as plentiful.

This steady exponential explosion in computer power, in turn, will spawn entire industries that have no counterpart in today's market. When the price of a computer chip is just one penny, the financial incentive to include them everywhere, from our appliances to our furniture, our cars, and our factories, will be enormous. *In fact, companies that don't include a few computer chips in their products will be at a severe competitive disadvantage.* (Already, for example, musical greeting cards, which contain disposable music-making chips, have more computer power than the computers that existed before 1950.) In the same way that practically every product on the planet contains writing on it, in the third phase of computers every product may contain a penny microprocessor.

As Andrew Grove, CEO of the giant Intel Corp. says, in the future computing power will be "practically free and practically infinite."

But to understand the dynamics and limits of Moore's law, one must understand the power of the quantum theory—the most fundamental physical theory of the universe.

## What Drives Moore's Law?

The secret behind the success of Moore's law lies in the transistor—how it behaves and the way it is manufactured. The transistor is basically a valve which controls the flow of electricity. In the same way that firemen can control huge torrents of water flowing in a fire hose by turning a valve, tiny voltages on a transistor can control the flow of large currents of electricity. The dynamics of semiconductor transistors, in turn, is governed by the quantum theory. (According to the quantum theory, the absence of an electron within a semiconductor acts like an electron of opposite charge, i.e. a "hole." The quantum theory dictates how these electrons and holes move in the transistor.)

What drives the success of Moore's law has been the struggle to miniaturize these transistors. The original transistors were crude electrical components, about the size of a dime, and were connected by wires. Transistors were originally built by hand. Today, transistors are made by using beams of light to make microscopic grooves and lines on silicon wafers (a process called "photolithography").

This process can be compared to making colorful T-shirts. The old-fashioned way is to paint each T-shirt by hand. But a more efficient method is to place a stencil over each T-shirt and then spray it with ink. In this way, one can repeatedly imprint images on T-shirts and mass-produce them in unlimited quantity. (Similarly light is projected through a special stencil called a "mask," containing the desired pattern of complex lines and circuitry, which is placed over a silicon wafer. The light beam focused through the mask imprints the pattern on the wafer, which is photosensitive. The wafer is then treated with special gases, which etch the circuitry into the wafer where it was exposed to light. The basic skeleton of the circuit is carved out this way. The transistors are created on these grooves by spraying the wafer with special ions. This process is repeated about twenty times, fashioning a multilayer system of silicon wafers containing wires and transistors.)

Philosophers used to debate how many angels could dance on the head of a pin. Computer experts today debate how many transistors can be crammed into a microprocessor by means of this etching process. The Motorola Power PC 620, for example, has almost seven million transistors squeezed into silicon chips smaller than a postage stamp. This miniaturization process, however, cannot continue indefinitely. There is a limit to how many wires can be etched on a wafer. That limit is the result, in part, of the wavelength of the light beam.

Typically, the etching of silicon wafers is done by light beams from a mercury lamp, which have wavelengths measured in microns (a micron is a millionth of a meter). Over the last few decades, Moore's law has been driven by using increasingly smaller and smaller wavelengths of mercury light to manufacture microprocessors. Mercury lamps emit light of wavelength .436 micron (in the visible range) and .365 micron (in the ultraviolet range). These distances are about 300 times thinner than a human hair.

The technology that may dominate the first few years of the next century, perhaps until 2005, is based on the pulsed excimer laser, which can push the wavelength down to .193 micron (in the deep ultraviolet range). But beyond 2020, this process will end and entirely new technologies will be required, which will be discussed in Chapter 5.

## Sensors and the Invisible Computer

The idea of ubiquitous computing has been amplified and embellished by many of the key thinkers in the computer field. Paul Saffo, director of the

Institute for the Future and one of the leading futurists in the country, is one of many computer experts who feel that some form of ubiquitous computing is inevitable, given the proliferation of cheap microchip technology. His particular version of this future is called the "electronic ecology."

When we analyze the ecology of a forest, we treat it as a collection of animals and plants which exist in harmony and interact dynamically with each other. To Saffo, every ten years or so there is a key technological advance which changes the relationship between the creatures in what he terms the electronic ecology.

The driving force behind the PC revolution in the 1980s, for example, was the microchip. In the 1990s, by contrast, the explosive growth in the Internet (which I will turn to in more detail in the next chapter) was driven by marrying the power of microprocessors with cheap lasers, which can carry trillions of bits of data at the speed of light along glass fibers.

In the twenty-first century, he thinks, the next revolution will be driven by cheap sensors coupled to microprocessors and lasers.

In Saffo's version of the third phase, we will be surrounded by tiny, invisible microprocessors sensing our presence, anticipating our wishes, even reading our emotions. And these microprocessors will be connected to the Internet. Equipped with these sensors, the "animals" of his "electronic forest" will be able to do what most computers cannot: sense our presence and even our mood. He points out wryly that toilets can now recognize our presence (via infrared sensors). But even the most advanced Cray supercomputer is totally unaware of who, what, or where the person using the computer is. Saffo says, "If a meteor smashed into my house and struck me while I was sitting next to my PC, it wouldn't have the slightest clue as to what happened. It will still be awaiting my next instruction!"

In Saffo's third phase, we will interact with our invisible computers by using our gestures, our voice, our body heat and electric field, and our body motions. Invisible computers will sense the world around them via two invisible media, sound and the electromagnetic spectrum. Different invisible media will be used for different purposes. For example, sensors will pick up our voice commands to carry out our wishes. Using hidden video cameras, computers will be able to locate our presence and even recognize our facial gestures. The location of our hands and body can be detected by measuring their electric fields. Smart cars will use radar to sense the presence of other cars. Infrared light sensors will be able to locate where we are by the heat we give off. Computers will communicate with each other and with the Internet via radio and microwaves.

## The Smart Office and Home of the Future

The first step in the long but exciting journey toward ubiquitous computing is to create marketable computer devices called tabs, pads, and boards, which are roughly an inch, a foot, and a yard in size. Perhaps 100 tabs, 10 to 20 pads, and one or two boards per room will likely be a fixture of the office of the future.

*Tabs* are tiny, inch-size clip-on badges that employees will wear—similar to an employee's ID badge, except that they will carry an infrared transmitter and have the power of a PC. Prototypes have already been built by Olivetti Cambridge.

As an employee moves within a building, the tabs can keep track of his or her precise location. Doors magically open up when they are approached, lights come on as people enter a room (and turn off automatically as they leave). Receptionists can actually locate anyone in the building. An intercom badge allows employees to verbally communicate orders or make inquiries to the master computer.

There are endless possibilities for computer tabs. They may be able to scan the Internet for important news and alert the wearer to crucial developments in industry or the stock market, to important calls, to family emergencies, and so on. Tabs will be able to communicate with other tabs as well, silently exchanging business information. Eventually, they may become so tiny they can be concealed in our cuff links or tie clasps.

The larger, foot-size *pads* are the counterparts of disposable scrap paper that we scribble on. In appearance, they will resemble extremely thin computer monitors, and will eventually become almost as thin as paper itself. Instead of employees lugging a heavy workstation from room to room, each room will have such "disposable" pads. These pads will have no individual identity at all. As with sheets of scrap paper, we might have scores of them scattered around our desk. Unlike ordinary sheets of scrap paper, however, each of these pads will be fully operational PCs connected to the main computer. It is, in a sense, the beginning of smart paper.

When we scribble on such smart paper, the graphics program inside will be able to convert our idle doodles into beautiful graphics or use editing capabilities to convert our notes into grammatically correct text. And after we are finished with it and have saved our work on the main computer, we simply toss it in the stack of pads on our desk.

Yard-long *boards* are huge computerized screens which are hung on a wall. At home, such boards can function as wall-size video screens for interactive TV or the Web. At the office, they serve as bulletin boards and "white boards" on which we can scribble notes and post notices or which we can use as a full-fledged PC connected to the Internet. They will also be used for teleconferencing. Instead of spending thousands of dollars flying a staff from distant offices, a manager will be able to simply convert the board into a large wall screen and teleconference with his or her staff. Or doctors will be able to use them to supervise surgery from distant locations.

Historically, one of the first devices in our office or home to become "smart" was the typewriter. When a chip was first inserted into the typewriter, it was christened the word processor. Today, although a few primitive chips may be scattered throughout the house, they are still not connected to each other. In the future, if a big storm is predicted, your house will pick up the weather forecast from the Internet and make the proper preparations, raising the temperature, alerting family members, and providing the latest updates. The smart bathroom will monitor family members' health. In Japan, a computerized toilet is being marketed which can diagnose simple medical problems. A person's pulse can be taken by the toilet seat by sensing the tiny pulse in the thighs. Urine can be chemically analyzed for diabetes.

Although these medical diagnostic tools are still primitive, in the future scientists expect them to blossom into sophisticated medical analyzers, acting as an electrocardiogram for the heart and detecting proteins emitted by precancerous tissues. In the distant future, the smart home may serve as a computerized nurse, carefully analyzing a person's state of health and sending the information silently and automatically to his or her doctor.

## The MIT Media Lab

Perhaps the institute most dedicated to bringing about the unification of media, art, and technology is the MIT Media Laboratory, founded by Nicholas Negroponte. Hidden among the austere, faceless buildings that make up the MIT campus, the Media Lab is housed in an ultramodern, white-tiled building designed by architect I. M. Pei. (Because of its distinctive design, the locals fondly refer to the building as the "Pei toilet.")

The director of what is probably its most ambitious and provocative enterprise inspired by ubiquitous computing, the Things That Think project, is physicist Neil Gershenfeld, who envisions a day when most inanimate things around us will think.

A young man in a hurry, Gershenfeld is tall, lanky, with a light beard and curly brown hair; he is a lively, intense man who has several irons in the fire at any one time. He can speak more rapidly on three subjects than most of us can on one.

Gershenfeld made a significant breakthrough when he found an entirely new way for computers to sense our presence. The space around our bodies is filled by an invisible electric field, like a spiderweb. This electric field is generated by electrons which accumulate on our skin like static electricity. When our bodies move, this electric-field "aura" moves with it.

In the past, this aura was considered useless from a commercial point of view. It was Gershenfeld who reasoned that if we have a sensor that can detect electric fields in the space around our bodies, they can be used to locate our arms and fingers.

As a result, the "smart table" was born. Gershenfeld likes to give demonstrations of this new piece of technology. He waves his hands over the computerized table like a symphony conductor. Nearby, a computer screen shows a ghostlike hand moving inside a cube, giving the precise coordinates of his hand in three dimensions. Gershenfeld calls this "electric-field sensing."

The phenomenon can have immediate applications, as it is a more versatile and powerful way to communicate with a computer than the two-dimensional mouse commonly found on PCs. It can also be used to enhance virtual reality, so that people don't have to wear clumsy gloves to locate the position of their hands. (The illusion of virtual reality is enhanced, in fact, without having to be wired up like a Christmas tree. Cyber shoppers of the future may be able to wave their fingers to navigate through virtual shopping malls on their computer screens.)

Gershenfeld's strategy by which he approaches computers of the next century is to ask himself, "Where can I find empty space that is not being used and how can I animate it?" One location, which has been overlooked for years, is the shoes we wear, which have valuable, unused workspace just waiting to be made intelligent.

In the future, our shoes may replace the computer batteries we are likely to need. Carrying bulky batteries around whenever we need to energize the computer in our tie clasp would be a nuisance. But the human body, Gershenfeld points out, generates about 80 watts of usable energy by its motions; about 1 watt of that can easily be drawn from the movements of the shoe alone.

Gershenfeld has found another use for our shoe. It may be possible in the future to put an electrode in one's shoe which will be able to transmit biographical data to others. Rather than exchange business cards, all one would have to do is shake a person's hands. Because skin is salty and conducts electricity, a résumé can travel electrically from shoe to hands and then to one's acquaintance's hand and shoe. This may ultimately prove to be a convenient way to exchange large computer files with someone on the street.

One motto of the Things That Think lab, not surprisingly, is:

> In the past, shoes could stink.
> In the present, shoes can blink.
> In the future, shoes will think.

## Wearable Computers

Another essential element of the Things That Think concept is the glasses many of us wear. The MIT Media Lab has already perfected a way to include a miniature computer screen over one's glasses. They do this by placing over the glasses a strange eyepiece resembling a jeweler's lens which contains a complete PC screen, illuminated by tiny LEDs (light emitting diodes). Peering into this tiny screen, barely half an inch wide, one can clearly see bright symbols appearing on a full-sized PC screen.

On nice days in Cambridge, one can sometimes see MIT students from the Media Lab dressed up like cyborgs, complete with helmets, goggles, special eyepieces, and a tangle of electrodes in their clothes. They carry a simplified keyboard which allows them to input data into their computer screens, which are located in their eyepieces.

These crude beginnings that make up part of the Wearable Computers project of the Media Lab will ultimately make any individual a walking node of the World Wide Web. Steve Mann of the Media Lab has connected the video images on his eyepiece to the Internet, so others too can instantly view exactly what he is seeing, even thousands of miles away. In the future, people in distant locations might be able to instantly share what we see through our glasses in this manner.

Wearable computers in many respects represent a merger of cellular phones with the laptop computer. Rocketing sales of laptops, which now account for almost a quarter of all PC sales, prove that mobile computers are no longer a niche market, but are an essential part of the computer landscape. As the costs continue to plunge, many of these users would likely leap at the opportunity to replace their cellular phones and laptops with an invisible smart device with the power of a supercomputer.

This could prove to be immensely liberating for people who ride in cabs, shop at the mall, or travel by airplane. Some of those who may require wearable computers are doctors who need access to emergency medical records, police who need access to files, reporters who need data to file reports, stockbrokers who need twenty-four-hour stock quotes, and so on.

Someday, wearable computers may also save lives. If you have a heart attack somewhere far from a hospital or phone, your wearable computer, by silently monitoring your heartbeat, will be able to recognize unusual patterns consistent with a heart attack and alert the EMS. After a car accident, a wearable computer could automatically call for an ambulance. By hooking up to the GPS satellite, which I'll discuss later, it will also be able to transmit your exact location. At present, tens of thousands die needlessly because there is no one around to alert the EMS when a heart attack or car accident occurs.

## The Smart Room

One long-range goal of the Media Lab is to be able to design machines that can identify and imitate the full range of ways people interact with each other. People don't use language alone; we employ a rich, complex body language to communicate with others consisting of a surprisingly wide variety of signals, including eye contact, facial gestures and grimaces, arm motions, voice intonations, and posture. One step in this direction is to design a "smart room" which can recognize not only people but also their signals and emotions.

The Media Lab's prototype smart room of the future is a very ordinary den with small cameras placed in the ceiling and a giant wall-sized screen on the floor.

"Imagine a house that always knows where your kids are and tells you if they are getting into trouble. Or an office that sees when you are having an important meeting and shields you from interruptions. Or a car that senses when you are tired and warns you to pull over," writes Alex Pentland of the MIT Media Lab.

Today's computers cannot reliably recognize a person's face from different angles. Faces are among the most difficult things to identify by computer. However, the Media Lab's computer takes a shortcut to this difficult problem. It already has a series of key faces stored in its memory. If the computer scans a stranger's face and matches it with a face already filed away in its memory, then it can correctly find a match 99 percent of the time in a group of several hundred people.

Computers at the Media Lab can also identify a person's mood by means of the face. Emotions are etched into our faces by the motions they induce in our facial features. By placing sensors on people's faces and having them smile, laugh, smirk, or scowl, the sensors are able to detect how much our facial muscles move. Scientists have found that emotions can be recognized by computers as a result of the well-defined stretching motions they cause in the face. A smile, for example, leads to a broad stretching of our mouth muscles. Surprise leads to rising eyebrows. Anger leads to a contorting forehead. Disgust creates motion throughout the entire face. Therefore when the computer focuses only on the parts of the face that are in motion, it has been able in tests to correctly identify the emotional state of the subject about 98 percent of the time.

## Smart Cards, Digital Money, and Cyber Cash

Money is already going digital.

As James Gleick of the *New York Times* commented, digital money "is money incarnated, finally, as pure information." For the big banks and international corporations, this is already a reality. Of the $4 trillion circulating in the U.S. money supply, only one-tenth is in the form of actual cash and coins stored in bank vaults and people's pockets. "People today do not put $5 billion in a truck and drive it from one bank to another—that's irrational," comments Kawika Daguio of the American Bankers Association. In the future, even that one-tenth will disappear into electronic bits.

In the years ahead, as microchip costs plunge to mere pennies, there will be enormous economic pressure for people to convert to smart cards and digital money. This is because maintaining a society based on cash is very expensive. According to Carol H. Fancher, who is researching smart cards for Motorola: "Counting, moving, storing, and safeguarding cash costs about 4 percent of the value of all transactions. The interest lost by holding cash instead of keeping money on deposit is also substantial."

"Money is the current liability of a bank," says Sholom Rosen of Citibank. "It's as simple as that. It's not gold, it's not silver." Cash sitting in a bank is money that is not collecting interest or appreciating in value and that has to be constantly guarded.

Europe has taken the lead in mass-producing primitive versions of smart cards which contain up to a few kilobytes of memory. The value of these smart cards to consumers, who have used them mainly as telephone cards, has already been demonstrated in France (which has over 20 million smart cards in use) and the rest of Europe, where most of the 250 million smart cards in circulation have been issued.

Germany has begun to issue a smart card that carries basic health information to all its citizens. The 1996 Olympics in Atlanta featured the largest trial of smart cards in the United States, with over a million smart cards issued that were honored by restaurants, shops, and the subway system.

In the future, smart cards will replace ATM cards, telephone cards, train and transit passes, credit cards, as well as cards for parking meters, petty cash transactions, and vending machines. They will also store your medical history, insurance records, passport information, and your entire family photo album. They will even connect to the Internet.

## Smart Cars

Even the automobile industry, which has remained largely unchanged for the last seventy years, is about to feel the effects of the computer revolution.

The automobile industry ranks as among the most lucrative and powerful industries of the twentieth century. There are presently 500 million cars on earth, or one car for every ten people. Sales of the automobile industry stand at about a trillion dollars, making it the world's biggest manufacturing industry.

The car, and the roads it travels on, will be revolutionized in the twenty-first century. The key to tomorrow's "smart cars" will be sensors. "We'll see vehicles and roads that see and hear and feel and smell and talk and act," predicts Bill Spreitzer, technical director of General Motors Corporation's ITS program, which is designing the smart car and road of the future.

Approximately 40,000 people are killed each year in the United States in traffic accidents. The number of people that are tragically killed or mangled in car accidents is so vast that we don't even bother to mention them in the newspapers anymore. Fully half of these fatalities come from drunk drivers, and many others from carelessness. A smart car could eliminate most of these car accidents. It can sense if a driver is drunk via electronic sensors that can pick up alcohol vapor in the air, and refuse to start up the engine. The car could also alert the police and provide its precise location if it is stolen.

Smart cars have already been built which can monitor one's driving, and the driving conditions nearby. Small radars hidden in the bumpers can scan for nearby cars. Should you make a serious driving mistake (e.g., change lanes when there is a car in your "blind spot") the computer would sound an immediate warning.

At the MIT Media Lab, a prototype is already being built which will determine how sleepy you are as you drive, which is especially important for long-distance truck drivers. The monotonous, almost hypnotic process of staring at the center divider for long hours is a grossly underestimated, life-threatening hazard. To eliminate this, a tiny camera hidden in the dashboard can be trained on a driver's face and eyes. If the driver's eyelids close for a certain length of time and his or her driving becomes erratic, a computer in the dashboard could alert the driver.

Two of the most frustrating things about driving a car are getting lost and getting stuck in traffic. While the computer revolution is unlikely to cure these problems, it will have a positive impact. Sensors in your car tuned to radio signals from orbiting satellites can locate your car precisely at any moment and warn of traffic jams. We already have twenty-four Navstar satellites orbiting the earth, making up what is called the Global Positioning System. They make it possible to determine your location on the earth to within about a hundred feet. At any given time, there are several GPS satellites orbiting overhead at a distance of about 11,000 miles. Each satellite contains four "atomic clocks," which vibrate at a precise frequency, according to the laws of the quantum theory.

As a satellite passes overhead, it sends out a radio signal that can be detected by a receiver in a car's computer. The car's computer can calculate how far the satellite is by measuring how long it took for the signal to arrive. Since the speed of light is well known, any delay in receiving the satellite's signal can be converted into a distance.

In Japan there are already over a million cars with some type of navigational capability. (Some of them locate a car's position by correlating the rotations in the steering wheel to its position on a map.)

With the price of microchips dropping so drastically, future applications of GPS into the next century are virtually limitless. "The commercial industry is poised to explode," says Randy Hoffman of Magellan Systems Corp., which manufactures navigational systems. Blind individuals could use GPS sensors in walking sticks, airplanes could land by remote control, hikers will be able to locate their position in the woods—the list of potential uses is endless.

GPS is actually but part of a larger movement, called "telematics," which will eventually attempt to put smart cars on smart highways. Prototypes of such highways already exist in Europe, and experiments are being made in California to mount computer chips, sensors, and radio transmitters on highways to alert cars to traffic jams and obstructions.

On an eight-mile stretch of Interstate 15 ten miles north of San Diego, traffic engineers are installing an MIT-designed system which will introduce the "automated driver." The plan calls for computers, aided by thousands of three-inch magnetic spikes buried in the highway, to take complete control of the driving of cars on heavily trafficked roads. Cars will be bunched into platoons of ten to twelve vehicles, only six feet apart, traveling in unison, and controlled by computer. By December 2001, engineers hope to have a full prototype system running.

Promoters of this computerized highway have great hopes for its future. By 2010, telematics may well be incorporated into one of the major highways in the United States. If successful, by 2020, as the price of microchips drops to below a penny a piece, telematics could be adopted in thousands of miles of highways in the United States. This could prove to be an environmental boon as well, saving fuel, reducing traffic jams, decreasing air pollution, and serving as an alternative to highway expansion.

## Virtual Reality and Cyber Science

Another technology that will become an integral part of the world of 2020 is virtual reality. Ubiquitous computing is, in some sense, the opposite of virtual reality, which seeks to re-create imaginary worlds that don't exist, rather than accentuate and magnify the world that does. Virtual reality tries to create a world inside the memory of the computer, using goggles and joysticks to simulate moving through space and time. But ubiquitous computing and virtual reality complement each other. While the invisible computer will infinitely enhance the world that does exist, putting intelligence into the inanimate objects which surround us, virtual reality, by contrast, puts us inside the computer.

Although virtual reality is still quite crude today, its technical flaws will be eliminated with time. The primitive joystick will be replaced by body suits and electric-field sensors, which will sense the location of every part of our body in three dimensions. The goggles will be replaced by lightweight LCD screens. Clumsy cables will be replaced by radio receivers hooked directly to the Internet.

Virtual reality is a powerful scientific tool as well as a training aid and a source of entertainment. It is creating a new type of science, called "cyber science," which gives us the ability to simulate complex physical systems (like black holes, exploding stars, the weather, and the surface of hypersonic jets).

For several centuries, science has advanced in two ways: experimentally and theoretically. Some scientists conducted experiments on the external world, while others tried to write down the mathematics and theory that explained the data that was collected. But, increasingly, a new, third form of science is appearing, based on computer simulations in virtual reality, opening up new areas of science.

Since Newton, nature has been described by "differential equations" which describe the tiny differences that occur in the shape or property of an object as it evolves in time. Differential equations have been able to provide surprisingly realistic descriptions of physical phenomena, from thunderstorms to rockets to subatomic particles. Computers are ideally suited to model differential equations because computers can calculate how an object changes every microsecond or nanosecond in time, giving us a sequence of snapshots which realistically predict its behavior.

Computer simulations are becoming so accurate that entire fields are already crucially dependent on them, which in turn will likely influence the development of multibillion-dollar commercial technologies. In many areas, computers are the *only* way in which to solve such differential equations. Here are but a few of the things that we can best study through cyber science:

**Exotic objects in space.** We depend upon computers to analyze supernovas, neutron stars, and black holes. "Computer simulation is our only hope of turning astronomy into an experimental science," says Bruce Fryxell of NASA.

**Protein folding.** When a protein cannot be crystallized, one cannot use X-ray crystallography to determine its structure. Scientists are forced to use the quantum theory and electrostatics to find the structure of a protein. The complex equations which determine the structure of these proteins can only be solved by using computers. Computers may be the only way to calculate the structure and hence the properties of a large class of proteins.

**Aerodynamics.** The airflow around everything from cars to hypersonic jets traveling at many times the speed of sound can now be simulated by using computers. This may be the key to cheap hypersonic flights in the future.

**The greenhouse effect.** At present, computers are the only way we have to determine if the buildup of carbon dioxide in the atmosphere (produced by the burning of fossil fuels) may cause temperatures to rise and set off global warming. If global warming becomes a reality early in the next century and the weather is disrupted, the economy of the entire planet could be adversely affected.

**Materials testing.** Testing stresses and strains on industrial materials can best be calculated by computer, saving millions of dollars in unnecessary testing.

### Doomed by the Point One Limit?

We have seen how the relentless march of Moore's law makes it possible to predict with reasonable accuracy when fascinating new computing devices will be within reach. Microprocessors, lasers, and sensors will be the tools which make the third phase of computing a reality. Moore's law should take us smoothly to about 2020, when the quantum theory forces scientists to adopt entirely new computer architectures.

But creating chips with light beams smaller than .1 micron is looming as a major roadblock. This is sometimes called the "point one" barrier (it is roughly the width of a DNA coil). Some computer specialists compare the difficulty of breaking the point one barrier to breaking the sound barrier. Below the point one barrier, chips can no longer be etched with ultraviolet light; scientists must resort to X-rays or electrons, which are much more difficult to control. Moreover, the ghostly, wavelike properties of electrons and atoms come into play at this scale, requiring scientists to abandon Newtonian physics altogether in creating smaller chips. Because of this, by 2020, we expect the point one barrier to end the gilded Age of Silicon.

Some of the futurists at the MIT Media Laboratory like to summarize the transition to an information economy as the difference between "atoms" and "bits." (The bit is the smallest unit of information, such as 0 or 1.) Since it is difficult and expensive to move atoms around, they claim that the future will be ruled by bits, which are effortlessly carried as digital signals along wires or cables at nearly the speed of light. The age of atoms will thus give way to cyberspace and the information age, they claim.

But this is only partially true. Ultimately, Moore's law, the driving force behind the information age, will yield to a force even more powerful than electricity, and that is the quantum theory. Ultimately, "atoms" will have their revenge over "bits." It was the quantum theory that first made the transistor possible, and the quantum theory ultimately will dictate when these technologies will fail. By 2020, the revolution begun by the microprocessor may be over and physicists will have to devise the next generation of computers.

But because Moore's law will continue unabated for the next twenty-five years or so as more and more transistors are jammed onto a silicon wafer, it is still possible to predict roughly when the marvelous inventions profiled in this and the next chapter will hit the market from now to approximately the year 2020.

Furthermore, by 2020, the Internet will likely create an entire universe in cyberspace, with electronic commerce, e-money, virtual on-line libraries and universities, cyber medicine, and so on. But what is even more fascinating is the world beyond 2020, when computers will become so powerful and widespread that the surface of the earth becomes a "living" membrane, endowed with a planetary "intelligence," creating the fabled Magic Mirror featured so often in fairy tales.

In the next chapter, I will investigate this fourth phase of computing, as we approach the "intelligent planet."

# 3

# The Intelligent Planet

"The Internet is like a twenty-foot tidal wave coming thousands of miles across the Pacific, and we are in kayaks. It's been coming across the Pacific for thousands of miles and gaining momentum, and it's going to lift you and drop you. . . . It affects everybody—the computer industry, telecommunications, the media, chip makers, and the software world. Some are more aware of this than others."
—ANDREW GROVE, CEO of Intel

"Mirror, mirror, on the wall, who is the fairest of them all?"
—EVIL QUEEN in *Snow White*

IN 1851, AMERICAN NOVELIST Nathaniel Hawthorne wrote prophetically in *The House of Seven Gables:*

> Is it a fact . . . that by means of electricity, the world of matter has become a great nerve, vibrating thousands of miles in a breathless point of time. Rather, the rough globe is a vast head, a brain, instinct with intelligence!

Hawthorne, surveying the near-miraculous advances made in his lifetime in connecting the great cities of the world via the telegraph, marveled that this mysterious substance called electricity could transmit signals across thousands of miles and make inert machines spring suddenly to life. He went beyond this to envision a wondrous day when electricity would endow the planet itself with a cosmic intelligence. Over a century later, this passage from Hawthorne would inspire Marshall McLuhan to coin the phrase "global village." And in the twenty-first century the telecommunications revolution, ignited by the microprocessor and the laser, will finally make Hawthorne's vision come to pass.

In the third phase of computing, invisible computers will converse with each other, eventually creating a vibrant electronic membrane girding the earth's surface. We can already catch fleeting glimpses of this powerful vision within the present Internet, which, like a dirt road waiting to be paved over into an information superhighway, is rapidly wiring up the computers of the world.

From now to 2020, computer scientists expect to see an entire world blossoming over the Internet: electronic commerce and banking, cyber malls, virtual universities and schools, cyber libraries, and so on. We will begin to have a glimpse of Hawthorne's vision when "intelligent agents" become part of this global network, capable of answering our inquiries in plain, conversational language. But the true fruition of Hawthorne's vision may not come until the period from 2020 to 2050, when true artificial intelligent (AI) programs will finally be added to the Net, capable of reason, common sense, and speech recognition. Some call this the "fourth phase" of computing, when we will be able to communicate with the Internet as if it were an intelligent being. Eventually, accessing the Internet may resemble talking to the Magic Mirror of children's fairy tales. Instead of typing arcane codes and symbols into a Web navigator and being flooded with fifty thousand incorrect answers, in the future we will simply talk to our wall screen or tie clasp and access the entire planet's formidable body of knowledge. This Magic Mirror, endowed with an intelligent system complete with common sense and reason, and, very possibly, a human face and a distinct personality, may act as an adviser, confidant, aide, secretary, and gofer all at the same time.

One computer analyst commented that the future could resemble a Disney movie, as inanimate objects come alive and talk to each other and to us, like the talking tea kettle, Mrs. Potts, in *Beauty and the Beast*.

## Why No Policeman on the Block?

To anyone who has ever cursed at a computer screen, railing at the utter chaos of the Internet, the idea that one day we will have an illuminating and revealing conversation with a Magic Mirror seems a remote reality. The promise of Hawthorne's "intelligent planet" is a far cry from the stark reality of today's Internet.

Any neophyte surfing the Net for the first time will be frustrated by the fact that it has no intelligence whatsoever; like a newborn baby, it is a blank slate. Worse, there are no rules or traffic cops, no regulations or even a directory of the Internet. Some computer enthusiasts are ecstatic over this, claiming that this is democracy at its purest and finest. Others flail futilely and stomp away in disgust.

Already, young computer nerds who have attempted to fill this curious vacuum by writing simple directories for the Internet have amassed fabulous fortunes overnight when their companies went public. Netscape's co-founder Jim Clark's fortune ballooned to an astonishing *half billion dollars* the day the company went public. (Clark reached billionaire status only eighteen months after starting his company; it took Bill Gates, co-founder of Microsoft, twelve years.) As the *New York Daily News* reported breathlessly: "Netscape's IPO was the most successful since God took earth public."

Why has the Internet, the first stage of the "intelligent planet," been born in this peculiar way, seemingly without any intelligence at all?

Many of the electronic marvels of today, including video conferencing, virtual reality, global positioning satellites, and the Internet, were largely developed in total secrecy by Pentagon scientists and kept hidden from the public. Some computer analysts feel this obsession with Cold War secrecy delayed the computer revolution for years and is responsible for the peculiar evolution of these technologies, leaving curious gaps which only now are being filled by software writers.

Only in the last decade, with the ending of the Cold War, have these technologies finally been fully released to the public domain. Free of military classification for the first time, they have now taken off, capturing the public fancy, generating new billion-dollar industries in the process, and paving the way to the twenty-first century.

Perhaps the lesson here is that science and technology advance and thrive in an open atmosphere, when scientists and engineers can freely interact with each other.

## How the Internet and Other Technologies Came About

In January 1977, a strange, madcap incident took place in the White House which helps to explain the dark atmosphere in which the Internet was born. The incident would be hilarious if it wasn't so serious.

As in a scene from *Dr. Strangelove,* President Jimmy Carter's National Security Adviser, Zbigniew Brzezinski, was being briefed by a junior officer about the elaborate plans to protect the nation's leadership in case of a full-scale nuclear war. The young officer explained at length that helicopters would land at the White House lawn, the Capitol, and the Pentagon to whisk the President and his advisers to carefully concealed sites, including secret bomb shelters near Culpeper, Virginia.

As the officer droned on, Brzezinski suddenly cut him short, and demanded a full-scale evacuation *immediately.*

"Right now?" the staff person asked incredulously.

"Yes, right now!!" Brzezinski barked back.

"The poor fellow's eyes . . . practically popped; he looked so surprised . . . he reached for the phone and could hardly speak coherently when he demanded that the helicopter immediately come for a drill," said Brzezinski.

Many agonizing hours later, after a series of embarrassing gaffes and mortifying blunders worthy of the Three Stooges, the helicopter carrying Brzezinski finally limped back to Washington for the return trip.

But the fiasco continued. The White House security guards, panicking when they saw an unauthorized and potentially hostile helicopter approaching the White House, immediately ordered an alert. They scrambled into position with automatic rifles, prepared to shoot down Brzezinski's helicopter.

This dismal failure was a sobering reality check for the Pentagon, pointing out the gross deficiencies in its grandiose plans to "win" a nuclear war.

To meet this challenge, the Pentagon's Advanced Research Projects Agency (ARPA) proposed several ingenious computer technologies and modified ones that already existed:

**Teleconferencing.** The Pentagon wanted to ensure that the leadership of the United States would survive to command our nuclear forces as the war progressed. While the rest of the planet was pounded into radioactive rubble, our leaders would command our nuclear fleet in the safety and comfort of high-flying jets and huge, air-conditioned underground vaults. Five top officials (including the President, the Vice President, the Chairman of the Joint Chiefs of Staff) would be dispersed to five different locations, from flying overhead in Air Force One to hiding out in hollow mountaintops or at SAC headquarters in Cheyenne, Wyoming. They would be patched into one another through TV monitors and computers. This plan marked the birth of teleconferencing.

**Virtual reality.** The Pentagon wanted to ensure that their pilots would be able to fly their jets and bombers in the most unpredictable, hostile environments, including the presence of huge winds whipped up by nuclear fireballs. To accomplish this, the Pentagon developed flight simulators, in what was the birth of virtual reality. The pilots would sit in chairs with goggles placed over their eyes and use joysticks to control the simulated computer image in their headset. Through their goggles, they could see an imaginary, computer-generated environment that simulated the conditions of warfare.

Tanks and submarines were easily simulated, since looking through goggles wasn't very much different from looking through a pair of binoculars or a periscope. Since the first head-mounted display was built for the Pentagon in 1968, primitive versions of virtual reality have since proliferated to video arcades around the country.

**GPS satellites.** The Pentagon wanted to make sure that its missiles were accurate. As a result, they launched a series of satellites around the world to guide the flight paths of these missiles in what became the Global Positioning System (GPS). It was so accurate that an ICBM launched from the United States could strike within 300 feet of a target several thousand miles away.

Because of this, the United States could destroy enemy missiles in their silos, submarines in their pens, and bombers on their airfields. The Pentagon realized it could also be used as a first strike weapon, to disarm the enemy before an opposing force had a chance to strike back.

These GPS satellites, once the backbone of a budding first strike capability, are now being used to guide passenger cars from Detroit.

**E-mail.** The Pentagon knew its technicians and scientists had to be able to communicate during and after a nuclear war. To facilitate this, a computer network would be necessary to rebuild the shattered cities and economy after the nuclear war was "won." Surviving scientists could plug into a telephone line to communicate with other scientists in order to begin the process of rebuilding modern civilization. Because most cities would no longer exist, messages would have to be broken up into pieces, scattered throughout the system, moved around cities that no longer existed, and then reassembled at the destination. ARPA combined these ideas with an existing system to create what is now called e-mail.

There was also a sense of urgency. The Pentagon was worried that the shattered remains of the Soviet Union might be rebuilt before the United States. Following a nuclear war, there would be a race to see who could rebuild their country first. In a scenario of two dazed boxers lying flat on their backs, slowly regaining consciousness, the winner of World War III would be that country which could stand on its feet first (and so go on to win World War IV). Therefore, the Pentagon's priority was to provide scientists with a way to rebuild the country as fast as possible, unimpeded by unnecessary restrictions.

It was clear that this meant the network would have to exist without a "policeman." Bureaucratic rules, censorship, and governmental meddling would only retard the rebuilding of America in the race with the Soviet Union (to fight World War IV!). It was partly for this reason that the Internet was built without censors, rules, and regulations.

ARPANET, which was conceived to link up the scientists and universities of the country, was modified to serve in this role. Eventually ARPANET became the Internet.

## The Mother of All Nets

In 1844, when Samuel Morse telegraphed the immortal words "What hath God wrought?" from Washington to Baltimore, he helped to usher in the age of electronic communication. On November 21, 1961, there were no prophets or sages invoking the wisdom of the information age when a half dozen scientists gathered at Boelter Hall, home of the computer science department at UCLA, to connect their computer with the computer at the Stanford Research Institute near Palo Alto.

"There wasn't a photographer present, and it didn't even occur to us that we should have one," recalled Steve Crocker, who was a graduate student at that time. No one, in fact, even remembers what was said in the first historic message linking two distant computers.

ARPANET initially connected only four sites (UCLA, the University of California at Santa Barbara, the Stanford Research Institute, and the University of Utah). ARPANET grew slowly, hindered at every turn because of the hush-hush nature of the project, and because computers back then were mostly incompatible. By 1971, there were only two dozen sites. By 1974, the ARPANET had grown to 62. By 1981, the number exceeded 200. Only in the mid-1980s did ARPANET finally reach critical mass among universities and scientific laboratories.

When the ARPANET finally took off, it was so successful that it was formally discontinued in 1990, having completed its original mission. With the ending of the Cold War, the baton was passed from the military to the National Science Foundation. ARPANET, once the private province of physicists and computer scientists, finally exploded into the public domain as people got wind of this marvelous technology.

By 1994, more than 45,000 smaller networks had joined the Internet. That year, physicists finally brought some order to the unruly, wild and woolly Internet. With the Cold War over, there was no longer any incentive to keep the Internet as unregulated as possible. Tim Berners-Lee, a mathematician working at CERN, the sprawling European physics research center in Geneva, Switzerland, created the World Wide Web in 1991, which made multimedia accessible on the Internet. Like

ARPANET, which could hook up physicists and technicians together during and after a nuclear war, the Web was originally designed to hook up particle physicists to keep track of their complex experiments and mountains of data pouring in from huge atom smashers.

Today, the Internet is growing at the phenomenal rate of 20 percent per quarter, almost doubling every year since 1988. At this rate, it is actually exceeding the growth rate of computers according to Moore's law. It is truly the "mother of all networks," with 10 million servers detectable on the Internet today. If we count those who dial up these servers from their home or work PC, the total number of Internet users in the world is roughly 40 million.

Vinton Cerf, one of the pioneers of the Internet, predicts that if the current rate of growth continues, by the year 2000, 160 million people will be on the Internet. Nicholas Negroponte of MIT estimates optimistically that perhaps up to a billion people could be surfing the Net by that date. Certainly the potential is there; in 1995, 65 million computers were shipped out of factories; by 1996, one-third of all U.S. households had computers, and about 10 to 15 percent of households are wired to the Net.

How big will the Internet become? Cerf says, "I'm not at all shy about predicting that by 2005 the Internet will be as big as the telephone system is today." (There are 600 million telephone lines installed worldwide.)

Because the landmark Federal Communications Commission ruling in 1996 will eventually pave the way for the merger of television with the Internet, and since 99 percent of all U.S. homes have TVs (more than have telephones, flush toilets, or computers), we may actually have 99 percent of the population wired to the Internet early in the next century.

The information stored on the Internet is also increasing at breakneck speed. In 1996, one could access about 70 million pages on the Internet. It is believed that by 2020 the Internet will access the sum total of the human experience on this planet, the collective knowledge and wisdom of the past 5,000 years of recorded history.

## The Historical Significance of the Internet

Visionaries, however, see the Internet as only the beginning; it's just a dusty dirt road that will pave the way for the true information highway of the twenty-first century. The "graphic jams" found on the Internet, which lead to many frustrating delays, will gradually be removed. (In 1996, for example, the switching station in San Jose was almost overwhelmed when Internet traffic reached 95 megabits per second, which was near the system's capacity of 100 megabits per second.)

Vice President Al Gore believes the Internet will be replaced by the National Research and Education Network (NREN), which will be a hundred times faster than the Internet and may cost $390 billion in federal funds over five years.

In many ways, the impact of the Internet can be compared to that of Gutenberg's movable type in the 1450s, when it became possible for large numbers of books to reach a mass audience in Europe. (China and Korea already had a version of movable type.) Before Gutenberg, there were only 30,000 or so books in all of Europe. Literacy and books were a luxury (and tool) of a tiny educated elite, which jealously guarded this precious resource.

By 1500, Europe was flooded with more than 9 million books, stimulating the intellectual ferment which paved the way for the Renaissance.

But the detractors of the Internet claim that it's a passing fad that will slowly fade away, as people get tired of being "flamed" and wading nose-deep in a pile of cyber junk.

They remind us of the fate of the Picturephone, the sensation of the 1964 New York World's Fair. Millions of visitors to the Fair were told that the ordinary telephone would soon be relegated to musty museums as people scrambled to buy Picturephones for their homes. AT&T spent a staggering $500 million in the 1960s to perfect this device—yet it sold only several hundred (which works out to about a million dollars per phone!). This was one of the great telecommunications blunders of all time.

Why did it fail? There were technical problems (phone lines and computers were not powerful enough to carry high-quality video images). But there were personal problems as well. Most people wanted to look at the person they were talking to, but not be looked at. One wag said, "After all, do you really want to comb your hair each time you use the phone?" Ultimately, we are reminded, the final arbiter of high tech is the consumer.

Perhaps the most consistent critic of the Internet is computer expert Clifford Stoll, author of the antimanifesto *Silicon Snake Oil*. Stoll pooh-poohs the claims that the Internet will one day swallow up all forms of human interaction. "Few aspects of daily life require computers, digital networks, or massive connectivity," says Stoll. "They're irrelevant to cooking, driving, visiting, negotiating, eating, hiking, dancing, speaking, and gossiping. You don't need a keyboard to bake bread, play touch football, piece a quilt, build a stone wall, recite a poem, or say a prayer."

He cites other products that became fads but later fizzled, such as CB radios, which in the 1970s grew in popularity until at their peak about 25 million people were using them. But by 1980 the novelty had worn off, and the CB radio market collapsed.

But here lies the difference. CB radios appealed only to people driving on a particular road trying to avoid a patrol car. Its range (a few miles) and audience (a handful of people on the highway in front of or behind you) were limited and as a result never attained critical mass. It never fulfilled the Law of Increasing Returns (which says that, after a certain level, the more people who use a certain technology, the more people will want to use that technology, thereby creating critical mass). On the Internet, however, the range is the planet itself, its subject matter is the sum total of human knowledge, and its audience is anyone with a computer and a modem, an audience which will number in the hundreds of millions to even billions before long. And when computers become invisible, we will be able to bake bread, hike, and drive while conversing with the Magic Mirror.

## To 2020: How the Internet Will Shape Our Lives

Larry Tesler, chief scientist at Apple Computer, and part of the original Xerox PARC team that perfected ALTO and the graphics-based system that eventually became the Macintosh and Windows, is one of the visionaries who left Xerox PARC and is now focusing his efforts to predict the impact that the Internet will have on our lives. He concurs with many of the criticisms of the Internet: yes, there is too much trash on the Internet; yes, there is too much hype. But the good far outweighs the bad. The bottom line, he says, is that the Internet is here to stay.

Of course, he agrees that the explosive growth of the Internet will ultimately level off, as Clifford Stoll suggests, and the Internet craze will subside as people get bored being flamed, but by then it will have become an indispensable part of modern civilization, essential for business, commerce, science, the arts, and entertainment. Tesler rattles off the many ways in which the Internet will change and enrich our lives for the better, from being able to work at home to bringing together specialized hobbyists around the world, to enjoying the "cyber marketplace" that will change the way we shop.

On-line travel agencies will be able to offer complete selections of thousands of travel packages over the Internet. "It's a match made in Maui," crows the *Wall Street Journal.* On-line brokerage firms, which currently account for only 1 percent of the stocks currently bought and sold, will skyrocket because they charge as little as one-tenth the usual fees and can provide instantaneous financial analysis. "Look out, Merrill Lynch," warns the *Wall Street Journal.* On-line bookstores will be able to offer millions of titles of books, the equivalent of several libraries. As much as 15 percent of the $400 billion grocery business will be done electronically by the year 2000, claims Mohsen Maozami of Kurt Salmon Assoc., a retail consulting firm. The future of banking might be seen in the Security First Network Bank of Pineville, Kentucky, which already conducts *all* of its business on the Internet.

"Slowly but surely, real commerce is going on-line," says the *Wall Street Journal*, because "the allure is overwhelming. No store ever closes on the Internet, and no location is isolated from the rest of the planet. Merchants that hang out an electronic shingle in cyberspace don't have to worry about shelf space and can target their marketing to interested customers at a fraction of the cost. And the sheer size of some on-line stores is far beyond anything that can be done with brick and mortar."

The Internet can also offer customers "mass customization." In the future, you will be able to pick out the precise style or pattern you want and have it sent via the Internet to the factory, which will then manufacture the product custom-made. Already, Technology/Clothing Technology Corp. is building a $8.5 million scanner which will provide three-dimensional full-body scans within two seconds. The customer first puts on a skintight body suit, and a scanner uses six projectors and six video cameras to photograph a series of horizontal lines cast on the body suit. The computer then calculates the precise three-dimensional coordinates of every curve in your body. When you select the type of garment you want, the computer sends your coordinates to the factory, which then feeds this information directly into a cutting device.

## Bottlenecks on the Internet

Such a vision of the Internet is truly breathtaking. But it is also one that is filled with potholes, ruts, and detours. In order to fulfill the promise of the information highway, several problems have to be solved and several milestones have to be reached between now and 2020: (1) resolving bandwidth bottlenecks; (2) designing better interfaces; (3) creating personalized agents and filters.

Microsoft CEO Bill Gates identifies "bandwidth bottlenecks" as the most immediate obstacle to this dream. Bandwidth is roughly the amount of information (or bits) that can be transmitted per second. The gold standard for bandwidths is 4 gigabytes, the amount of information contained in a feature-length movie video. Many consider the transmission of movies-on-demand to be the "killerapp" (killer application) which will energize the market for the Internet, in the same way that movies were the killerapp for the VCR market, or spreadsheets became the killerapp for the PC in business.

The question that generates feverish speculation on Wall Street is: which medium can best send 4 gigabytes of information into individual homes in the shortest amount of time?

Virtually everyone who has surfed the Net feels like pulling his or her hair while waiting for pictures to show up on the screen. Even with a fast 28.8 kilobit modem, for example, it can take from fifteen to thirty seconds for a single picture to appear. (With an ISDN link, with a transmission rate of up to 144 kilobaud, the time can be reduced to a second or so.) To see full-length videos on the screen, you have to be able to project about 30 frames per second, many orders of magnitude faster than the fastest modem. Moreover, an analog signal moves through the phone lines, the primary Internet link at present, at about 64,000 bits per second. At that snail's pace, it would take over a hundred hours to transmit *The Silence of the Lambs*.

Because of this, it was once thought that videos could never be transmitted over copper wires by the telephone companies. However, if the video signal is digitalized, then it is possible to compress the digital signal so that it can be transmitted over copper wires. Compression loses a small part of the information, but gains many times over in increasing the speed of transmission.

Some attractive alternatives to the phone lines that are being explored are satellites as well as cable. Each has its strengths and weaknesses. Transmitting from outer space, satellites have the advantage that companies need not spend billions of dollars laying millions of miles of wires. The drawback is that one has to launch hundreds of communication satellites to cover all parts of the earth at any given time. Similarly, cable wires are convenient because they already broadcast videos into households around the country at high speeds. Cable companies are beginning to offer fast Internet access along with their usual cable service. Cable has some problems too, however. Cable wires, for example, need costly "boosters" to amplify the signal over long distances.

If the dirt road of the Internet is made up of copper wires, then the paved information highway will probably be made of laser fiber optics. Lasers are the perfect quantum device, an instrument which creates beams of coherent light (light beams which vibrate in exact synchronization with each other). This exotic form of light, which does not occur naturally in the universe, is made possible by manipulating the electrons making quantum jumps between orbits within an atom.

Light beams traveling inside thin transparent glass fibers are trapped inside, even if the fiber is wrapped in circles. The beam simply bounces off the interior walls of the fiber, an effect called total internal reflection. (This is the same effect used to create spectacular water fountain light shows in Italy and elsewhere. If powerful lamps are placed at the base of the fountain, the light shooting skyward is captured by the streams of water, giving the illusion that the water is on fire.)

The laser will become the main medium for the Internet because 10 to 100 times more information can be carried on laser beams than on copper wires. The frequency of red laser light from a common helium-neon gas laser, for example, is on the order of 100 trillion cycles per second. The faster the vibrations, the more information you can pack into the signal.

Moreover, in terms of congestion, there is a crucial difference between ordinary highways and the Internet. The more highways that are built, the more space they take up, until they begin to cause congestion and eat up valuable land. That's caused a backlash against some highways. The Internet, however, as Tesler points out, is *limitless.* You can always string more optical fibers, increase the speed of switches, expand the bandwidth via new lasers. Nor is there any limit to what kinds of information can be transferred on the Internet. In fact, the only physical limits to the growth of optical fibers appears to be the bottlenecks at the ends—i.e., the switches and cables at the receiving end.

Fiber cables have now been manufactured which can carry a staggering 100 billion bits of information per second, which is equivalent to sending the Encyclopaedia Britannica over a glass fiber in a fraction of a second. This appears to be the upper limit attainable with present-day technology, but it is probably more than sufficient to handle the exploding volume of traffic on the Internet. This speed is so great, in fact, that standard electronic switches are too slow to handle this influx of information. Eventually, the switches and components will also have to be made out of lasers and other optical devices.

Already, thousands of miles of copper wires are being replaced by thin flexible glass fibers—as thin as an eyelash—which can carry up to millions of messages. The $6 billion fiber optic cable industry is growing at an astonishing rate of 20 percent per year. Annual installation of fiber optic cable has doubled since 1993, to an estimated 16.25 million miles in 1996 alone. Unlike the microprocessor, which will likely begin to be phased out by the year 2020, the power of the laser seems boundless, limited only by the crude technology at either end.

The second problem that the intelligent planet faces is interface bottlenecks—i.e., screens and voice inputs. In order to have a genuine Magic Mirror, one has to have digital TV—wall screens with very fine resolution—and an intelligent agent behind the screen capable of understanding English and common sense.

## The Merger of TV and the Internet

One bottleneck to the Internet of the future has been the incessant and fierce infighting between the computer and TV industries over who will dominate the future of electronic media. Since 99 percent of American homes have one or more TVs, many feel the commercial direction of the Internet eventually lies with a merger with TV.

After a decade of bickering, a long-awaited agreement was reached in late 1996 which will set the course of electronic communications well into the next century. Already, the agreement is being heralded as the most important of the last few decades. Gary Shapiro, head of the Consumer Electronics Manufacturers Association, describes what will result from it this way: "People like to say this transition is going to be like moving from black-and-white television to color. But I think it's more fundamental, like the change from radio to TV."

The Federal Communications Commission and the giants of the television and computer industries finally agreed to adopt digital as the standard mode of transmission, which will further the merger of the TV with the computer, making TV interactive.

Previously, the TV images in the United States were based on the cathode-ray tube, which scans 525 lines across the picture screen, producing 30 images every second. The signal was sent in "analog" format—i.e., in a continuous wave pattern—and could not be easily modified. (Most waves that we usually encounter in daily life, such as sound, light, radio or TV waves, are analog signals. When they are amplified, static builds up and information loss occurs. That's why long-distance phone calls, which must be amplified many times, sound so scratchy.)

The new agreement changes all of this. The TV of the future will have double the resolution (1,080 lines per screen, which approaches the quality of a 35 mm photograph), and will be digital. Instead of being square, the screen of the future will be shaped very much like the wide-screen image found in the movies.

(The key word is "digital." When a binary signal is sent in discrete packets of ones and zeros, the signal can be manipulated in thousands of ways to clean it up and modify it. Error-correcting programs can provide nearly error-free transmission, reducing the distortion and fuzziness commonly found on standard TV screens. The signal will always be picture-perfect and free of interference, no matter where it originated from. A signal from halfway around the world will be just as crisp as if it were broadcast from next door. The signal can also be enhanced and magnified [as is done to the faint images sent by distant NASA space probes]. The signal can also be fragmented, so that it can carry the World Wide Web and the stock market as well as TV.)

The first commercial models of these hybrid TV/computers are expected to be shipped in late 1998. In 1997, the FCC ruled that all analog TV signals will be phased out by 2006. Consumers then may have no choice: they will buy either a digital TV or a converter. The standard big-box TV found in most TV stores will be relegated to the museum.

More important, they will have a box on top which will connect it to the Internet, making digital TV fully interactive. Instead of vegetating passively in front of the TV screen, viewers in the future will be able to engage and interact with the images on the TV screen.

Already, six major companies are marketing TVs hooked up directly to the Internet. Rick Doherty of the Envisioneering Group estimates that a third of U.S. households will have them by 2002. Once the digital transmission becomes mandatory, the Internet may eventually become a standard feature for 99 percent of the U.S. population.

But by 2010, even the wide-screen digital TV may be phased out as a new generation of paper-thin wall screens are introduced.

## Wall Screens

Eventually, as another consequence of Moore's law, computer and TV screens will be flat enough to hang on a wall like a picture, or small enough to fit on your wristwatch.

The cathode-ray tube (CRT), which has been the workhorse of television screens since their inception, currently makes up two-thirds of the computer monitor market. The CRT is a large glass vacuum chamber in which electron guns fire several beams across a large phosphorescent screen, which glows when hit by the beams. (Color CRTs use three electron beams, one for each of the primary colors: red, blue, and green, out of which all the other colors can be made.) The advantage of the CRT is that it creates a brilliant image. Early attempts at flat panels have been notoriously hard to read. However, the CRT has numerous defects. Because electron beams move only in a vacuum chamber, CRTs will always be heavy and thick—too bulky and heavy to make them easily transportable.

The CRT will eventually be replaced by either liquid crystal displays (LCD) or plasma screens. LCDs contain special liquid crystal chemicals which flow like a liquid but have molecules arranged in crystalline order. Such crystals have been known to science for about a century. In fact, they are quite common, appearing in everything from cell membranes to soap scum. Normally, LCDs are transparent. However, when a small electric current is sent through them, they immediately become opaque. By controlling the flow of electricity through the LCDs, we can make blinking letters flash on and off on the screen.

Although LCDs are cheap, consume little power, and can be made notebook-thin, in the past they have had an Achilles' heel. Because they generate no light of their own, they are difficult to read in dim light. But this problem has been solved with the most advanced form of LCD screen, the active-matrix screen, which is able to create a brilliant image because each pixel on the screen is controlled by its own thin-film transistor. The miniaturization of transistors is so advanced that each dot on a computer screen corresponds to one transistor.

The active-matrix screen, already commercially available, will dominate the market in the coming years. Active-matrix screens are now being developed which measure twenty-two inches (measured diagonally), larger than the standard seventeen-inch CRT computer monitor.

With the falling cost of transistors and mass production, the cost of flat panel displays is now dropping rapidly. The engineers at Stanford Resources in Silicon Valley predict that by the year 2000 they will overtake CRTs in sales. "For the first time, it is possible to conceive of the end of the era of the desktop CRT," declares Carry Lu, a computer expert.

Another possible alternative to the CRT in the future is the plasma screen, which uses thousands of tiny chambers containing an ionized mixture of neon and xenon gas that lights up in various colors and intensities. Plasma screens can be compared to stacking together thousands of tiny neon lights, each smaller than a pinhead, to make a screen. Their advantage is that they can be quite large, with a wide viewing angle: plasma screens have already been made which measure forty-two inches. Sixty-inch plasma displays suitable for wall screens are already in the experimental stage. Their drawback, however, is that they consume a fair amount of power and appear a bit misty.

The market for LCDs and plasma screens is projected to soar to $19.23 billion and $5.11 billion, respectively, by 2003.

By 2020, the flat panel displays will likely come in a variety of forms. They will be miniaturized to work as wristwatch screens and may be added to eyeglasses or key chains. Eventually, they will become so cheap they will be everywhere: on the backs of airplane seats, in photo albums, in elevators, on notepads, on billboards, on the sides of buses and trains. They may one day be as common as paper.

## Speech Recognition

In fairy tales, characters don't type in instructions to a Magic Mirror on a keyboard; they talk to it. But speech recognition today is yet another interface bottleneck on the information highway. Remarkable progress has been made in designing computers which can take dictation. The problem is that these machines can recognize human speech but cannot understand what they are hearing.

In principle, voice recognition should be easy to solve. In casual conversation we use perhaps only 2,000 words. An educated person can reasonably be expected to know 10,000 to 20,000 words. A vocabulary of this size is easily stored in a computer. These words, in turn, can be broken down into the various phonemes which have long been cataloged by linguists.

Computers can identify phonemes by breaking them up into two quantities: the frequency and the intensity of the sound. Measuring these two quantities, a computer can obtain a visual "voice print" of each phoneme, consisting of a series of vertical squiggly lines. (For example, the larger the squiggles, the louder the volume. The faster the squiggles, the higher the pitch.) This can be demonstrated at most science museums, where you speak into a microphone and then see your voice pattern displayed as waves oscillating on a screen.

At present, there are voice recognition software programs on the market which can take dictation with over 95 percent accuracy. A typical voice recognition machine is capable of recognizing 40,000 words spoken by a person the computer has never heard before. But the programs are imperfect. One has to talk with a slight hesitation between each word in order for the computer to distinguish various words. But many expect that by 2005 even these problems will be solved. They are merely technical problems, and do not involve overcoming any new scientific hurdles; they just require more computer power.

What is more difficult is to design machines which not only can hear human voices but can understand what is said. Computers can read and hear, but they do not comprehend. To have a true Magic Mirror, in essence, involves perfecting artificial intelligence, the most difficult problem of all in computer technology. Ultimately, it is a problem that goes to the heart of the age-old question: what makes us human?

The first tiny step is to develop "intelligent agents," programs which can make primitive decisions and act as filters. However, the true resolution of this problem will have to wait until the fourth phase of computing, which will likely take place from 2020 to 2050, when scientists expect intelligent agents to be replaced by true artificial intelligence.

## From the Present to 2020: Intelligent Agents

An intelligent agent should be able to act as a filter on the Internet for the user, distinguishing between junk and valuable material. As anyone who has cruised the Net knows, much of the information on it is cyber junk and cyber babble, including everything from someone's five-year-old wedding pictures to the rantings of would-be prophets. Intelligent agents will have to make complex value judgments about what a user wants.

One person who is working hard to make this vision of the future come true is Pattie Maes of the MIT Media Laboratory, one of the pioneers of the intelligent agent, a computer software program which combines aspects of a secretary, personal planner, and even companion.

After getting her Ph.D. in computer science, Maes started in the field of artificial intelligence, working with Rodney Brooks of MIT, helping to build Cog, a humanoid robot which would learn like a child. But like Larry Tesler before her, she gradually became disillusioned with artificial intelligence. "I am not convinced that if I can build a robot with the intelligence of a two-year-old, it will teach us much about adults," she concluded. "It's easier to simply build a two-year-old the biological way!"

When she became pregnant, she made a bet with Brooks over who would attain the intelligence of a two-year-old first, Brooks's robot Cog or her baby. (She won the bet.)

She decided that if we can't achieve artificial intelligence at this stage, why not at least try to augment our own intelligence by writing software programs for intelligent agents which could perform herculean feats of information gathering and decision making?

On the lowest level, such an agent should be able to sift through one's e-mail, prioritizing letters, throwing away junk e-mail, and putting them in order. At a higher level, the agent should be able to update one's schedule and route important calls through, informing one of new appointments, and even blocking annoying requests. In emergencies, it can contact a user wherever he or she may be.

These agents of the future will act as filters, preventing us from drowning in an ocean of trivia and junk from the Internet as well as enabling us to search the Internet for items we might need. Already, Maes and her colleagues have developed an intelligent agent which can search various scientific databases and retrieve selected articles that may be of interest to a scientist. The most successful agents are allowed to combine or "mate" and pass on their "genetic information" (i.e., your likes and dislikes) to the next generation of agents. In this way, each generation of evolving agents could become more "adapted" to the wishes of the programmer. "In my vision of the ultimate intelligent software agent system," she says, "there are all these 'life' forms evolving by themselves and specializing toward whatever you happen to be interested in." She adds: "Each succeeding generation matches its owner's interest better."

Such agents will be invaluable for people who want continual updates on sports events, news items, hobbies, or human-interest stories. Even as we sleep, our computers will be able to silently collect information we might need. Other uses of intelligent agents include acting as personal intermediaries or go-betweens to other people. Singles could use them to create a database from the entire planet. Job seekers could scan the world's want ads. Companies could locate consultants even in obscure disciplines. Hobbyists could contact people with similar interests.

Maes has felt that the most effective intelligent agent would be a personal one, with a humanlike face and personality. Programs have already been written for prototype intelligent agents resembling a kind of Happy Face, with a range of perhaps ten to twenty different emotions.

"Rather than manipulating a keyboard and mouse, people will speak to agents or gesture at things that need doing. In response, agents will appear as 'living' entities on the screen, conveying their current state and behavior with animated facial expressions or body language rather than windows with text, graphs, and figures," she has written. This means talking directly to a humanlike face, which can smile, grimace, frown, even crack jokes.

## 2020–2050: Games and Expert Systems

In the world beyond 2020, scientists expect genuine artificial intelligence to begin to permeate the Internet. The next step beyond intelligent agents is a branch of artificial intelligence called *heuristics*, which tries codify logic and intelligence with a series of rules. Ideally, heuristics would enable us to speak to a computerized doctor, lawyer, or technician who could answer detailed, technical questions about diagnostics or treatment. One of the earliest branches of heuristics which has actually exceeded human abilities is the chess-playing machine. Heuristic machines excel in chess games because they are based on simple, well-defined rules; millions of moves can be analyzed at the speed of light, making the most advanced programs able to beat all but the greatest chess grand masters.

In 1996, world chess champion Gary Kasparov accepted the challenge of a computer, IBM's Deep Blue chess-playing program. Kasparov was shaken to the core. With 32 microprocessors, Deep Blue could analyze *200 million* positions per second.

"I could feel—I could smell—a new kind of intelligence across the table," Kasparov admitted. "I got my first glimpse of artificial intelligence . . . when in the first game of my match with Deep Blue, the computer nudged a pawn forward to a square where it could easily be captured." It dawned on Kasparov that for the first time he was facing a machine that could see ahead in novel ways. "I was stunned by this pawn sacrifice," he admitted.

In the first match, although Deep Blue took the first game in the series, eventually Kasparov found its Achilles' heel and trounced the computer, 4 to 2, claiming the $400,000 prize money offered by the Association for Computing Machinery. Kasparov found the weak spot of the computer: chess-playing machines pursue a set strategy. If you force the computer to deviate from that strategy, it becomes helpless, flailing like an overturned turtle on its shell. "If it can't find a way to win material, attack the king or fulfill one of its other programmed priorities, the computer drifts planlessly and gets into trouble," Kasparov said. "So although I think I did see some signs of intelligence, it's a weird kind, an inefficient, inflexible kind that makes me think I have a few years left."

He was too optimistic. Just one year later, an improved version of Deep Blue soundly trounced Kasparov, sending shockwaves around the world. The media asked the question, "Can machines now think?"

But Douglas Hofstadter, a computer scientist at Indiana University, echoed the thoughts of many when he said, "My God, I used to think chess required thought. Now, I realize it doesn't. It doesn't mean Kasparov isn't a deep thinker, just that you can bypass deep thinking in playing chess, the way you can fly without flapping your wings."

Hyped-up statements in the press about machines being able to outthink humans are wildly premature. After all, handheld calculators already compute much faster than any human alive, yet people do not have nervous breakdowns and identity complexes contemplating that fact. Chess-playing machines, in a sense, are glorified handheld calculators.

But the area of heuristics which may have the biggest impact on everyday life is "expert systems." These heuristic programs contain the accumulated knowledge of human experts and can dissect problems like a human.

This branch of AI is based on listing all possible "if . . . then" propositions—i.e., if something breaks down, then you do this. Since computers are good at rapidly analyzing a well-defined set of rules and outcomes, there is commercial gold to be had in an AI system that can incorporate vast amounts of "rules of thumb." For example, when you are sick, the doctor asks you a series of questions about your symptoms. Then the doctor tells you what might be wrong. This "if . . . then" type of questioning is easily duplicated by computers, because they can store the thousands of rules necessary to diagnose an illness. (The gruff holographic doctor featured on *Star Trek: Voyager* is based on a heuristic program. So was the murderous HAL 9000 in the movie *2001*, where the "H" stood for "heuristics.") Not only will computer heuristic health programs help to reduce health costs, they may actually be more accurate than a human doctor for most simple problems since the computer program will be comprehensive and up to date.

In 1975, an expert system called Mycin surpassed the average doctor's ability to diagnose meningitis in patients. As long as the program stayed within carefully defined boundaries, it performed remarkably well. (However, as computer expert Douglas Lenat quips: "Ask a medical program about a rusty car and it might blithely diagnose measles!")

Certain heavy industries are keen on expert systems since they can be used to replace expert factory engineers and chemical technicians as they retire, taking their valuable experience with them. In the 1980s, General Electric had only one engineer who knew how to repair all of GE's electric locomotives. Over a lifetime, he had accumulated a vast trove of detailed knowledge about the idiosyncrasies of these large locomotives. He was getting old, however, and his esoteric knowledge, worth tens of millions of dollars, would be lost when he retired. With the transferal of his knowledge into an AI program, called the Diesel Electric Locomotive Troubleshooting Aid (DELTA), however, computers may be able to diagnose 80 percent of the breakdowns.

As early as 1985, 150 companies spent a staggering $1 billion on AI—primarily on expert systems. The fundamental problem with expert systems, however, has been that they lack common sense. No matter how many rules they contain, they make glaring errors because they lack even a child's intuitive understanding of the world. The reason why expert systems ultimately collapsed in the marketplace can be summarized in one popular phrase: "It's easier to simulate a geologist than a five-year-old"—i.e., an expert system can do a reasonably good job of handling the facts necessary to do geology, but it cannot simulate the common sense that even a five-year-old has.

## Common Sense Is Not So Common

The problem with computers is that stripped of their mystique and dazzling accessories, at present they are nothing more than glorified adding machines, or "idiot savants." While these adding machines can be modified to become word processors, at their core they are still adding machines. They can manipulate vast amounts of data millions of time faster than humans, but they do not understand what they are doing and have no independent thought. Nor can they program themselves.

One of the principal problems in the era between 2020 and 2050 will be to build intelligent systems with common sense. Like the huge concealed portion of an iceberg hidden beneath the waves, common sense is so embedded in our brains at such an unconscious level that we don't even ponder how we use it in our daily lives. Only the tiniest fraction of our thinking is devoted to conscious thought. Most of our thinking is actually unconscious thought, including common sense.

Ironically, our brains never evolved the remarkably simple neural circuits it takes to do arithmetic. Being able to multiply five-digit numbers, which is effortlessly performed by handheld calculators, was of no use in escaping a hungry saber-toothed tiger hundreds of thousands of years ago. To perform arithmetic requires surprisingly few neural circuits, but because they were not needed in our evolution, we never developed them. Our brains did, however, evolve the sophisticated mental apparatus that enables us to understand common sense without thinking about it and survive in a hostile world.

Computer systems are the opposite; they are marvelous at abstract mathematical logic, but in general they do not grasp the simplest concepts of physics or biology. They have difficulty, for example, solving the following problem:

> Susan and Jane are twins. If Susan is now twenty years old, then how old is Jane?

The concept of "time" (that all objects age at the same rate, that a son is younger than his father, etc.) is easily grasped by children, but not by computers. It's a law of physics, not mathematical logic. The computer must be told that time progresses uniformly.

Computers have trouble with "obvious" biological facts about living things. For example, computers make the following mistake:

> Human: All ducks can fly. Charlie is a duck.
> Robot: Then Charlie can fly.
> Human: But Charlie is dead.
> Robot: Oh. Then Charlie is dead and he can fly.

Computers have to be told that once something is dead, it cannot move. This is not obvious from the laws of logic.

The problem is that computers are mathematically logical, whereas common sense is not. Biological and physical laws of nature are not necessarily inherent in the laws of logic.

### The Encyclopedia of Common Sense

Douglas Lenat has devoted a lifetime to conquering the mysteries of common sense. He feels that the problem is that artificial intelligence (AI) researchers have tiptoed around the periphery of the real problem. What is needed, he says, is nothing less than an "AI Manhattan Project," a full frontal assault on common sense. The challenge is to create an Encyclopedia of Common Sense—i.e., a nearly complete set of common-sense rules. In other words, instead of analyzing isolated pieces of logic, he is advocating a brute force, take-no-prisoners approach.

Beginning in 1984, Lenat began to create Cyc (short for encyclopedia), a $25 million project which was funded by a consortium of companies including Xerox, Digital Equipment, Kodak, and Apple. While previous programs could barely achieve the common-sense logic of a three-year-old, Cyc's goal was to achieve the common-sense knowledge of an adult. "No one in 2015 would dream of buying a machine without common sense," Lenat claims, "any more than anyone today would buy a personal computer that couldn't run spreadsheets [or] word processing programs."

Lenat believes that in the future everyone will load common-sense programs into their computers, allowing them to have intelligent conversations with their computers, which will be capable of interpreting and carrying out people's commands. Lenat's goal is to write down a complete list of all common-sense rules. Some "obvious" rules include:

- Nothing can be in two places at the same time.
- When humans die, they are not born again.
- Dying is undesirable.
- Animals do not like pain.
- Time advances at the same rate for everyone.
- When it rains, people get wet.
- Sweet things taste good.

Each "obvious" statement can take his crew many weeks or months to break down into its logical components. After ten years of work, he has accumulated 10 million such assertions, requiring a billion bytes of information. He ultimately hopes to accumulate a staggering 100 million "obvious" assertions.

At times Lenat despairs of compiling all the ambiguities hidden within the English language, ambiguities that are only resolved by a person's knowledge of the real world. Take, for example, the statement: "Mary saw a bicycle in the store window. She wanted it." Lenat says, "How do we know that she wanted the bicycle, and not the store, or the window?" The actual resolution of this simple problem requires that Cyc understand the nearly complete set of likes and dislikes of human beings.

The difficulty of the problem is underscored by the fact that it took Lenat three months for him to program Cyc to understand the following:

"Napoleon died on St. Helena. Wellington was saddened." Unraveling these two deceptively simple sentences was complicated because Cyc had to untangle a chain of "obvious" statements. First, Cyc had to figure out that Napoleon was a person; that persons have the unfortunate habit of dying; that death is irreversible and undesirable; that death, in turn, often triggers emotion; and that sadness is one of these emotions.

Lenat and his staff get ideas for Cyc from a most unusual source: by reading scandalous supermarket tabloids and asking themselves what Cyc needs to know to understand (or refute) them. Lenat asks the question: can Cyc spot the errors in the tabloids? (If Cyc succeeds in seeing through the misconceptions in the supermarket tabloids, it probably will have already exceeded the common-sense abilities of a great many Americans!)

One of Lenat's intermediate goals is to hit the "break even" point, where the computer will be able to learn faster by simply "reading" new material than by having an army of private tutors with Ph.D.s. Like a young bird taking off on its maiden voyage, Cyc will then be able to soar on its own power. At that point, it can dispense with human teachers and, like a ten-year-old child, read and learn on its own.

Lenat sums up his philosophy with the statement: "Intelligence is 10 million rules." This is the opposite of the approach taken in physics, where physicists try to reduce vast amounts of material to the simplest equations. According to Lenat, this is, in fact, the problem with AI research. Like AI founder Marvin Minsky, Lenat believes that AI researchers have fallen victim to "physics envy." Impressed with how successful physicists have been in representing the physical world with a handful of equations, they have mistakenly thought that artificial intelligence could also be reduced to a few lines of logic.

But to Lenat, common sense and intelligence are the sum total of millions of lines of code. They cannot be reduced to a few lines of logic. This is why Lenat feels the Cyc program is so important. Beyond 2020, if Cyc can be successfully incorporated into an expert system, it could give us computerized doctors, industrial chemists and engineers, lawyers, and so on.

Not everyone in the AI community believes in Lenat's work. Maes, for example, thinks that a truly intelligent agent must learn from and interact with its environment. Randall Davis, another AI researcher, gives it an "outside chance" of working, but he concedes that "Cyc is not a rocket ship that is going to make it to the moon or not. It is a vast experiment in absolutely hard-core empirical AI. Something important will come out of it." Perhaps we are being too critical, he says. After all, he reminds us, "you can look around and see that the planet is populated by semi-intelligent systems [i.e., us] who have only the barest theory about time, space, causality, and so forth."

## A "Week in the Life" in 2020

So how might these scientific revolutions affect our lives? Although I will discuss the next 100 years and beyond in *Visions*, scientists can guess with reasonable accuracy about life in the year 2020 because many of the prototypes of the inventions and technologies contained in the following story already exist in the laboratory. Far from being science fiction, many of the technologies I profile are already beginning to prove their worth. As Paul Saffo of the Institute for the Future has said: "The future is already here. It's just distributed unevenly."

What follows is a "Week in the Life" scenario of what life might be like in the year 2020 if you were an executive dealing with the latest technology.

*6:30 A.M. June 1, 2020*

A gentle ring wakes you up in the morning. A wall-sized picture of the seashore hanging silently on the wall suddenly springs to life, replaced by a warm, friendly face you have named Molly, who cheerily announces: "It's time to wake up!"

As you walk into the kitchen, the appliances sense your presence. The coffeepot turns itself on. Bread is toasted to the setting you prefer. Your favorite music gently fills the air. The intelligent house is coming to life.

On the coffee table, Molly has printed out a personalized edition of the newspaper by scanning the Net. As you leave the kitchen, the refrigerator scans its contents and announces: "You're out of milk. And the yogurt is sour." Molly adds: "We're low on computers. Pick up a dozen more at the market while you're at it."

Most of your friends have bought "intelligent agent" programs without faces or personalities. Some claim they get in the way; others prefer not to speak to their appliances. But you like the convenience of voice commands.

Before you leave, you instruct the robot vacuum cleaner to vacuum the carpet. It springs to life and, sensing the wire tracks hidden beneath the carpet, begins its job.

As you drive off to work in your electric/hybrid car, Molly has tapped into the Global Positioning System satellite orbiting overhead. "There is a major delay due to construction on Highway 1," she informs you. "Here is an alternate route." A map appears ghostlike on the windshield.

As you start driving along the smart highway, the traffic lights, sensing no other cars on this highway, all turn green. You whiz by the toll booths, which register your vehicle PIN number with their laser sensors and electronically charge your account. Molly's radar quietly monitors the cars around you. Her computer, suddenly detecting danger, blurts out, "Watch out! There's a car behind you!" You narrowly miss a car in your blind spot. Once again, Molly may have saved your life. (Next time, you remind yourself, you will consider taking mass transit.)

At your office at Computer Genetics, a giant firm specializing in personalized DNA sequencing, you scan some video mail. A few bills. You insert your smart wallet card into the computer in the wall. A laser beam checks the iris of your eye for identification, and the transaction is done. Then at ten o'clock two staff members "meet" with you via the wall screen.

*4 P.M.*

Molly informs you that it is time for your doctor's appointment. As Molly makes the connection, your virtual doctor appears on the wall screen. "We picked up trace amounts of a certain protein in your urine. There is a microscopic cancer colony growing in your colon," he says.

"Is that serious?" you ask anxiously.

"Probably not. No more than a few hundred cancer cells. We'll zap them with a few smart molecules."

"And just out of curiosity, what would have happened before protein testing and smart molecules?" you ask.

"Well, in ten years, you would have developed a small tumor; at that point there would have been several billion cancer cells growing in your body, and your chances of survival would be about five percent."

The virtual doctor frowns and says, "We also used the new MRI machine to take a peek inside your arteries. At the present rate of plaque buildup, the computer calculates that within eight years, you will have an eighty percent increased risk of a heart attack. I'm video-mailing a strict program of exercise, relaxation, meditation, and yoga."

Oh great, Molly will have one more function: that of your personal trainer.

*Evening*

That evening, you attend a company cocktail party. As you wander among the guests, the video camera in your glasses scans the faces in the crowd and Molly matches the faces with the computer profiles in her memory.

Molly whispers in your ear who each person is from a special miniature transmitter in your glasses.

By the end of the party, you've drunk a bit too much. Molly whispers, "If you drink any more, the breath analyzer in the dashboard won't allow you to start the car."

*Midnight, Wednesday*

You decide to do some last-minute shopping. "Molly, put the virtual mall on the screen; I need to buy a new sweater."

The wall screen flashes an image of a town mall. You wave your hands above the coffee table, and the video image changes, as if you are walking through the mall.

You pick out the sweater you want from the racks. You like the design, but the size is wrong. Fortunately, Molly maintains your precise 3-D measurements.

"Molly, I want a red sweater, not a blue one, but without those frills. Send the order, and put it on my smart card."

Then you decide to house-hunt some apartments in the city and several beach houses in Europe. Pictures of apartments and beach houses in the price range you specified appear on the wall screen. You walk through them with your fingers.

*Thursday night*

You have no date for that weekend. On a whim, you tell Molly to scan the names of all the eligible single people in the area, matching them to your tastes and hobbies.

A list of faces appears on the screen, with a brief description beneath each picture.

"Well, Molly, whom do you think I should contact?"

"Well, I think numbers three and five look rather promising. They're an eighty-five percent match to your interests." Molly then scans the facial features of each person and performs some computations on their facial measurements. "Plus, I think numbers three and six are rather attractive, don't you?" Molly says. "And don't forget number ten. Good parents."

Molly has picked out the most austere, conservative-looking people in the group. Molly is beginning to sound just like your mother!

*Saturday night*

One of the people you picked from the list has agreed to go out with you.

You and your date go to a romantic restaurant, but just as you are about to eat, Molly scans your meal for its nutritional content. "There's too much cholesterol in that food."

You suddenly wonder if you can turn off Molly.

Afterward, the two of you decide to go back to your apartment to watch an old movie.

"Molly, I'd like to see *Casablanca.* But this time, could you replace Ingrid Bergman's and Humphrey Bogart's faces with ours?"

Molly downloads the movie off the Net and begins to reprogram all the faces in the movie.

Soon you see yourselves transported on the screen back to war-torn Morocco. You can't help but smile at the end of the movie as you see yourselves in the final scene at the airport, staring into each other's eyes.

"Here's looking at you, kid."

## Conclusion

In the period from 2020 to 2050, we might interact daily with expert systems and common-sense programs in our Magic Mirror, which, in turn, could revolutionize the way certain professions are organized. Although specialized information and services will necessarily be provided by human experts, many everyday questions may well be answered by intelligent expert systems.

Of course, such computer assistance raises questions. What is it that makes us human? How do we think? In the next two chapters, I will explore the culmination of artificial intelligence, the creation of an artificial mind.

Unlike the quantum or biochemical revolution, the study of human consciousness is still in its infancy. The Newton or Einstein of artificial intelligence probably has not been born. However, there is a revolution taking place within that field which is upsetting previous thought, provoking entirely new discussions about what it means to be human.

# 4

# Machines That Think

"Sometime in the next thirty years, very quietly one day we will cease to be the brightest things on earth."
—JAMES MCALEAR

## Creating the Future

A VISIT to MIT's famed Artificial Intelligence Laboratory is full of delightful twists and surprises. The AI Lab is located on the eighth and ninth floors of a modern building in Technology Square just off the main MIT campus, looking very much like an ordinary office building.

But when one opens the door, he or she sees a strange spectacle: the world's most expensive toy factory, an elaborate mechanical playpen for brilliant engineers who have never grown up. With teams of intense grad students slumped over their benches, carefully assembling dangling legs, arms, bodies, and heads with their tools, the place looks very much like a high-tech version of Santa's workshop.

Wandering around the lab, one sees a menagerie that would light up any child's eyes: a toy battlefield with realistic-looking miniature tanks, large plastic dinosaurs, a huge Plexiglas-enclosed sandbox, complete with a two-foot mechanical ant, and a ten-inch-long mechanical cockroach. These mechanical denizens, far from being futuristic children's toys appearing beneath a Christmas tree, may one day evolve into an army of automatons which will walk on the surface of Mars, explore the solar system, and even enter our homes.

Everywhere in the lab, there is a playful atmosphere. The blackboards are covered with silly rhymes and syllogisms; there is even a yellow brick road painted on the floor that leads to the computer nicknamed Oz.

In one corner of the lab sits Odie, a contraption about two feet tall resembling a mechanical dog, equipped with video cameras for eyes and a smart bow tie, resembling the dog Odie from the comic strip Garfield. Odie responds to motion; wherever one's hand moves, Odie's video eyes lock on precisely, following every twist and turn. Unlike his namesake, however, Odie is no slowpoke: drop a book without warning and Odie's video eyes can track it instantaneously as it falls to the floor.

In another corner lies WAM, a large mechanical arm attached to a TV camera. Throw a Day-Glo red plastic ball at WAM and the camera locks onto the ball in midair, the computer plots its future trajectory, and WAM's arm lunges out and snatches it. Not bad for a one-armed bandit.

In the basement lies Trudy, an elaborately crafted four-foot mechanical dinosaur named after the Troodon, a sleek chickenlike dinosaur that once walked the earth. Trudy is designed to walk, run, and one day hop like its namesake. It is one of several walking robots at MIT, some of which can hop, skip, even flip upside down in midair, and do everything except break-dance.

Touring this bizarre room, one gradually realizes that the AI Lab is a romper room for geniuses, what might have happened if Peter Pan's Lost Boys all turned into computer whizzes and hackers. The future, it seems, is being invented by a team of mischievous, overgrown children with Ph.D.'s.

Amidst the chaotic jumble of oversized mechanical toys is the devilishly simple creation of Rodney Brooks named Attila. Attila has a face only a mother (or creator) could love. Weighing in at 3.6 pounds, it looks like a gangly, six-legged oversized cockroach made out of rods, complete with ten computers and 150 sensors. It spends most of the day crawling like a bug at the brisk speed of 1.5 miles per hour, successfully avoiding whatever obstacles are placed in its path.

Like a proud, beaming parent, Brooks boasts, "Ounce for ounce, Attila is the world's most complex robot."

To Brooks, the future of artificial intelligence does not belong to giant computers which fill up entire floors, romanticized in countless Hollywood films. It belongs instead to tiny but remarkably agile mechanical bugs like Attila and a fresh, entirely new approach to artificial intelligence and robotics.

Unlike traditional mobile robots, which must be fed huge computer programs before they can move, Attila learns everything from scratch. It even has to learn how to walk. When it is first turned on, its feet flail in all directions, like a drunken cockroach. But gradually, after much trial and

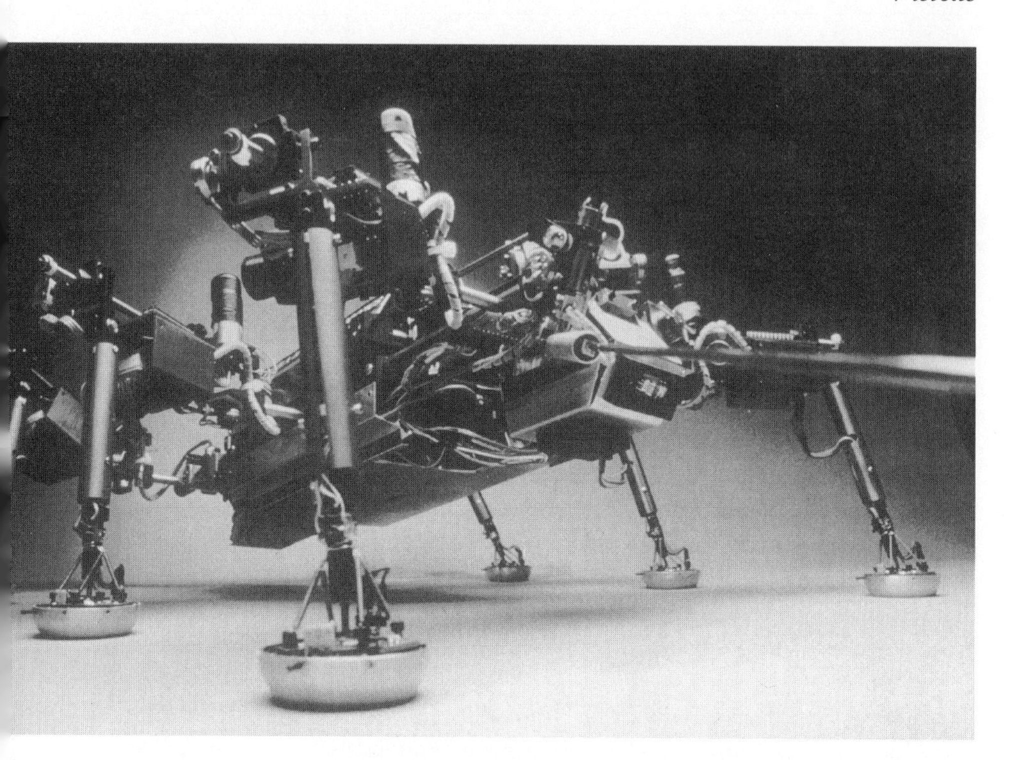

Attila and Hannibal (shown here) are insectoids, representing a new approach to artificial intelligence called the "bottom-up" approach—using paradigms found in nature to simulate intelligence. Unlike preprogrammed robots, these are true automatons that can make their own decisions. (© Bruce Frisch)

error, it learns how to move its six legs with the proper coordination, like a real insect. A simple feedback mechanism is all that is necessary for Attila to learn how to crawl all over the AI Lab.

The robots of this new generation are affectionately called "insectoids" and "bugbots."

"Insects have immensely slow computers with just a few hundred thousand neurons, and yet they fly around in real time and avoid stuff," Brooks notes. "Insects must organize their intelligence in some better way that allows them to get around so well, and that started me thinking about how to reorganize a robot's computations so it could get around in the real world in real time."

Evolution has built insects with less brainpower than a standard computer, yet they can outperform all of their mechanical rivals at MIT. Compared with the tiny, fleet-footed insects that have taken over the earth's surface, traditional AI robots are oafish mechanical stumblebums.

Brooks has little use for the monstrously long computer programs that try to mimic the process of "reason" and human thought; his creations, by contrast, have tiny brains and slim, streamlined circuits which learn to do what bugs do best in the real world, such as poke around and bump into their environment.

Already, the pioneering creations of Brooks and his colleagues are sailing into outer space to invade the planet Mars. NASA was so impressed with Brooks's insectoids that it patterned its first Mars Rover (named Sojourner) after Attila.

The Sojourner, sent to Mars aboard a Delta II rocket in December 1996, weighs twenty-two pounds, has six wheels, and can crawl and maneuver over steep craters and rough boulders, with little assistance from ground control. As part of the Mars Pathfinder mission to Mars, the Rover will be the first autonomous land vehicle to roam freely over that desert planet. (Since it takes roughly ten minutes for a radio signal to reach Mars, when Mars is in range at all, guiding a Mars Rover by remote control was out of the question.) Five similar robots are being planned for the future space station.

Brooks's papers, with provocative titles like "Intelligence Without Reason" and "Elephants Don't Play Chess," have ruffled more than a few feathers in the close-knit field of AI. But decades of intense effort at writing chess-playing programs have not given us the slightest insight into why animals like elephants, who cannot play chess, are so successful in the wilderness. In contrast, Brooks's tiny robots are machines that can walk and maneuver in the real world, not the carefully controlled, sterile environments of the standard mobile robot. He makes no pretense that his machines have anything close to "reasoning" ability.

## Cross-Fertilization of the Three Revolutions

This biology-based approach to artificial intelligence is called the *bottom-up* school. The inspiration for this comes not only from insects but also the rich variety of simple structures found throughout biology and physics—e.g., frog eyes, neurons and neural networks, DNA, evolution, and animal brains. And perhaps one of the most bizarre (and promising) approaches comes from the quantum physics of atoms.

The many bottom-up approaches share one feature: they let the machines learn from scratch, the way biological organisms do. Like a newborn baby, they learn from their own experience. This philosophy can be summarized roughly in one phrase: learning is everything; logic and programming are nothing. First, you create a machine that can learn; later, it learns the laws of logic and physics by itself by bumping into the real world.

As I stated earlier, scientific progress in the future will be propelled up by the intense interplay between quantum physics, molecular biology, and computers. After years of stagnation in the field of artificial intelligence, the biomolecular revolution and the quantum revolution are beginning to provide a flood of rich, new models for research.

One of the strangest consequences of this tight interplay between the three revolutions is a sociological one: the migration of theoretical physicists (who normally work on arcane subjects like superstring theory, trying to unify the laws of the physical universe) into brain research. Several of my colleagues with established reputations in quantum gravity and superstring theory are now applying their formidable knowledge of quantum physics to understanding how the brain functions, treating neurons like atoms.

This interplay between the three scientific revolutions is one of the most important factors driving the science of the future, as we will see throughout this book.

Although researchers of the two schools of artificial intelligence sit side by side in the same building, the lines between them are clearly drawn. On one side of the debate are the distinguished founders of artificial intelligence who have spent a lifetime programming mammoth computers to model human intelligence. Their inspiration for a thinking machine was a powerful digital computer—the bigger, the better. Their strategy was dubbed the *top-down* approach; they believed they could program in the logic and reasoning ability necessary for a machine to think. They assumed that thinking machines—like Minerva, the Roman goddess of wisdom, who sprang from Jupiter's forehead fully grown—would emerge fully developed from a computer.

Their recipe for building a thinking machine was simple: First pour the complex rules and programming into a digital computer in order to reproduce logic and intelligence, then sprinkle on a few subroutines for speech and vision, attach mechanical hands, legs, and eyes . . . *Voilà!* You would have an intelligent robot. Inside that robot's brain would be a complete representation of the outside world, a detailed manual that described the rules for living in the real world.

Their philosophy was based on the idea that intelligence can be simulated by a "Turing machine," which forms the basis of all digital computers. But the traditionalists soon slipped into a quagmire; they profoundly underestimated the enormity of writing down the complete road map of human intelligence. Their computer-based machines turned out to be pathetic, feeble creatures. The mobile robots built on their approach consumed vast amounts of computer power, yet they were supremely inept: agonizingly slow and timid, they frequently got lost. They were useless in the real world.

Thomas Dean of Brown University admits that the lumbering mobile robots built along these lines are quite primitive. His machines, he says, are "just at the stage where they're robust enough to walk down the hall without leaving huge gouges in the plaster."

Furthermore, the practitioners of the top-down approach to AI, according to AI pioneer Herbert Simon, have often shot themselves in the foot by making outrageous claims. MIT's Berthold Horn tells the story of an AI conference in Boston where reporters were buzzing around one scientist who was claiming that in five years robots will pick up things left on the floor. He recalled that he dragged the scientist into a corner and told him, " 'Don't make these predictions! People have done this before and gotten into trouble. You're underestimating the time it will take.' He said, 'I don't care. Notice that all the dates I've chosen were after my retirement date!' I said, 'Well, I won't be retired and people will come back and ask me why they don't have robots picking up socks in their bedrooms!' "

Amidst the ruins of the top-down approach, many felt the time was ripe to start all over from scratch. Brooks's insect machines, based on a bottom-up approach, in comparison are downright retarded, but, after a period of trial and error, manage to crawl successfully across rugged landscapes, effortlessly avoiding obstacles and zipping past the competition.

The researchers in the bottom-up school see their creations as upstart mammals: fast, nimble creatures which can take over when the lumbering computer dinosaurs perish. While the top-down school drowns in millions of lines of computer codes, they boast that the sleek, efficient brains of the bottom-up school will conquer the world.

Although relations between the two schools are cordial, this does not disguise the fact that Brooks and his colleagues in the bottom-up school are considered heretics by some in the AI community. They have thumbed their noses, rhetorically speaking, at computer-based machines by adopting paradigms borrowed from biology and evolution.

Marvin Minsky, co-founder of the lab, issuing a broadside against the bottom-up school, says, "Why bother building a robot that's capable of getting from here to there, if once it gets there it can't tell the difference between a table and a cup of coffee?"

Brooks fires back, "I get very frustrated when people say to me, Yeah, but your robots don't do such and such. Well, of course they don't. Chess-playing programs don't climb mountains, either."

With these disagreements, one would think that the lab would be paralyzed. Actually, the diversity is tolerated, even encouraged.

"I think it's great that everyone is fighting and disagreeing," muses lab director Patrick Winston. "They're making things very interesting again, just like it was in the early days."

Tomas Lozano-Perez, the lab's associate director, agrees. "Complete agreement is a sign of rigor mortis," he notes.

Ultimately, the final resolution of this split may come from a merger of these two schools in the twenty-first century. AI pioneers like Hans Moravec of Carnegie-Mellon University believe that the final step in artificial intelligence may ultimately lie in a sophisticated synthesis or blend of both schools. "Fully intelligent machines will result when the metaphorical golden spike is driven uniting the two efforts," he says, predicting that this union will take place in about forty years.

An eventual merger between these two opposing schools in the middle of the next century is probably the most reasonable estimate of the future of AI. Both schools have distinct advantages and disadvantages. Humans, after all, combine the best of both schools; not only do we learn from bumping into the real world, we also absorb certain data by sheer memorization, as well as having certain circuits "hard-wired" into our brains. Whether we are learning music, a foreign language, a new dance step, or higher mathematics, our brains use a combination of trial-and-error learning as well as memorization of rules.

## Preprogrammed Robots

Given the rather primitive state of artificial intelligence, it may be twenty-five years or more before we see any of the creations of the MIT AI Laboratory enter the marketplace. Instead, from now to the year 2020, what may gradually gain acceptance in the market are increasingly sophisticated industrial robots which are either preprogrammed or remote-controlled.

In the period from 2020 to 2050 we are likely to enter the "fourth phase" of computing, when intelligent automatons begin to walk the earth and to populate the Internet. During this period we may finally see the synthesis of the top-down and bottom-up schools, giving us true robots with common sense which can learn, move, and interact intelligently with humans. Beyond the year 2050 we are likely to enter the "fifth phase" of computing, with the beginnings of robots with consciousness and self-awareness.

To more fully appreciate the importance of these developments, it is necessary to distinguish between MIT's robots, which are true *automatons* which can act independently, and the industrial robots on Detroit's automobile assembly lines, which are preprogrammed. Preprogrammed robots possess the "intelligence" of simple windup toys, music boxes, and mechanical pianos. These industrial robots obey instructions written on computer disks and chips; otherwise these robots are largely identical to overgrown toys. Every movement has to be tediously scripted and spelled out. Disney Studios, for example, has produced a series of remarkable robots that can sing, dance, gesture, even tell jokes, often significantly better than the average human. But although they can execute sophisticated humanlike movements, they are in essence just clever preprogrammed windup toys which are carefully scripted ahead of time.

Preprogrammed and remote-controlled robots are already being used to carry out extremely dangerous missions. Rover 1 was used to repair the damaged Three Mile Island reactor in 1979 after it came within thirty minutes of a full-scale meltdown. Jason Jr., a robot submarine, took historic photos of the wreckage of the ocean liner *Titanic* rusting on the bottom of the Atlantic Ocean in 1986. Lunokhod, a Russian "dune buggy," landed on the moon and roamed over its craters under remote control.

Because of the explosion in computer power, by 2020 we should see increasingly sophisticated preprogrammed robots become commercially available and entering our homes, hospitals, and offices. Some robots already on the market include HelpMate, a four-foot medical robot used at the Danbury Hospital in Connecticut to fetch drugs and equipment for doctors and nurses, following a map of the hospital lodged in its memory. It is operated by punching commands on a keypad. Eventually, medical robots could reduce the skyrocketing cost of health care for the aged. Robo-Surgeon, a medical robot at the Long Beach, California, Memorial Medical Center used in brain surgery, can drill a precision hole in the human skull within a thousandth of an inch. It resembles a large mechanical arm, with a removable scalpel or needle at the end. Sentry was a 485-pound robot which acted as a security guard. Denning Mobile Robotics, Inc., used to sell Sentry for $50,000. Looking like R2D2 in the movie *Star Wars*, it resembled a fifty-five-gallon drum on wheels. As long as it conducted its patrols repetitively along the same path, moving at five miles per hour, it worked fine, and even thwarted a burglary at Boston's Bayside Exposition Center.

Hans Moravec believes that these clumsy robots will eventually evolve into more sophisticated robots, roughly according to the following time schedule.

From 2000 to 2010, these robots will increasingly develop into reliable helpers, able to navigate in factories, hospitals, and the home and perform well-defined functions. He calls such robots "Volks-robots." They will mow lawns, act as butlers, perform car tune-ups, perhaps even cook gourmet meals.

From 2010 to 2020, these robots will begin to be replaced by machines that can learn from their mistakes. Although clumsy at first, they will learn from their constant interactions with humans. They may even possess a primitive "pain" and "pleasure" system to reinforce certain positive acts and prohibit others.

## From 2020 to 2050: Robotics and the Brain

An area where the top-down approach has been less successful is the field of *robotics*, which studies mobile robots that can recognize obstacles and move around them. The first mobile robot, called Shakey, was built in 1969 at the Stanford Research Institute, and resembled a large tin can sitting on wheels. On top of the can were TV cameras, range finders, and a radio antenna to connect it to a remote computer. It could only recognize geometric objects in a carefully controlled environment, and even then it took hours to move across a room. Unfortunately, over the intervening thirty years, not much progress has been made beyond Shakey.

One difficulty that has dogged these mobile robots is the notorious problem of *pattern recognition.* These primitive mobile robots can see, but they can't understand what they are seeing. When their cameras scan a room, they break up the image into thousands of tiny dots, which they have to tediously compare, dot for dot, with the images stored in their memory, which can take anywhere from hours to days. Driving a car, which requires recognizing an ever-changing landscape, is out of the question for the most powerful robot. Recognizing faces is a particularly difficult problem. Computers have great difficulty recognizing a familiar human face if it is rotated by even a few degrees.

Yet our human brain can recognize new surroundings and can identify a single face out of thousands, all within a fraction of a second.

The three and a half pounds of neurons sitting on our shoulders is perhaps the most complex object within the solar system, perhaps even in this sector of the galaxy. Although we can hold the brain in our hands, take it apart neuron by neuron, we have only the most primitive understanding of how it works.

Scientists have been fascinated by the fact that the brain consists of several layers, reflecting the gradual progression of our evolution.

Since nature is frugal, usually recycling lower forms into higher ones, rather than destroying the older form, our own brain serves somewhat as a museum preserving its own evolutionary history. As a consequence, our brain consists of several distinct and concentric layers, starting with the most primitive layer and having successive and more advanced layers surrounding previous ones.

The first and deepest layer of the brain is what biologist Paul MacLean has called the "neural chassis," which controls the basic life functions, such as respiration, the heartbeat, and blood circulation. It consists of the spinal cord, brain stem (medulla and pons), and midbrain. In fish, the neural chassis makes up most of the brain.

Surrounding the neural chassis is the R-complex (olfactostriatum, corpus striatum, and globus pallidus), which controls aggressive behavior, territoriality, social hierarchies. This layer is found in reptiles, and is sometimes called the "reptilian brain."

Surrounding this is the limbic system (thalamus, hypothalamus, amygdala, pituitary, hippocampus), which is found in mammals. It controls mainly emotions and social behavior, but also smell and memories. As mammals evolved complex social relations for survival, a greater part of the brain was required to handle the problems and dynamics of living in a cohesive group.

And lastly, surrounding all the previous layers, is the neocortex (frontal, parietal, temporal, occipital lobes), which controls reason, language, spatial perception, among other functions. In contrast to other animal brains, which are quite smooth, we have pronounced wrinkles on the surface of our brain, which increases the surface area of the cerebral cortex.

From this perspective, we can see that our present-day robots are still in the most primitive phase, possessing only the neural chassis. Our robots have yet to evolve any social hierarchies, emotions, socialization skills, or cognitive skills that typify animals more complex than fish. But from this, we can appreciate the complexity of the animal brain and how far we must still advance before we approach the abilities of the human brain.

One quantum physicist fascinated by the architecture of the human brain is Miguel Virasoro, recently named director of the famed International Center for Theoretical Physics at Trieste, Italy, operated in part by the United Nations. Originally, Virasoro made an international reputation for himself in superstring theory—the fundamental symmetry of strings is called the Virasoro algebra, in his honor. Virasoro, however, is one of many quantum physicists whose fascination with artificial intelligence has taken him to neural networks and brain theory.

Virasoro believes that the power of microchips will one day approach the raw computing power of the human brain. But does that mean that the brain is a computer? he asks. Our computers have already exceeded or matched the computational power of certain animal brains. A typical

SUN-4 computer can process information at the rate of about 200 million bits per second. In speed alone, that matches the ability of a snail's brain, which contains 100,000 neurons. The Cray-3, one of the fastest computers on earth, can process information at the rate of 100 billion bits per second, which is comparable to the brain of a rat, which contains about 65 million neurons.

By comparison, some scientists estimate that the human brain can calculate at the rate of 100 trillion bits per second, or about a thousand times faster than the Cray-3. Since computing power doubles every eighteen months, it is possible, barring an interruption as the Silicon Age comes to a close, to derive a mathematical estimate of the time when computers will overtake the raw calculational power of the human brain. If current trends continue, we should be able to build computers which are as fast as the human brain and contain as much information by early in the next century, perhaps between 2010 and 2030. By 2040, even desktop computers will have the computing ability of a human brain.

In 1996 the Department of Energy awarded a $93 million contract to IBM to build the world's fastest computer by 1998, a computer that will handle 3 trillion operations per second and will process 2.5 trillion bytes of information—within striking distance of the power of the human brain.

Virasoro's fundamental objection to this top-down approach to the brain, however, is that *the brain is not a Turing machine;* in fact, it's not a computer at all. Creating faster and faster computers in the hopes of duplicating the human brain is a wild-goose chase.

To see this, it is necessary to understand how the brain is wired up. There are about 200 billion neurons in the brain—or about the number of stars in the Milky Way galaxy. They fire perhaps 10 million billion times per second. Although nerve impulses travel at an excruciatingly slow rate of 300 feet per second (or 200 miles per hour), the brain makes up for this by the vast complexity of its parallel connections.

Virasoro points out that each neuron is connected to about 10,000 other neurons, and hence the brain functions as a *parallel processor*, carrying out trillions of operations simultaneously per second. Yet it only consumes about the energy of an ordinary lightbulb. To appreciate its efficiency: if one could somehow build a standard computer as powerful as the human brain, it would consume about 100 megawatts, enough to power an entire town.

Although computers can calculate at nearly the speed of light, they perform calculations one at a time. The brain, in comparison, calculates at a snail's pace, but makes up for this by performing trillions of operations simultaneously. As a result of the way it functions, large portions of the brain can be destroyed by a stroke and yet the brain still can function and even regain some lost function. By contrast, a Turing machine can be completely destroyed by the loss of even a single transistor. The brain is thus very fault-tolerant. To Virasoro, the brain is actually an extremely complex neural network, which is one of the foundations of the bottom-up school of AI.

## Talking Robots

Aoooeeehiooaaaaa! A low, almost inhuman howl fills the room.

Like a proud father listening to his child say "papa" for the first time, Terry Sejnowski, a young professor working on neural network theory, smiles with deep satisfaction. The eerie guttural sound, almost a wail, comes from his machine, NETalk, a neural network he created one summer at Johns Hopkins University that has made history, a neural network that can learn to pronounce the English language almost from scratch.

Sejnowski rejected the usual top-down approach to reproducing human speech. He threw out the fat dictionaries of pronunciation and programs brimming with the rules of phonetics and the tedious list of exceptions to all the previous rules, which had no rhyme or reason. Instead, he replaced all this with a surprisingly simple neural circuit. Miraculously, NETalk learns to speak English the way we do, from trial and error alone. No programs, no dictionaries, no rules, no rules for exceptions—just the ability to learn from its mistakes.

Sejnowski begins a typical demonstration by giving NETalk a tape recording of a text (usually a child's essay of about 100 words). NETalk begins by randomly trying to read the text. Then it applies "Hebb's rule." Each time it "reads" the text, it compares its almost pathetic effort with the text and makes small adjustments in its neural net. Each neural connection which comes closer to the correct pronunciation is strengthened. With each adjustment, NETalk gets closer to the text.

In this way, NETalk mimics the way children learn how to pronounce words. Psychologists, who have placed tape recorders next to infants when they are alone at night before they go to sleep, have long known that they will endlessly repeat the sound of certain words to themselves, until they slowly perfect each word. With each trial, the child gets a bit closer to the correct pronunciation.

Sejnowski explains how NETalk begins to learn: "The first thing it discovers is the distinction between vowels and consonants. But it doesn't know which is which, so it just puts in any vowel or consonant. It babbles."

A neural network such as NETalk is a collection of electronic neurons which mimic the behavior of the brain. Each time a neural network makes a correct choice, the circuits are reinforced by changing the "weights" of each neuron. Each time it makes an error, the connections are deemphasized. After a few hours of this painfully slow process, one can detect unmistakable progress toward the correct pronunciation.

"Hear the difference?" Sejnowski says excitedly. "Now it has discovered spaces, the distinction between words. So it speaks in bursts of sounds, pseudowords."

After about a day, the progress is astonishing. Overnight, NETalk can read the text with 98 percent accuracy at the level of a third grader. After sixteen hours, with uncanny accuracy it was able to read the words "I walk home with some friends from school. I like to go to my grandmother's house. Because she gives us candy."

Of course, neural networks still have a long way to go before they can model the human brain. As physicist Heinz Pagels has said: "The difference between a real neuron and the model neurons . . . is like the difference between a human hand and a pair of pliers." But the fact that a simple neural network can speak at all is remarkable, indicating that perhaps human abilities can be simulated by electronics.

## Robotics Meets Quantum Physics

Sejnowski is part of a recent migration of quantum physicists who have found a rich new field of investigation: using the laws of the quantum theory to probe the secrets of the brain.

Brain research, of course, is vastly different from pure theoretical physics. In physics, the goal is to find the simplest, most elegant solution to the most fundamental problems, such as the Big Bang and the unified field theory. Biology, however, is messy, inelegant, full of dead ends, and the brain represents the end product of all these detours. While physics is based on "universal" laws, the only universal law recognized in biology is the law of evolution, with all its twists and accidents.

Sejnowski remarks, "A lot of the details and organizational decisions in biology are historical accidents. You can't just assume that nature took the simplest and most direct route to do something. Some features are remnants of some earlier stage of evolution, or it may be that some genes that happen to be around are commandeered for some other purpose."

In designing NETalk, Sejnowski is following in the footsteps of John Hopfield, the quantum physicist who helped to break open the field of neural networks in 1982 and the current tidal wave of interest in neural network theory after decades of neglect.

Tall, handsome, and nattily dressed, John Hopfield looks more like a distinguished college president or CEO than a solid-state physicist with his head buried in a mountain of arcane tables listing the properties of crystals, metals, magnets, and semiconductors.

In the late 1970s, Hopfield began to attend seminars on neuroscience twice a year at MIT. After a while, he began to realize that the field of artificial intelligence had few, if any, organizing principles. It was a loose hodgepodge of interesting but disjointed tidbits of knowledge. Hopfield began to ask himself if there are any fundamental principles behind AI, as there are in physics.

In solid-state physics, where the atoms are tightly bound in a lattice structure, there are simple organizing principles given by the quantum theory. Hopfield, for example, was studying the properties of spin glasses, which are composed of arrays of spinning atoms. Hopfield asked himself whether the array of atoms found in a solid are similar to the neurons in the brain. *Can a neuron in the brain be treated like an atom in a lattice?* This led to the publication of a celebrated 1982 paper, "Neural Networks and Physical Systems with Emergent Collective Computational Abilities."

This was a truly revolutionary idea, representing a leap of logic that caught both the world of AI and quantum physics by surprise. Previously, the top-down school held that the "mind" was an incredibly complicated program inserted into a large computer. Hopfield was suggesting that intelligence might arise from the quantum theory of mindless atoms, without any programs whatsoever!

"One of the side effects of Hopfield's work was that many theoretical physicists working on spin glasses became overnight experts on the properties of neural nets. Some of them, like Hopfield, switched fields," noted physicist Heinz Pagels.

The idea, Hopfield pointed out in his pathbreaking paper, is not as preposterous as it sounds. Each atom in a solid is spinning and can, for example, exist in a few discrete states, such as spin up or down. Similarly, the neuron also exists in discrete states: it can fire or not fire. In a quantum solid, there is a universal principle that determines which state the system prefers—i.e., the atoms arrange themselves so that the energy is minimized. Hopfield's idea boiled down to this: like the quantum solid which minimizes its energy, a neural network circuit must also minimize its "energy."

This was Hopfield's breakthrough. Before Hopfield, there was no unifying principle which allowed one to understand neural networks. Hopfield, using the broad principles of the quantum theory, found the unifying principle behind the neural network: all the neurons in the brain would fire in such a way as to minimize the "energy" of the net. "Learning" is the process of finding the lowest energy.

As Brown University's Jim Anderson says: "We always knew that neural nets worked, but Hopfield showed why they work. That was really important, because it gave us legitimacy."

As a result, an entirely new world of research opened up, and physicists became part of a new vanguard of neural network research. Like British mathematician Alan Turing, who captured the mathematical essence of the universal computing machine, Hopfield had discovered one of the universal laws behind neural nets. This in turn helped to lead to the current revival in neural network theory.

The essential idea behind Hopfield's breakthrough is easy to visualize. Consider a ball rolling down a hilly terrain, full of rifts, valleys, and mountains. The ball, of course, will spiral down into one of the valleys. The ball, in other words, seeks the state of the lowest gravitational energy (a valley). Now imagine that the hilly terrain represents all the possible states of the neurons in a brain. Each point in the terrain represents a certain setting of the weights of the neural net. (The terrain exists in N-dimensional space.) Each time the ball rolls, the weights of the neural net change, such that the ball rolls toward a state of minimum energy. A rolling ball is therefore a metaphor for the complex process of learning. Although the mathematics of a neural network can be fiendishly difficult, Hopfield showed that the essential mathematical picture was no more difficult than a ball rolling down a hill!

Hopfield went on to find that his neural nets exhibited unexpected behavior which mimicked actual brain functions. He found, for example, that even after the removal of many neurons, the neural network behaved pretty much the same; the geometry of the valleys did not change. In other words, the valleys corresponded to "memories." Like actual memories in the brain, which can persist even after the loss of millions of brain cells, these valleys within the neural net were quite stable even after being partly destroyed. These valleys or memories, instead of being localized in one place in the brain, were spread out over the entire system.

Another by-product of this model was that it gave an interpretation for obsessions. Sometimes, if you weren't careful in preparing a neural net, a particular valley might become so large that it ate up all the neighboring valleys. Then the ball would inevitably fall into this gaping hole. This may be just what happens in the case of an obsession.

But the strangest by-product of this simple but seminal idea was totally unexpected. He found that his neural nets began to dream!

### What Are Dreams?

What causes dreams? Mystics once thought they were omens foretelling future events. To Sigmund Freud, dreams were a window to the unconscious mind, representing fragments of repressed desires. Through dreams, Freud thought, he could probe the hidden recesses of the libido and the id.

Today, there are as many theories about dreaming as there are schools of psychology. Not one of them, however, can produce convincing empirical evidence in its favor.

Psychologists have found that dreams are essential to our emotional well-being; if we are interrupted each time we begin to dream, we become increasingly irrational and unstable, even if we are allowed to sleep for hours. (We can interrupt a sleeping individual precisely as he or she enters a dream state by monitoring the eyes and brain for REM—rapid eye movement—and alpha waves on an electroencephalogram. In this way, we have determined that certain mammals probably dream as well.)

To Hopfield, dreams are fluctuating energy states in a quantum mechanical system. Hopfield discovered that his neural networks reproduced many of the properties of dreams identified long ago by psychologists, who found that we need to sleep and dream after a series of exhausting experiences. He found that if he filled a neural net with too many memories (i.e., valleys), then the system began to malfunction from overload—i.e., the amount of time it took to access different memories began to become increasingly unequal. It began to malfunction in recalling previously learned memories. In fact, unwanted ripples began to form on the surface of the terrain that did not correspond to any real memories at all. These ripples are called "spurious memories" and correspond to dreams. Unlike real valleys, they do not represent real events, but are composed of fragments of existing memories.

In order to eliminate these spurious memories, he would add a small disturbance to the system, abruptly changing the terrain (so the ball would be thrown out of a valley and would roll once again). The system was then allowed to settle down again into a state of deep energy minimization. Hopfield says this corresponds to sleep.

After several episodes of dreaming and sleeping, the system "awakened" refreshed—i.e., it stopped malfunctioning and could recall all its memories at the same rate.

If Hopfield is right, then perhaps all highly developed neural nets, mechanical or organic, must dream in order to process their memories. Whenever a neural net is overloaded, it necessarily begins to act abnormally, creating memories that are not real—i.e., dreams consisting of random fragments of real memories. The system sleeps to cleanse itself of these fake ripples or dreams.

Hopfield thinks that these spurious memories may be intimately tied to the creative process of the brain. He notes, "If you want to have a new behavior, what you'd call *originality*, this is a way to generate it."

Yet another quantum physicist who has jumped onto the neural network bandwagon is Nobel Laureate Leon Cooper of Brown University, who is the founder of Nestor, Inc., a Rhode Island business that markets neural network devices. Cooper points out that the usual rule-based top-down approach is too clumsy for performing tasks like recognizing handwritten numbers on credit card receipts. "It's not that you couldn't build it. But it would be like building a car that runs on four feet—it just wouldn't make sense," he claims.

One of the first commercial applications of neural network theory is a bomb detector for airlines that can seek out certain chemicals, like plastic explosives, which are usually invisible to X-rays. Luggage is first flooded with neutron radiation, which is absorbed by the explosive. When the explosive then emits a distinctive gamma ray, the neural network machine can recognize that pattern and sound an alarm.

In contrast to the traditional top-down computers, you do not program these machines. "You train the system rather than program it," sums up Barbara Yoon, program manager for artificial neural network technology at the Defense Advanced Research Projects Agency.

Other promising applications being analyzed for future commercial use include:

- identifying handwriting
- spotting fraudulent credit card charges by knowing your spending habits
- recognizing patterns appearing on sonar and radar screens
- analyzing mortgage risks
- identifying patterns in blood cells (already being used to authenticate the pedigrees of horses)

Neural nets are also giving us a new way to attack the stubborn problem of pattern recognition, which is necessary for vision. The current strategy of the bottom-up approach uses simple nature-based models, such as modeling the eye of animals. Instead of tediously comparing the millions of dots contained in a picture with every picture stored in its memory bank, an animal sees by focusing on simpler cues, such as motion, edges, colors, shading, etc.

Frog eyes, for example, are especially keen for detecting abrupt motion, such as that of a fly. It's said that you can capture a frog by standing motionless directly in front of it and then slowly moving your hand toward it, thereby evading the motion-detecting sensors of the frog's brain. What is remarkable about the frog's eye is that the retina alone has the ability to recognize moving objects. The cells in the frog retina have a built-in "bug detector."

Caltech's Carver Mead has scored some impressive successes modeling the frog's retina with "silicon retina," a neural net with photoreceptors that can detect motion, just like frogs. Mead was, in fact, the first person to put a Hopfield neural net on a silicon chip. Using transistors and standard chip-making devices, he crafted a 22 neuron chip that demonstrated Hopfield's ideas. "You just put a lens on it," he says, "and it will 'see.' It can compute how objects are moving. Of course, that's only one of the things your real retina does, but it is an important thing. And it's one of the things you can't do with conventional computer vision systems. *You just can't do it!* There are people putting supercomputers behind television cameras to try to do stuff like this little chip does, and it doesn't work. That's why I did the motion first—it's something they [in the top-down approach] can't come close to."

Another achievement is the duplication of the pattern recognition of a bee's brain. Although its brain contains only a million neurons (about 100,000 times smaller than the human brain), it still can perform operations about a thousand times faster than most of today's computers. Biologists have determined that the bee's brain cells, called VUMmx1, have connections which can be stimulated when the bee encounters sugar or aromas. After foraging among flowering plants, the bee is left with a memory linking the scent of a flower with the reward: nectar. In this way, the bee learns how to determine which flowers yield the most rewards. Terry Sejnowski was able to successfully create a neural network which carried out the same function as the bee's brain. In fact, he found that in a bed of flowers the preferences of his artificial bee were identical to those of real bees.

### Cog

Rodney Brooks of MIT, whom we encountered earlier in this chapter with his insectoid robot Attila, has begun to build his first humanoid machine, an android called Cog, a robot that actually looks a bit like a human.

At first sight, Cog resembles some of the androids appearing in science fiction movies, such as the murderous robot played by Arnold Schwarzenegger in *The Terminator* (after its outer skin is burned off near the end of the movie). Without its skin, one can see all of Cog's delicate mechanical parts. It has miniature motors instead of muscles, metal bars instead of bones, and video cameras instead of eyes. It has one long arm with a large pincers at the end by means of which it can interact with its environment.

About four feet tall, Cog has no feet. "It's a paraplegic," Brooks admits. Although it lacks legs, it can execute most of the physical motions of the human trunk, head, and arms.

When you turn Cog on in the morning, it moves its head and arm around, as if it were yawning. (Actually, it is merely locating the position of its head and arms.)

The "brain" behind Cog is a collection of eight 32-bit 16MHz Motorola 68332 microprocessors modified to form a neural net, patterned after the way the neurons in our own brain are wired. Ultimately, Cog will have considerable brainpower when his circuits are augmented to include 239 microprocessors. Because it is not a Turing machine in the usual sense, Cog is not programmed. Like all bottom-up machines, Cog learns the way a child learns.

Newborn babies, for example, are such a tabula rasa that only by biting and bumping into things do they finally realize that their limbs are actually connected to their body. By flailing about, babies slowly begin to be aware of the three-dimensional world that lies in front of them. Then, later, once the babies understand the objects in the surrounding world, they learn through interactions with humans.

Similarly, in the first phase, Cog is trained to grasp objects, which is one of the first responses of a baby. By tedious trial and error, Cog learns to move its arms until it can reach out and touch objects. Eventually, it learns how to grasp and hold on to them. In this way, it develops its own "world map," rather than having it programmed in from the start.

Cog interacts with a human in the same way that a baby learns from its mother. Thus Cog has to be taught to recognize a human. Cog also has to learn to make eye contact. (In fact, Cog's eyes were designed so that humans can easily make eye contact with Cog.) Via eye contact, the "mother" can teach Cog increasingly difficult tasks. For example, Cog will eventually learn by "taking turns." After the mother performs a task, she will make eye contact with Cog, so Cog will know that it's his turn. After Cog performs the act, Cog will make eye contact with the mother. This exchange of eye contact will then be repeated, until Cog learns the task. So far, Cog is still at the experimental stage. It does not even have the capability of a two-year-old.

Conceptually, Cog is the exact opposite of Douglas Lenat's Cyc. While

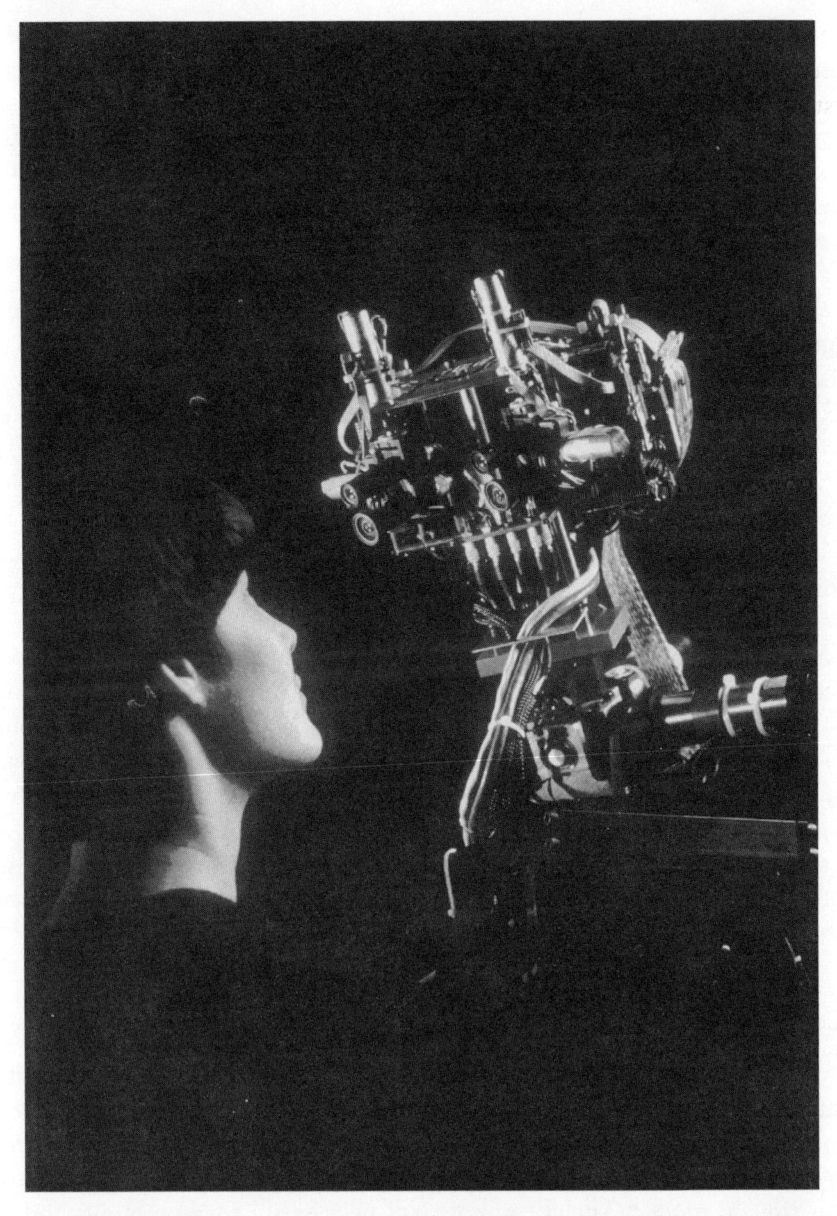

Cog, the android, learns the way a baby does. Instead of having information fed into its brain, it learns by interacting with its environment and its "mother." (© Sam Ogden / Boston)

Cog is the ultimate in bottom-up androids, Cyc is the ultimate in top-down preprogrammed common-sense machines. (There is also some friendly rivalry between these two diametrically opposite approaches.

Brooks even toyed with the idea of calling his android Psych! [pronounced the same as Cyc] just to get Lenat's goat.)

It is likely that androids such as Cog and top-down common-sense programs such as Cyc will be primarily experimental in nature until sometime in the middle of the twenty-first century. Ultimately, there will be a merger of the top-down and bottom-up schools, perhaps within forty years, before we arrive at a genuine free-thinking robot.

This process, according to Hans Moravec, may take place in stages.

From 2020 to 2030, robots will excel at imagery—i.e., being able to simulate a task in its head before carrying it out. These robots will be able to model the world and anticipate the consequences of their actions. For example, before cooking a meal or crossing the street, a robot will be able to simulate the future possibilities many times in its head before acting. To accomplish this imagery, the robot must combine the strengths of the bottom-up approach, which is good at interacting with the real world, with the top-down school, which is good at creating abstract models of the world. By this time, robots should have the intelligence of a monkey.

From 2030 to 2040, a true synthesis may occur, the culmination of both schools. Robots with true reasoning capabilities may well be achieved by this time, according to Moravec. Given the astronomical rate at which computer power increases, the top-down school should be able to create robots which far surpass human reasoning abilities. The final unification will come when scientists can combine this superhuman reasoning capability with the ability to navigate and function in a real or a simulated world. By finally merging these two powerful functions, Moravec believes, "the combination will create a being that in some ways resembles us, but in others is like nothing the world has seen before."

## Can Robots Feel?

It is reasonable to assume that by 2050 we may have robots that can interface intelligently with humans, machines with primitive emotions, speech recognition, and common sense. In other words, we will be able to talk to them and have relatively interesting conversations. In order to function in modern society, robots will necessarily have emotions and a certain amount of common sense, if only because humans will find it easier to interact with them; this may increase "bonding" with a robot.

At the very least, however, robots must be able to understand and cope with the emotions of their bosses and clients. Mechanical butlers must be able to screen pesky visitors and annoying clients for the owner and politely refuse their requests, or make up white lies; mechanical secretaries must be able to spot which appointments are critical and which are not;

mechanical servants must be able to determine when their masters are acting irrationally or overreacting to a situation. Not only will robots have to anticipate the idiosyncrasies of their masters; they will also have to make value judgments as to what is best for them.

This, of course, flies in the face of the typical stereotype of the robot in Hollywood movies, robots with flat, monotone voices, incapable of feeling the joy of first love, the beauty of a blazing sunset, or the wonder upon gazing at the infinite heavens. Some people say that robots are, after all, clumps of wires and steel. It is emotions which distinguish us from creations of metal. That is why the Tin Man always wanted a heart.

From the point of view of AI, however, while it is difficult to reproduce emotions in a robot, it is not impossible to do so. One purpose of our emotions, from the evolutionary point of view, is to increase our chances of survival by preparing us for action. This was done via focusing certain forms of behavior and by censoring.

A "focuser," for example, is an emotion such as "like." When we say, "I like apples," this narrows the infinite universe of possibilities down to a few choices. It focuses our attention on a few desirable alternatives which increase our chances of survival. Not surprisingly, humans usually "like" a small handful of things which are good for them and enhance their survivability. As Minsky has said: "Liking's job is shutting off alternatives; we ought to understand its role since, unconstrained, it narrows down our universe." "Love" is an even more powerful focuser, since it is important in forming pair bonds between humans, which likely increased our reproductive success.

Moravec envisions providing a robot with the capacity to "love" its master, which would increase its commercial success and acceptability by the owner. He says that "when you bring one into your house, it will understand that you're the person it's there for, and that it had better keep you happy. . . . It will care how you feel about its actions. It will try to please you in an apparently selfless manner because it will get a thrill out of this positive reinforcement. You can interpret that as a kind of love."

"Jealousy" is another human focuser because it directs our attention to potential rivals for our mates. "Anger" is valuable because it warns others of our kind that we really do not "like" something.

"Fear" is a focuser which channels our behavior in a specified (and beneficial) direction. A robot can be programmed to experience "fear" as soon as its batteries begin to run low and there is no power source in sight. Moravec says, "It can't let its batteries run down to nothing . . . it would express agitation, or even panic, with signals that humans can recognize. It would go to the neighbors and ask them to use their plug, saying, 'Please! Please! I need this! It's so important, it's such a small cost! We'll reimburse you!' "

"Laughter" is an entirely different type of emotional reaction which is evolutionarily desirable. Instead of focusing our attention on a few alternatives, laughter acts as a "censor," defining the limits of acceptable behavior, helping to rule out dangerous or forbidden acts. Certain dirty jokes are funny because the punch line surprises us with forbidden outcomes. Laughter is the mechanism which tries to integrate these new, forbidden outcomes into the body of known "censored" forms of behavior. Sex, for example, is vitally important for the survival of the species. Yet because society has developed so many constraints and taboos to control and regulate volatile sexual emotions, there are a great many censored forms of sexual activity that have to be learned by everyone over many years. That's why naive teenagers are among the most avid listeners to dirty jokes.

Once the forbidden form of behavior is incorporated into our body of censored activities, we are no longer surprised by the message of the joke. That's why a joke is no longer funny the second time around.

Even "fun" has an important evolutionary role. Anyone who has ever watched children play has noticed that their games imitate complex adult social interactions. Adult society's rules of acceptable behavior are quite complex, developed over millennia; games capsulize one tiny facet of human society and make it digestible to children. That's why they play "cops and robbers," "doctor," "teacher," and so on.

We are, of course, unconscious of all this. I once asked a child why she was having "fun" playing a game of "teacher," suggesting that perhaps it helped to explain the complicated process of learning at school. She stared at me, as if I was from Mars, and replied authoritatively, "Fun is fun. I'm having fun because it's fun." She looked pleased with herself, as if she had just given me the definitive explanation of fun.

To program a robot to have emotions is difficult, but not impossible. How can it be done? Scientists might assign "weights" or numbers to certain behaviors. When faced with danger, the robot must assign a negative number to the situation and, as a result, avoid it. When faced with a pleasant alternative (e.g., ample sources of power), the robot would assign a positive number and thus pursue it. Then the response must be programmed in (as it is for humans): facial muscles contract for laughter, legs move for flight, arms flex for fight, eyebrows rise for surprise or lower for anger.

Anthropologists, studying the possible emotional states of primates, have discovered that they too use complex gestures, facial expressions, and hand motions to convey their feelings. For example, I once visited a science museum that had a very sophisticated model of the head of a gorilla used in a recent Hollywood movie. By twisting certain levers, I could move specific facial muscles, which would evoke unmistakable facial expressions corresponding to surprise, anger, or happiness. As I moved the levers, the gorilla head sprang to life, as if it were a living, breathing creature expressing genuine feelings. The children who were gathering around squealed and roared with delight as I made the gorilla laugh, giggle, and look goofy. Then I suddenly made the gorilla look murderous with rage, with its teeth bared, eyes narrowed, and nostrils flared; the children all instinctively screamed and fled in terror.

I was shocked that it was so easy to evoke realistic emotions from a piece of synthetic rubber, plastic, and wires—that just the slightest movement of a few facial muscles could generate terror in others.

We have limited control over our emotions because evolution has hard-wired them deep into our brain's limbic system; we respond to emotions viscerally and unconsciously, without thinking. Language may be only a few hundred thousand years old, but body language, especially facial expressions, go back before the dawn of the apes. Facial expressions were one of the dominant modes of communication millions of years before our vocal cords could express language. Creating facial expressions—the outward manifestation of emotions—in robots will not be very difficult. One may still argue that robots with sophisticated facial expressions do not really "feel" or "understand" the emotion they are exhibiting. Their emotions are hollow. So can robots be "aware" of their own being?

## Beyond 2050: Robot Consciousness

By 2050, we expect AI systems to have a modest range of emotions. Intelligent systems will then be truly ubiquitous, animating many of the objects around us and even sharing some of our feelings. By then, the Internet will have evolved into a true Magic Mirror, not only capable of accessing the entire database of human knowledge, but also capable of gossiping or joking with us. (Some AI experts have written that this may inadvertently create a resurgence of interest in magic and superstition. To many, a world populated by intelligent systems may seem, as in medieval times, to be animated by mysterious spirits.)

But the questions arise: Are they "aware" of what they are? Can they set their own goals and plans? Are they "conscious"? Such predictions are, of course, quite controversial, since up to now no one has even given a compelling definition of what consciousness is. Indeed, it seems as if everyone has their own definition of consciousness.

Christian theologians have sometimes defined the "soul" as something independent of the material world which even exists after death. Christian theology, with its elaborate rewards and punishments for sin and promises of an afterlife, is predicated on separating the flesh from the spirit.

Eastern philosophers have raised the "mind" to a state of spiritual awareness. Here, for example, is the fable of three Zen monks viewing a flying banner above a temple.

> The first monk says, "The banner is moving."
> The second monk says, "No, it is the wind which is moving."
> Finally, the third monk says, "It is the mind which is moving."

Eastern religions, in other words, do not seek to separate the body from the mind, but to raise the harmony and unity of the two, to achieve a higher state of consciousness within the material world.

But many of the scientists who have dedicated their lives to building machines that think feel it's only a matter of time before some form of consciousness is captured in the laboratory.

To the scientists in the AI community, it is an article of faith that thinking machines already exist, and they are called "human beings." Some of them believe that neural networks have already produced consciousness, and they point out the human brain as their prime example. Most people who work with neural networks believe that consciousness is an "emergent" phenomenon—i.e., it happens naturally when a system becomes complex enough. In other words, the whole is no longer just the sum of its parts. But to say that consciousness is something that springs out of complexity begs the question. Even the most ardent advocates of this emergent theory admit that the theory says everything and says nothing—it's such a sweeping, grandiose concept that it's of little use in guiding specific new areas of research, generating new ideas, or creating new avenues of investigation. This "emergent" theory of consciousness is more a matter of faith than a strategy for success.

And then there are scientists who claim that the question of consciousness has already been solved. Philosopher Daniel Dennet of Tufts University, in fact, wrote a book with the (perhaps premature) title *Consciousness Explained*.

To Herbert Simon, who won a Nobel Prize in economics but is also an expert on artificial intelligence, thinking is little more than the rules that computer programmers place into their robots. "Is human thinking just heuristics?" asks Simon. "I'd say yes, it is."

To Marvin Minsky, the mystery of consciousness is "trivial," because he feels he's solved it. In his book *The Society of Mind*, he argues that the mind is built up from interactions among many smaller parts, each mindless by itself. In this framework, there is no "seat of consciousness" as was once thought. There is no "little man" hidden in the brain somewhere in which all conscious activity takes place. Consciousness simply arises out of the complex interactions of many nonconscious systems. Minsky adds, "Freud had the best theories so far, next to mine." But he admits, "As far as I know, nobody read the book."

PET scans of the living brain seem to bear Minsky out. By tracing the flickering flashes of light within the brain, corresponding to the consumption of glucose and the release of energy, scientists have seen that consciousness is an ephemeral thing, spread out over many structures within the brain. Consciousness seems more and more like a dance between different competing parts of the brain, but without a master conductor orchestrating the whole process. With all these thoughts and sensations rippling past our brain, we are only left with the illusion that there is a "place" where our soul and consciousness resides.

Others believe that various parts of the brain simultaneously generate different "thoughts" which compete with each other for the brain's attention. Only one thought then "wins" in this competition. Consciousness, in this sense, is not continuous, but just the succession of thoughts that win this contest.

At the other extreme, there are some philosophers who claim that robots will never become conscious. Some of them, such as Colin McGinn of Rutgers University, are dubbed the New Mysterians, who argue that consciousness will never be explained. McGinn claims this feat "is like slugs trying to do Freudian psychoanalysis. They just don't have the conceptual equipment." Roger Penrose, the noted Oxford relativist, uses philosophical arguments taken from the quantum theory to bolster his argument against the possibility of creating consciousness in machines.

The problem with these criticisms is that trying to prove that machines can never become conscious is like trying to prove the nonexistence of unicorns. It can never be done to everyone's satisfaction or rigor. Even if one could show that unicorns, for example, do not exist in most parts of the world, there is always the possibility of finding one in unexpected or unexplored areas. Therefore, to say that thinking machines can never be built has, to me, no scientific content.

Ultimately, whether machines can think can only be resolved when someone builds a thinking machine. Until then, the question is undecidable.

This dilemma was exemplified recently when the Dalai Lama met with scientists at the New York Academy of Sciences to explore the link between science and religion. He was asked if he was familiar with work on artificial intelligence. When he said he was, he was asked if an artificial being was a reincarnated being.

Realizing he had been tricked, the Dalai Lama roared with laughter. He then said, "There, there! When you have such a machine and put it there before me, then we will have this discussion again!"

Physicist Heinz Pagels recalled, "In other words, it was put up or shut up. I was secretly pleased, however, that he shared my view of strict constructivism—you have got to design and build, not just talk about your philosophical fantasies." In other words, the only way to settle the question is to build one.

Many critics of AI, like John Searle, concede that robots may one day successfully simulate thinking but they will still be unaware of what they are thinking. They may exhibit emotions, but really do not "feel" them, in the same way that a CD of Bill Cosby telling a joke will not understand what was so funny. To Searle, robots cannot be conscious, just as simulated thunderstorms can never make anyone wet.

But as Turing stressed decades ago, it is possible to give a perfectly reasonable operational definition of intelligence without opening the Turing box. By analogy, if a robot performs in a way which is indistinguishable from that of a conscious being, then, for all intents and purposes, it is conscious. What is actually happening inside the robot's brain is, to a large degree, irrelevant.

There are probably many degrees of consciousness. In the coming decades, AI scientists will almost certainly, slowly and inexorably, be able to create increasingly sophisticated versions of "conscious" machines. These levels of consciousness will probably be developed in much the same way that evolution produced sentient beings on earth over billions of years. Although there are major gaps in the animal kingdom, there is probably a rough continuum of consciousness, starting with even simple one-celled organisms that later evolved into increasingly more complex ones, including humans. Since humans evolved from less complex forms, it seems reasonable to conclude that there are many levels of consciousness.

Contrary to science fiction stories where a robot suddenly "wakes up" and becomes conscious, in reality scientists will probably create robots over the coming decades which have increasing levels of consciousness.

## Degrees of Consciousness

The lowest level of consciousness is the ability of an organism to monitor its body and its environment. By this definition, even a lowly thermostat has some "consciousness" since it monitors the surrounding temperature. Computers that perform self-diagnostics and that print error messages also fall into this category. Higher up on this same level of consciousness are plants. Even without nervous systems, they have to be aware of numerous shifts in the environment and react to them in sophisticated ways. Machines with vision are on this scale, since they are programmed to recognize various patterns in their immediate environment. Animals at rest function at this level of consciousness. Even relaxing, animals are constantly scanning the environment and identifying patterns for danger, food, mates, etc.

At the second level is the ability to carry out well-defined goals, like survival and reproduction. The future Mars probes scheduled into the next century fall into this category, since they will be mobile and able to scout out unknown terrain, detect danger, seek out interesting formations, all without human commands.

Higher up on this second level lies the entire animal kingdom. Once primary goals (e.g., finding food and mates) are fixed or preprogrammed into the animal brain, they determine the complex plans that the animal must carry out in order to fulfill them. For foxes, it means planning how to hunt and capture rabbits. For rabbits, it means planning how to avoid foxes. These animals have only a limited understanding or awareness of what they are doing when they hunt or flee. Most of their behavior is hard-wired into their brain.

(Remember, this level of consciousness is probably the dominant one for most human activity. Most of us do not spend inordinate amounts of time asking philosophical questions about self-awareness and pondering the paradoxes of the meaning of existence. Although we are reluctant to admit it, we spend most of our time thinking about survival and reproduction, much like the animals. And when we are not thinking about survival and reproduction, we are usually thinking about entertainment and fun. So we shouldn't get carried away about the esoteric and mythical nature of human consciousness.)

The more sophisticated the goal and subsequently the plans necessary to carry them out, the higher the level of consciousness. In other words, there may be thousands of subcategories of consciousness within this broad level, depending on the complexity of the plans that the robot can generate to pursue a well-defined goal.

Predators, such as foxes, for example, are probably more "intelligent" than prey. Foxes have to devise complex hunting strategies to capture rabbits; they have to learn how to hunt with stealth, how to stalk, how to ambush, how to deceive, and they also have to learn the behavior of rabbits. Foxes therefore probably have more developed cognitive skills than rabbits, whose main strategy is to flee. It may take until the middle part of the next century before we have robots that possess the level of consciousness consistent with, say, dogs, who can devise sophisticated strategies for hunting.

The third and highest level of consciousness is the ability to set one's own goals, whatever they may be. Robots able to function at this level are "self-aware." Some scientists believe that we will have a class of robots which can set their own goals, rather than having their goals predetermined, sometime after the year 2050.

But such ability raises other questions: What happens when the goals of our machines and our own goals do not match? What happens when they are superior to us intellectually and physically? These are rather delicate questions that I will address in Chapter 6.

Although pattern recognition and common sense are beyond the capabilities of present-day computers, we can now see the vague outlines of a solution emerging from two fronts: the ever-increasing power of neural nets and conventional computers. A combination of the top-down and bottom-up approaches may one day crack these problems.

Within the next forty years or so, the top-down and bottom-up approaches will likely meet somewhere in the middle, giving us the best of both worlds, a machine that can learn by bumping into its environment and also possesses the expert knowledge of a professional engineer, chemist, doctor, or lawyer. And sometime after 2050, we will likely enter the fifth phase of computing, when we see the arrival of automatons which are conscious and self-aware.

The potential stumbling block to this dream is the wall that computer chip makers will hit as they reach the physical limits of silicon technology. Before we can begin to rival the computer power and memory of the human brain, scientists will have to discover a new architecture for computers. It is a search that has physicists, computer scientists, and engineers scrambling for solutions.

# 5

# Beyond Silicon

## *Cyborgs and the Ultimate Computer*

"All things must pass."
—GEORGE HARRISON

ALEXANDER THE GREAT conquered most of the known world by the time he was twenty-five, organizing a group of isolated Greek settlements and, through fierce military campaigns, carved out an empire. Before his greatest battle, he visited the famed oracle of Amon, which foretold that he would become a world conqueror and attain the power of a god. He died, however, at the age of thirty-three and his empire did not long survive him, falling apart as his generals bickered among themselves.

The microchip conquered the information age in twenty-five years, compressing the power of mainframe computers and enabling that processing power to be placed on everyone's desk. In a few decades, it would become the new engine driving business, industry, science and technology, spawning a lucrative $150 billion semiconductor industry, with 170 million microprocessors stamped out every year.

The question on the minds of physicists and engineers is whether the computer empire created by the microchip will survive its demise. Like the spectacular but short-lived empire of Alexander the Great, the microchip industry could eventually collapse, reduced to bickering among competing designs vying to propel computer processing power forward.

The iron laws of quantum physics are clear: the principle of Moore's law, which, like an oracle, has successfully predicted the growth of microprocessing power, cannot last much longer. Like Alexander, the microchip too shall pass. And relatively quickly. This realization sends a shudder through most computer scientists, some of whom have amassed fabulous fortunes riding the coattails of the microchip.

As we saw in Chapter 2, physicists will soon be pushing the famous "point one" barrier: silicon components cannot be shrunk much below .1 micron in size. Once we reach that scientific limit, entirely new technologies must be introduced to etch ever tinier transistors onto silicon wafers. Components on a microchip will have to be made as small as the coil of a DNA molecule. Sooner or later, the elements of the microchip will become so small that they will reach the size of molecules, where the bizarre laws of quantum physics prevail.

Furthermore, the speed of electricity will be too slow for the computers of the next century. Supercomputers, like the Cray T90, can already perform calculations at the rate of 60 billion calculations per second (60 gigaFLOP per second). In the previous chapter, I mentioned that in 1996 the Department of Energy awarded a $93 million contract to IBM to build the world's fastest supercomputer by 1998, capable of 3 trillion calculations per second (3 terraFLOP per second, with 2.5 trillion bytes of memory). In comparison, it is believed that our brain routinely calculates at 10 terraFLOP speeds or faster, which will be surpassed early in the next century by supercomputers. But this may be getting close to the ultimate limit for ultrafast computers. In a trillionth of a second, electric signals can travel only a tiny fraction of a millimeter, which is too short to reach other components of the computer.

We are able to make rational predictions about the evolution of computer science and technologies through the year 2020 on the basis of Moore's law. In this chapter, I will look at the world beyond 2020, when an entirely new architecture will be required. Some visionaries have written about optical computers, which compute on dancing beams of laser light, and molecular computers, which perform calculations on the atoms themselves. Remarkably, DNA computers have already been built which can solve problems in mathematics faster than supercomputers. Other visionaries talk of the "quantum computer," perhaps the ultimate computing machine.

Still others dream of the distant day when cyborgs will walk the earth, the ultimate merger of humans with their electronic creation. Marvin Minsky of MIT even believes that cyborgs may represent the next stage in human evolution! Then we would achieve true immortality, replacing flesh with steel and silicon.

This debate between competing designs is not an academic one. The future of a multibillion-dollar industry, the jobs of millions of people, the economic fate of entire nations, and the machines driving our future will ultimately depend on the answer.

## Into the Third Dimension

Even by the year 2005, scientists will begin bumping up against the point one barrier. Given the enormous stakes involved, a variety of incremental measures have been exploited to squeeze new life out of the microchip.

Perhaps the simplest way to modify the microprocessor and extend its life is to stack microprocessors into a cube, etching layers of transistors on top of each other. Not only does such a chip have the advantage of packing more transistors into a tiny volume, but the distance electrons must travel is also reduced.

But there are also problems with replacing chips with cubes. Foremost is the enormous heat they generate. In a supercomputer, the heat generated by the surface of a microchip can approach that of a hot skillet, intense enough to melt the chip. Elaborate cooling systems are needed to remove the excess heat.

In a standard microchip, the heat dissipates through the surface. But by stacking microchips on top of each other, the heat dissipation is reduced significantly, as there is less surface area for cooling for a given number of transistors. (This is the famous surface-to-volume problem. If we double the size of a 3-D microchip, the heat generated is proportional to the volume, which goes up by a factor of eight, but the ability to cool the microchip is proportional to the surface area, which goes up only by a factor of four. Thus, it is twice as hard to cool down a 3-D chip if we double its size.)

Heat is generated by these microchips as a result of the electrical resistance of the components. In supercomputers, this heating problem can be partially solved by cooling components with liquid nitrogen or helium. But they are quite expensive and require elaborate cooling systems.

If the heat generated by cubical microprocessors becomes excessive, requiring the use of sophisticated refrigeration systems, then the chips would likely be too clumsy for use in desktop and laptop computers (unless scientists can perfect a room temperature superconductor, as discussed in Chapter 13). Such cubical microprocessors would be confined to supercomputers if heating problems became too severe.

In addition to 3-D "cubelets," here are a few other solutions that have been suggested to squeeze new life out of silicon chip technology:

- Replace silicon with gallium arsenide, which can make circuits up to ten times faster because its crystal lattice structure impedes electrons less than silicon. Such a switch may give a few more years to the microprocessor. Others scientists have proposed using silicon-germanium to replace standard silicon technology.

• Replace laser light beams (which are used to etch components onto silicon wafers) with X-rays, which have smaller wavelengths. One problem is that X-rays are quite energetic. According to Planck's law, the smaller the wavelength of a beam of light, the more energy is packed into that beam. But unlike laser light, X-rays are very penetrating, difficult to work with, and cannot be easily focused. In other words, X-rays can distort the silicon wafer they are supposed to etch. No commercial chip has yet been made with X-rays.

• Use electron beams to etch the silicon chip. But although electron beams can probe increasingly small distances, a fact which is exploited every day in electron microscopes in biology laboratories, they are slow. While light beams can scan entire chips in a flash of light, electron beams must draw each line separately, a process that requires hours, making them uneconomical. Computer experts, however, believe that some form of ingenious X-ray/electron beam technology will be developed around 2005, squeezing new life out of silicon chips and extending their viability until about 2020. IBM, for example, already is experimenting with generating X-rays from an atom smasher (a synchrotron) at its facility in New York.

But as the wires in silicon chips become thinner and thinner, at some point another problem surfaces. With such tiny distances between wires, electrons can leak or "tunnel" across the wire barrier, ruining the logic circuit. There is a limit to silicon technology that, due to the laws of physics, cannot be breached.

This sense of impending doom, although it is still years away, is already creating anxiety among computer experts. "I won't quote anybody, but I was in a meeting where people said that when we get out of optics, we're out of the business," says Karen H. Brown, director of lithography for Sematech, the U.S. consortium for research and development.

## Beyond 2020: Optical Computers

Imagine what New York City or Los Angeles would be like if cars could pass right through each other. Traffic snarls, gridlocks, and pileups would disappear instantly. Rush-hour driving would become a pleasure rather than a medieval torture. That is the potential of optical computers, in which light beams may eventually crisscross each other in an optical cube carrying digital information.

Such optical messages would also be incredibly fast, traveling at the speed of light. And the fact that they generate less heat solves one of the persistent problems with cubical microchips.

In 1990, the scientists at Bell Labs, where the original transistor was invented, created the first prototype of an optical computer. It eliminated wires and transistors in favor of lenses, mirrors, and laser beams. The key to building the optical computer is to find the optical counterpart of the transistor, the heart of any computer. The transistor is simply a valve which regulates the flow of electrons; the scientists at Bell Labs created an optical transistor which regulates the flow of light. It works on the same signaling principle used by navies around the world, sending pulses of light beams by rapidly covering and uncovering a powerful lamp. The optical transistor is called the "S-Seed" (short for "symmetric self-electro-optic effect"); it works on the simple property that light may or may not pass through a filter. (When a voltage is applied to the S-Seed, the filter becomes transparent and the laser light passes through. This is equivalent to 1 in binary code. But if another laser beam is directed at the switch, the S-Seed becomes opaque and shuts off the main laser beam. This is equivalent to 0 in binary code. Thus, a binary message of 1s and 0s on a laser beam, consisting of short pulses of laser light, can be generated by changing the voltage on the S-Seed.)

The original optical computer was embarrassingly crude. Whereas silicon microchips have millions of transistors etched onto a silicon wafer the size of a fingernail, the first optical computer had just 128 optical transistors on a tabletop about three feet across. But one must remember that John Von Neumann's original electronic computers filled up entire rooms with vacuum tubes.

"This work is very significant, because eventually these devices will become the transistors of the twenty-first century," says John Moussouris, a Silicon Valley designer.

The next step in optical computers will be to replace the cables entirely, so they pass freely across each other in three dimensions, carrying millions to billions of instructions per second. And to store the fabulous amounts of data that will be carried by light beams, scientists are contemplating exploiting the power of the most dazzling display of laser light: the hologram.

## Holographic Memory

Holograms are well known for creating remarkably realistic three-dimensional images. One day, TV images in home living rooms may be holographic and three-dimensional. But a much more immediate and important use of holograms may be to store vast amounts of computer data.

A typical CD, for example, can store 640 million bytes of information (equivalent to about 300,000 pages of double-spaced type). Multilevel CDs, which stack several CDs on top of each other, may reach tens of billions of bytes before the year 2000, sufficient to store entire 35 mm motion pictures. But a holographic memory system could store hundreds of billions of bytes of information. The reason has to do with the fact that the wavelength of light is so small. When two beams of laser light are made to interfere with each other, they create tiny whorls in a web of interference lines on photographic emulsion. Astonishing amounts of information can be stored in these interference lines. In fact, the total information presently stored in all the world's computers may one day be stored in a single holographic cube.

Optical computers with holographic memory would be an ideal successor to silicon: they are faster, more powerful, easier to cool, and can store nearly unlimited amounts of information. But optical computers have their disadvantages as well. The issue of miniaturization has to be solved before optical transistors can be made competitive with silicon-based computers.

The key to reducing the size of the next generation of optical computers will be to create truly microscopic lasers and S-Seeds, which can be packed by the millions in a tiny cubical volume. This technology is not far away: the etching process used to carve transistors out of silicon can be used to carve S-Seeds out of gallium arsenide, thereby achieving significantly faster switching speeds, as we will see in Chapter 13. If etching technology can ultimately be adapted to create microscopic lasers, the optical computer will be a strong candidate to replace the silicon microprocessor.

## DNA Computers

One of the most original and unexpected discoveries in recent years is the DNA computer, which may eventually outperform silicon computers on difficult mathematical problems. The DNA computer represents the combined power of the biomolecular and computer revolutions. Leonard Adelman of the University of Southern California has showed that even a tiny test tube of DNA might be able to crack problems that would choke a supercomputer.

DNA molecules are the ideal material for a molecular computer. They are efficient and compact, making up only 0.3 percent of the volume of the nucleus of the cell. And DNA packs over a *hundred trillion times* the information stored in current sophisticated computer devices. In a DNA computer, an astronomical number of DNA molecules stored in a typical test tube (about $10^{20}$ molecules) can all be performing calculations simultaneously.

While silicon chip computers are very fast, they calculate one number at a time and generate lots of heat. DNA computers, on the other hand, while slower, can calculate simultaneously on an astronomical number of molecules, and are a billion times more energy-efficient.

An important point of similarity between silicon and DNA computers is that they are both digital—they are both based on *information.* For computers, this information is encoded in binary code, a series of zeros and ones, which may look like this:

0001110010101001001011110101001001

For DNA, the code is written with four symbols, A, T, C, and G, corresponding to the four nucleic acids which make up DNA. To the naked eye, the DNA code for a human being, if written out, would consist of 3 billion letters appearing as a continuous strand of nonsensical letters:

ATTTCCCGAATCGGTCTGTGAGAGCGCGAAAAAA . . .

Because the DNA code is digital, the information can be manipulated much like a Turing machine. A Turing machine takes an input code, consisting of a string of 0s and 1s, such as 1011100101010000, and performs four operations on it in order to produce an output. One can change a 1 into a 0, a 0 into a 1, move backward or forward one step on the tape. All serial digital computers, no matter how fast or complicated, can be reduced to a humble Turing machine.

Similarly, the DNA molecule consists of a series of four nucleic acids arranged like AACCGTTCCC. One can convert this to standard binary. For example, one can set ATTCG = 1, TCGGA = 0, GATTC = 1. By using a series of complex chemical processes (i.e., using restriction enzymes to cut DNA and the polymerase chain reaction to reproduce DNA sequences) one can duplicate, step for step, all operations of a Turing machine. Starting with a sequence such as AACCGTTCCC, one can perform manipulations to convert it to another DNA sequence. In this way, a DNA Turing machine can be created. A pound of DNA molecules (suspended in about 1,000 quarts of liquid, which would take up about a cubic yard) could store more memory than all the computers ever made. It would have 100 trillion times the capacity of the human brain. Furthermore, just an ounce of DNA could be 100,000 times faster than the nation's fastest supercomputer.

"The floodgates have started to open," says Richard Lipton of Princeton University. "I have never seen a field move so fast."

Ronald Graham of AT&T Bell Laboratories says that it is as if a door opened "to a whole new toy shop."

DNA computers have already proven their worth. A DNA computer was constructed by Adelman which solved a version of the famous Traveling Salesman problem (i.e., calculate the shortest path a salesman must travel in order to connect N cities, such that he visits each city only once; this deceptively simple problem becomes exceedingly difficult as the value of N increases). The DNA computer solved one version of this problem in one week; it would have taken a standard serial computer several years to solve.

One measure of computer power is the ability to break the DES (data encryption standard) code, which was devised by the National Security Agency to safeguard government transactions and is also used by big banks. Hundreds of billions of dollars in corporate records are routinely sent through communication lines via the DES system.

Because a considerable amount of the nation's commerce and military operations is based on the DES, the government has long been curious to know whether the code can be cracked. The DES's sophisticated code is based on a 56-bit number called a "key." (A key is the set of logical instructions used to scramble a message.) The trick is to find the correct key among $2^{56}$ possible keys. A standard computer would take 10,000 years to try each of these keys. The government once thought that this DES code would be secure for the next 10,000 years. DNA computers, however, could change all this. Lipton thinks that it would take as little as "a few months of biological computing" to crack the DES. Dan Boneh of Princeton University agrees; he calculates it would take 907 biological steps to crack the DES, roughly four months of DNA computing time. (International finance will not be plunged into a catastrophic collapse when a DNA computer breaks the DES code. Banks often run their most secret data through a second or even a third DES.)

DNA computers, alas, have their own drawbacks. One is that DNA molecules eventually decay. As a result, one cannot store vast amounts of data on DNA computers for long periods of time. One must transfer the memory ultimately into standard computers.

Second, they are not exceptionally versatile. At present, each problem requires setting up a unique sequence of chemical reactions. Doing another mathematical problem involves preparing an entirely new sequence of chemical reactions. Silicon-based computers, by contrast, are all-purpose devices; the same computer can solve millions of different problems without having to rewire the computer each time.

DNA computers are not likely to replace laptop computers or PCs—they are too bulky and just not versatile enough. Silicon chip technology is much more useful for most everyday applications. However, DNA computers will become superior to mainframe computers in the heavy-duty number crunching when an organization needs sheer firepower to crack a problem.

At present, most computer analysts believe that the DNA computer (and other organic computers, such as protein computers) will be useful for solving specific classes of computer problems now solved by huge mainframe supercomputers. But no matter how powerful these DNA computers become, they always pale beside the ultimate transistor (the quantum transistor) and the ultimate computer (the quantum computer). In fact, the smallest transistors and components are not molecules, but the electrons themselves.

## Beyond 2020: Quantum Transistors

Eventually, all electronic circuits come up against the laws of quantum physics. One of the essential postulates of the quantum theory is that matter can exhibit both wavelike and particlelike characteristics. Electrons at low energies, for example, behave very much like a wave, while high-energy electrons behave like pointlike particles. Because of this dual nature, electrons exhibit bizarre wavelike properties which are counterintuitive. While particles may be blocked by tall barriers, for example, waves can ooze around them. (More precisely, the quantum theory says that the electron is a point particle, but the probability of finding it is given by the square of the Schrödinger wave function. As the electron speeds up, the wavelength of the Schrödinger wave gets smaller, so the probability of finding it peaks around a point. As the electron slows down, the wavelength expands, and the probability of finding it smears out over space. As a result, we cannot precisely locate the position and velocity of the electron, which is encoded in the Heisenberg Uncertainty Principle.)

One of the deepest principles of the quantum theory—the Heisenberg Uncertainty Principle—states that there is a finite probability that seemingly implausible events can happen. Imagine being trapped in a maximum-security prison. Normally, hitting one's head against the massive brick walls will only give one a headache. However, there is a finite probability the atoms of one's head will slip right through the atoms of the brick wall, allowing one to escape the prison. (The probability of such an event is calculable and is so small that the event will not occur within the lifetime of the universe, so the quantum theory is a not practical way to break out of prisons.)

Similarly, electrons are trapped in a prison of their own, a wire. Like the prisoner, they are constantly butting against the walls of the wire, but there is a crucial difference. Both the number of electrons and the number of times they hit against the walls are truly astronomical. Therefore, there is a nonnegligible probability that some will tunnel through the wire, especially if the wire is exceedingly thin. In other words, because wires are beginning to approach the scale of atomic distances, and because of the large number of electrons hitting the walls of these wires, a fraction of the electrons will leak through barriers, thereby making standard logic circuits impossible.

Progress in quantum electronics is now advancing so rapidly that devices are now being manufactured that were once considered impossible a few years ago, devices which manipulate *single electrons*, which may make possible "quantum transistors." So far, scientists have been able to make a "quantum well," a single electron sandwiched between two flat layers. A "quantum line" consists of a single electron confined to a line. And a "quantum dot" consists of a single electron confined to one point in space (usually about 20 nanometers across, about the size of five to ten atoms).

Inside these quantum devices, the single electron can vibrate at distinct frequencies, exhibiting the wavelike property of "resonance." When a violin string vibrates, for example, only certain frequencies (e.g., A, B, C, G, etc.) are allowed to resonate. (When singing in the shower, even someone with a tinny or squeaky voice is able to sing in a voice of operatic proportions because certain frequencies are amplified and resonate between the two shower walls.)

Likewise, a single electron trapped inside a quantum dot will resonate, just like a violin string or your voice in the shower. But certain allowed frequencies can vibrate inside the quantum dot. By changing the voltage on the quantum dot slightly, one can make electrons flow through the dot. This corresponds to the bit 1. If the voltage is raised a bit more, the resonance is destroyed and the current stops flowing. This corresponds to the bit 0. But if the voltage rises again, you hit the next resonance, and current flows once more. In this way, a quantum dot is equivalent to several transistors. By controlling the voltage on the quantum dot, you can create a series of binary messages.

In other words, the world's smallest transistor consists of a single electron trapped within a dot little bigger than an atom, which can mimic the action of not one but many transistors.

Such quantum transistors are no longer dreams of quantum physicists. They have actually been constructed. But because they are so sensitive and difficult to work with, they exist only in the laboratory stage. They will not hit the marketplace for years to come.

Gary Frazier of Texas Instruments says, "No one is ready to provide the million-circuit quantum transistor yet, but the concepts are crystallizing."

This, however, hasn't stopped scientists from speculating about the next and final step: the "ultimate computer," the quantum computer.

## The Ultimate Computer

Quantum computers differ from quantum transistors in that they are totally quantum mechanical devices. While quantum transistors still use conventional wires and circuitry, the quantum computer will replace all this with quantum waves.

One of the first to ponder the possibility of a quantum computer was Nobel Laureate Richard Feynman. In an article in 1981, Feynman asked himself how small computers could become. When computers reached the size of atoms, he reasoned, then they would respond to an entirely new set of laws totally alien to ordinary experience. Feynman was frustrated that many of the fundamental problems of the quantum theory could not be solved by ordinary Turing machines. Many of the objects found in quantum physics require an infinite number of computations, and hence are beyond the capability of ordinary computers. It would take an infinite amount of time on a computer to calculate the interesting questions in quantum physics, such as what happens in a liquid as it begins to boil or what happens when two subatomic particles collide.

Feynman's solution was simple: why not use a quantum computer to solve a quantum problem? His ideas were finally put into concrete form in a paper by David Deutch of Oxford University in 1985. Deutch realized that quantum processes are like gigantic adding machines. The only difference is that quantum computers regularly handle infinite quantities in a blink of an eye. Quantum computers are an entirely different animal from Turing machines. The essential point is this: calculations which take an infinite amount of time on a computer can be processed rapidly on a quantum computer.

To give an example: Imagine walking across Central Park in New York City. In quantum mechanics, to calculate the probability of reaching the other side of the park, you must first add up the contribution of *all* possible paths from one point to the next in Central Park—including the paths which take us to Mars, Jupiter, even past the Andromeda galaxy to the quasars. When all these incredible journeys to the outer reaches of the universe are added up, we obtain the probability that we will walk across Central Park. In other words, the quantum theory is the most ridiculous theory ever proposed in the history of science, flying in the face of all common sense and intuition. The quantum theory opens the door to all sorts of sticky paradoxes which defy all our notions about the universe. The quantum theory has only one thing going for it: it is unquestionably correct. It has survived every experimental challenge hurled at it.

Because the quantum theory sums over all paths between two points, including paths which take us to distant stars, it follows that a quantum computer is one gigantic adding machine, adding an infinite number of paths in a twinkling of an eye.

In this respect, a quantum computer is not a Turing machine—it is fundamentally different from a DNA or a molecular computer (which can process an enormous but only finite amount of information using vast numbers of molecules acting in parallel).

In 1994, there was a flurry of excitement when Peter Shor of AT&T Labs made a breakthrough in quantum computing, showing that if a quantum computer could be built, it could rapidly factorize any number, no matter how long. A quantum computer would have an immediate impact on the worlds of commerce, banking, and espionage. Some of their secret transactions are based on the difficult problem of factorizing a number which can have up to a hundred digits. Since computers factorize large numbers mainly by trial and error, it would normally take decades to solve this problem. But a quantum computer, Shor showed, can easily crack this difficult problem.

To put this into perspective, realize that it took eight months for 1,600 computers from around the world wired up via the Internet to factor a 129-digit number. It would take centuries for this armada of computers to factor a 250-digit number—written out, the reasoning would take up $10^{500}$ lines of paper. To get a sense of how big a number this is, note that there are only $10^{80}$ atoms in the visible universe. In other words, *there are not enough atoms in the visible universe* to allow us to write down the steps necessary to factor a 250-digit number. Yet a quantum computer could perform even this monstrous calculation.

### Beyond 2050

In principle, a quantum computer would be a simple device. Normally, a Turing machine processes a series of bits, given by 1 or 0, written on a tape. A quantum computer replaces this tape with a sequence of atoms. Assume that the atoms in the array are spinning like tops, arranged such that the axis of spin can point in either the "up" or the "down" direction. Scientists say that the atom can be in two states, either spin "up" or spin "down." This gives us a convenient binary code: 0 = spin down, and 1 = spin up. This quantum bit is called a "qubit."

The heart of quantum computing lies in the qubits, which are quite different from bits. In a Turing machine, a bit is either a 1 or a 0. There is no in-between. In a quantum computer, by contrast, the spin of an atom is actually not well defined, but can actually exist as the *sum* of a spin up and spin down state. Thus, a qubit is neither a 1 nor a 0, but a superposition of both simultaneously. (This bizarre feature, that a qubit can exist simultaneously in this never-never land between 1 and 0, means that a quantum computer can perform infinitely more complicated operations than a standard Turing machine.)

When a photon of light is shined on this array, the photon, as it bounces off the atoms, can flip the orientation of a particular atom from spin up to spin down. Now measure the spin of the photon after it has bounced off the array. In principle, the quantum theory has added up all possible paths that the photon could traverse and all possible spin states. But the number of possible states of an array of 1,000 atoms is $2^{1000}$ or roughly 1 with 300 zeros after it. Again, this is much larger than the number of atoms in the visible universe. Thus, a quantum computer can easily manipulate astronomical numbers which would choke a standard Turing machine.

If quantum computers are infinitely more powerful than the largest supercomputers, and if they can crack encryption codes worth hundreds of billions of dollars, why isn't there a crash program to build one?

The problem is that the slightest impurity or contamination from the outside world could disrupt a quantum computer. The computer would have to be isolated from all possible interactions with the outside world, an exceedingly difficult task. In principle, even a single cosmic ray piercing the quantum computer could interfere with the infinite number of calculations it performs. Space probes require "clean rooms" so that even dust particles do not disrupt the delicate gyroscopes. Quantum computers, by comparison, would have to be isolated from even stray subatomic particles.

Progress in this direction is slow, but it is accelerating. David Deutch adds: "Technological progress in this area has absolutely amazed me in the last couple of years. When people asked me this question three or four years ago, I used to say this is a matter of centuries. Now, I'm much more optimistic."

As Seth Lloyd of MIT has said: "It's just hard to string a lot of atoms together. I mean, these things are wickedly small. They're sensitive little buggers too. But people are getting to the point where they can control these things. It's a big technological crapshoot. In the not so distant future, people might be able to do full-blown quantum computation."

Their optimism is based on key developments in two laboratories, which are building some of the components for a quantum computer. The work is being done by Jeff Kimble at Caltech and David Wineland and Chris Monroe at the National Institute of Standards and Technology in Boulder, Colorado. In the NIST experiment, they begin with a series of mercury atoms aligned in a row. Each mercury atom is spinning up or down. As laser light is shined on the row, it can flip a mercury atom spinning down into one spinning up. In principle, the laser light which has bounced off the line of mercury atoms has within it the information for all possible states of the array. The problem is that no one at present knows how to extract usable information from it.

It may be well into the middle of the next century before any substantial progress is made experimentally. But the quantum computer continues to captivate the imagination of computer scientists. It represents, in some sense, the ultimate frontier. Given the rapid advances being made in quantum computers, they may become a reality in the latter half of the twenty-first century.

However, there is another way to approach petaFLOP capability without having to use the bizarre properties of the quantum theory. And that is to exploit a device which already comes close to petaFLOP speeds: our own brain.

## Bionics

Is it possible to interface directly with the brain, to harness its fantastic capability?

Scientists are proceeding to explore this possibility with remarkable speed. The first step in attempting to exploit the human brain is to show that individual neurons can grow and thrive on silicon chips. Then the next step would be to connect silicon chips directly to a living neuron inside an animal, such as a worm. One then has to show that *human* neurons can be connected to a silicon chip. Last (and this is by far the most difficult part), in order to interface directly with the brain, scientists would have to decode the millions of neurons which make up our spinal cord.

In 1995, a big step was taken by a team of biophysicists led by Peter Fromherz at the Max Planck Institute of Biochemistry just outside Munich. They announced that they had successfully created a juncture between a living leech neuron and a silicon chip. In a dramatic breakthrough, scientists have been able to weld "hardware" with "wetware." Their remarkable research has demonstrated that a neuron can fire and send a signal to a silicon chip, and that a silicon chip can make the neuron fire. Their methods should work for human neurons as well.

Of course, neurons are frustratingly thin and delicate, much thinner than a human hair. And the voltages used in experiments would often damage or kill the neurons. To solve the first problem, Fromherz used the neurons from leech ganglia (nerve bundles), which are quite large, about 50 microns across (half the diameter of a human hair). To solve the voltage problem, he brought the leech neurons, using microscopes and computer-controlled micromanipulators, to within 30 microns of a transistor on a chip. By doing so, he was able to induce signals across this 30 micron gap without exchanging any charges whatsoever. (For example, if you vigorously rub a balloon and place it next to running water, the stream of water will bend away from the balloon without ever touching it. Likewise, the neuron never touches the silicon.)

This has paved the way to developing silicon chips that can control the firing of neurons at will, which in turn could control muscle movements.

So far, Fromherz has been able to make as many as sixteen contact points between a chip and a single neuron. His next step is to use the neurons from the hippocampus of rat brains. Although they are much thinner than leech neurons, they live for months, while leech neurons last only for a matter of weeks.

Another step in trying to grow neurons on silicon was achieved in 1996. Richard Potember at Johns Hopkins University succeeded in coaxing the neurons of baby rats to grow on a silicon surface which was painted with certain peptides. These neurons sprouted dendrites and axons, just like ordinary neurons.

The ultimate aim of his group is to grow neurons so their axons and dendrites follow predetermined paths that can create "living circuits" on the silicon surface. If successful, it might allow neurons to conform to the architecture of a logic circuit in a chip.

The doctors at the Harvard Medical School's Massachusetts Eye and Ear Infirmary have already begun taking the next step: getting a team together to build the "bionic eye." The group expects to conduct human studies with computer chips implanted into the human eye within five years. If successful, they may be able to restore vision for the blind in the twenty-first century.

"We have developed the electronics, we have learned how to put a device into the eye without hurting the eye, and we have demonstrated that the materials are biocompatible," says Joseph Rizzo. They are designing an implant consisting of two chips, one of which contains a solar panel. Light striking the solar panel will start up a laser beam, which then hits the second panel and sends a message down the wire to the brain.

A bionic eye would be of enormous help for the blind who have a damaged retina but whose connection to the brain is still intact. Ten million Americans, for example, suffer from macular degeneration, the most common form of blindness among the elderly. Retinitis pigmentosa, an inherited form of blindness, affects another 1.2 million.

Already, studies have shown that damaged cones and rods in animal retinas can be electrically stimulated, creating signals in the visual cortex of the animal's brain. This means that, in principle, it may one day be possible to connect directly to the brain artificial eyes which have greater visual acuity and versatility than our own eye. Our eye is essentially the eye of an ape; it can see only certain colors that apes can see, and cannot see colors which are visible to other animals (for example, bees see ultraviolet radiation from the sun, which is used in their search for flowers). But an artificial eye could be constructed with superhuman capabilities, such as telescopic and microscopic vision, or the ability to see infrared and ultraviolet radiation. Thus at some point it may be possible to develop artificial eyesight that exceeds the capability of normal eyesight.

In the world beyond 2020, we may be able to connect silicon microprocessors with artificial arms, legs, and eyes directly to the human nervous system, which would be of enormous help in aiding people with disabilities. But although it may be possible to connect the human body to a powerful mechanical arm, the stunts we saw on the TV show *The Six Million Dollar Man* would place intolerable stresses on our skeletal system, rendering most superhuman feats impossible. To have superhuman strength would require superhuman skeletal systems that can absorb the shock and stress of such feats.

## Merging Mind and Machine

With all these successes, one can reasonably predict that by 2020 scientists will be able to connect a variety of organs to silicon chips, possibly reactivating paralyzed or inactive body organs. The reason for this optimism is that only a handful of neurons are involved in controlling many body organs, so it should be relatively simple to sort out the wiring of these organs. However, connecting directly to the brain itself poses a whole new set of problems.

The number of neurons in the spinal cord of the body is so large that it is impossible for the foreseeable future to connect even a portion of them to electrodes. It would be like trying to splice into the tangle of telephone lines connecting New York City to the rest of the world without a guide or manual. The brain's wiring is so complex and delicate that a bionic connection with a computer or neural net is something that is, at present, seemingly impossible without causing permanent damage.

At present, our understanding of the brain is quite primitive. We know only in the broadest terms (by analyzing brain-damaged individuals or using PET scans) which areas of the brain are connected to which parts of the body. Scientists only know at the structural level which parts of the brain are involved in which general function. At the cellular level, scientists have no understanding whatsoever of how the wiring is connected. By analogy, consider trying to understand a modern industrialized nation, with its arts, literature, science, commerce, and politics, if we are only given a map of its interstate highway system.

It may be the twenty-second century before scientists begin to understand how the wiring of the brain is connected, let alone be able to tamper with it.

Ralph Merkle of Xerox PARC has made some rudimentary calculations of the time and money it would take to determine how the brain is wired, neuron for neuron. He thinks that it would first be necessary to cut a human brain into millions of thin slices and then analyze each one by means of an electron microscope. Using future automated, computerized image recognition programs, computers could scan these photographs to yield a three-dimensional picture of the brain, revealing how each neuron is connected to all the others. Since there are roughly 200 billion neurons, each connected to 10,000 other neurons, this is a herculean feat. In a sense, it would take another Human Genome Project to determine precisely how the brain is wired. Merkle estimated that the cost, using current technology, would be a staggering $340 billion. But because of Moore's law, it is likely that prices will tumble over the years. He estimates that around 2010 the technology will finally be cheap enough to begin this awesome task. The actual sorting out of the neurons in the brain would then take three years at a cost of $120 million.

But even after possessing a detailed map of the wiring of the brain, one still has to determine how signals move inside the brain and how the various organs are connected to it.

Nonetheless, this has not prevented some individuals from making certain conjectures about mind/machine links, which properly belong in the far future.

## The Distant Future: Growing Cyborgs in the Lab

Some scientists feel the ultimate direction of scientific research would be the merger of all three scientific revolutions in the far future. The quantum theory would provide us with microscopic quantum transistors smaller than a neuron. The computer revolution would give us neural networks as powerful as those found in the brain. And the biomolecular revolution would give us the ability to replace the neural networks of our brain with synthetic ones, thereby giving us a form of immortality.

Evolution has always favored the organism with those adaptations which best enable it to survive. Perhaps a blend of human and mechanical properties could create a species with superior survival possibilities. Humans, according to this line of reasoning, may well be creating the bodies for the next stage of human evolution.

What happens in the distant future when we are able to manipulate individual neurons? Assume, for the moment, that Merkle's idea of mapping every neuron in the brain becomes a reality late in the twenty-first century or beyond. Can we then give our brains immortal bodies?

In his 1988 book, *Mind Children*, Hans Moravec imagines that a bionic merger of this sort between humans and machines will lead to "immortality" of sorts. He envisions humans in the distant future being able to gradually transfer their consciousness from their bodies to a robot, without ever losing consciousness. Each time a tiny clump of neurons is removed, a surgeon will connect it to a clump of neural nets in a metal hull which duplicates the precise firing of the original clump. Fully conscious, the brain could be gradually replaced, piece by piece, by a mechanical mass of electronic neurons. Upon completion, the robot brain will have all the memories and thought patterns of the original person, but will be housed in a mechanical body of silicon and steel which can potentially live on forever.

Of course, the technology necessary to manipulate individual neurons at will, let alone transfer their functions to a neural net, is far beyond anything possible within the next century. But the question is well posed, for if such a scenario is possible, then we may be laying the groundwork for the next step in human evolution.

One person who takes such wild and woolly ideas seriously is AI founder Marvin Minsky. Instead of natural selection providing us with the next step in evolution by trial and error, he believes, the next step will be "unnatural selection," as AI scientists deliberately try to duplicate the human brain, neuron for neuron.

But how will people react when they wake up one day and find that their bodies are made of steel and plastic? When he asked other scientists these questions, he found that they responded by saying, "There are countless things that I want to find out and so many problems I want to solve that I could use many centuries."

"Will robots inherit the earth?" he asks. "Yes, but they will be our children. We owe our minds to the deaths and lives of all the creatures that were ever engaged in the struggle called evolution. Our job is to see that all this work shall not end up in meaningless waste."

□ □ □ □

Clearly, the computer revolution will interact with society in a way that opens up new and exciting possibilities, from petaFLOP and DNA computers to cyborgs. But these are only possibilities, not actualities. In the final analysis, it is up to us to decide among these various choices, given their diverse impacts on our lives, our families, and our jobs. It is we who must decide how much authority we wish to give to our creations. Are we to be masters of the machines, or will the machines become our masters?

# 6

# Second Thoughts

## *Will Humans Become Obsolete?*

"To err is human, but to really foul things up requires a computer."
—*Farmer's Almanac*, 1978

"Computers can solve any problem in the world, except the unemployment they create."
—Anonymous

"[The postbiological world] is a world in which the human race has been swept away by the tide of cultural change, usurped by its own artificial progeny. . . . When that happens, our DNA will find itself out of a job, having lost the evolutionary race to a new kind of competition."
—HANS MORAVEC

THE COMPUTER REVOLUTION evokes two startlingly different visions of the future. The first is a future of prosperity and leisure, a world of instantaneous communication, unlimited knowledge, unparalleled conveniences, and unbounded entertainment. Vibrant new industries will be formed by the computer revolution as Moore's law inexorably increases the power and reach of computers. Scores of new high-tech jobs, called "cyber jobs," will sprout around the country. Already, computers are so vital to the economy that the airline industry, the banks, the insurance companies, and even the federal government would grind to a halt without them.

There is a darker vision, however, that computers could help make possible, as outlined in George Orwell's novel *1984*, a nightmarish world in which a totalitarian government controls and monitors every aspect of our lives. Electronic listening devices could be hidden anywhere, silently monitoring our activities and eavesdropping on our conversations. Society could be controlled by the harsh rule of Big Brother and an army of informers, censors, and spies. History could be rewritten at the whim of a cruel, self-serving bureaucracy which controls the flow of information.

Ironically, we have listening devices today infinitely more powerful and pervasive than anything envisioned by Orwell in his novel. Yet we still enjoy basic democratic freedoms. Rereading *1984*, one is surprised by how primitive the electronic methods described there were, compared with today's devices. Yet the influence of the computer and the Internet have arguably increased, rather than decreased, our freedom of expression and access to information. Many have hailed the Internet as an intrinsically democratic and decentralizing force, weakening the bonds of dictatorships and authoritarian regimes. Oppressive governments are at a disadvantage if information can be dispersed worldwide to a million people with a single keystroke. Nonetheless, there are real dangers. The first is the threat to civil liberties (privacy, censorship, and eavesdropping), which will only get worse in the next century. Each generation of secret codes will stimulate new efforts to crack them. The second danger is the real possibility that the computer revolution will throw tens of millions of people onto the breadlines, skewing the distribution of wealth on this planet. Society may increasingly become a nation of information "haves" and "have-nots." This is already happening on a small scale, and will accelerate into the next century. And late in the twenty-first century, perhaps from 2050 to 2100, there will be a danger that robots may gradually become "self-aware" and hence pose a threat to our existence. Although this notion is entirely speculative, scientists have devoted a fair amount of thinking to the question of how best to control robots as they gradually assume more and more humanlike characteristics.

## Eavesdropping on the Internet

The media is full of lurid stories of computer break-ins, mischief, and outright thievery. So is there an ultimate code that can never be broken, no matter how clever the hacker or government?

In 1918, Gilbert S. Vernam of AT&T proposed the celebrated Vernam cipher, which, in the 1940s, was mathematically proven to be unbreakable. Unfortunately, the Vernam cipher was extremely awkward, and is impractical for most use. (It requires that the sender and receiver both possess a long "key," a set of secret random numbers.) Simplified variations of the Vernam cipher are used today, but it is believed that some of them are, in fact, mathematically breakable.

Within the next twenty years, this situation could change significantly as a new encryption system is adopted, a system not from the world of mathematics, but from the quantum theory. A new area of the quantum theory, called *quantum cryptography*, promises by 2020 to revolutionize the entire concept of computer secrecy. It is here where James Bond meets Werner Heisenberg. How does it work? Whenever somebody listens in on another person's secret conversation, they necessarily disturb it a bit by making an observation. Because some of the information in the original message has been disturbed, the disturbance can be detected.

On one level, this means that observing an object changes its state. For example, one way to tell if a phone is being bugged is to check the voltage on the phone line. A bugged phone usually has reduced voltage because its energy is being siphoned off by a bug.

The quantum theory goes much deeper, however. The quantum theory says that no matter now sensitive the bugging device, there will *always* be some disturbance in the original signal.

Quantum cryptography uses the fact that light can be polarized—i.e., light is a wave which vibrates in a particular direction (perpendicular to its direction of motion). For example, if a light beam is moving toward you, its vibrations can be in the vertical or horizontal direction. (This has practical use in polarized sunglasses, which reduce glare by blocking all light which vibrates in the incorrect direction.) Quantum physicists use this fact to send messages along a polarized light beam by alternating the direction of the polarization. Since different pulses of light can have different polarizations, we can send a digital message along the light beam.

According to quantum mechanics, if a spy were to intercept the message and make an observation of the beam, that would disturb the beam and force it into an incorrect polarized state. The person at the receiving end would immediately know that someone was eavesdropping on the message.

Unlike quantum computers, which may be many decades in the future, prototypes of quantum cryptography have already been developed. The first prototype was demonstrated in 1989.

"In a few years, quantum cryptography has come to the point where it's really an engineering issue," claims James D. Franson of Johns Hopkins University, who has successfully performed his own experiments on quantum cryptography. "We now have demonstrated the ability to transmit secure messages between two buildings and over distances of roughly 500 feet in this way."

In 1996, a milestone was reached when scientists sent a secret message along an optical fiber that was 22.7 kilometers long. The message was carried by infrared light and was sent from Nyon, Switzerland, to Geneva, showing that an abstract principle of the quantum theory could have practical applications in the real world.

Given the explosion in computer power, it is only a matter of time before computers can crack most encryption codes. Because of this, the most important messages of the future will necessarily incorporate some form of quantum cryptography. Early in the next century, it is likely that we will see large corporations and institutions begin to use quantum cryptography for their sensitive data. So although the problem of computer privacy will be fiercely debated for decades to come, in principle there is an ultimate solution to the question of computer privacy.

### Roadkill on the Information Highway

By 2020, industries will rise and fall because of the information highway, just as the transcontinental railroad system in the nineteenth century made ghost towns out of the rural towns bypassed by the railroad, while townships located near railroad junctures became burgeoning cities.

The danger of the computer revolution is perhaps capsulized by the famous but probably apocryphal story of an exchange between Henry Ford and union leader Walter Reuther during the Great Depression. Henry Ford, pointing proudly to the rows of shiny new machines which were replacing union workers, chided his rival and asked, "Mr. Reuther, where are your workers?"

To which Mr. Reuther replied quietly, "Mr. Ford, where are your customers?"

Because the strengths and weaknesses of electronic computers are well known, it is possible to predict which kinds of jobs will be directly threatened by the computer revolution in the coming decades. They include three basic kinds of employment:

- Jobs which are repetitive (Factory workers involved in mass production are prime targets for the robotic revolution.)
- Jobs which involve keeping track of inventory
- Jobs which involve a middleman

The first kind of employment has been threatened for decades. More surprising is that many seemingly secure middle-class jobs requiring a college education involved with tracking inventory or acting as a middleman are being downsized in the 1990s. The Internet will only speed up this transition.

"The Internet is a rifle aimed at all middlemen: insurance salespeople, investment bankers, travel agents, car dealers. It's going to hit everybody," claims Jeffrey Christian, who runs an executive-search firm in Cleveland.

Andrew Grove, CEO of Intel, put it even more bluntly. For anyone whose business depends on large databases, Grove says, "I would view the Internet as a tidal wave that's going to wipe me out. I would be running as far as my feet go, redoing all my reservation systems, order systems, customer databases, so that masses of people would be able to reach them from their computer."

Businesses like travel agencies, banks, video stores, and stockbrokers are all ultimately threatened by the Internet. Security First Network Bank of Pineville, Kentucky, for example, is doing business entirely on the Internet today. No tellers. No lines. No waiting. And no branches either. "It means we don't need all these bodies," boasts James S. Mahan III, Security First's chief executive. The human teller, like the friendly telephone operator who used to chat with the customers as he or she connected their calls, is rapidly becoming a thing of the past.

*The Wall Street Journal* summed it up: "Electronic commerce promises to free workers to do more productive, higher paying work. But it's going to hurt a lot of employees along the way." The real question, however, is whether the computer revolution will create new jobs to make up for the old ones and make the economy more productive and prosperous.

In one sense, the elimination of the middlemen may increase the efficiency of the economy. In the last century, for example, there was a proliferation of private roads in the United States, where wagons had to pay a stiff toll for their use. There was a pike at each tollgate, and, after the toll was paid, the gatekeeper would turn the pike and allow the wagon to pass. This was the origin of the term "turnpike." This archaic system, which greatly hindered the growth of interstate commerce, was mercifully ended when states bought up most of the roads and created the modern system. The elimination of these middlemen greatly accelerated the free flow of trade and commerce, generating millions of new jobs and helping to create a modern industrial state.

Another example from the last century is the horse and buggy industry. Thousands of people were employed as blacksmiths, carriage repairmen, coachmen, stable managers, horse trainers, and breeders. Most of these jobs were destroyed with the coming of the automobile and the internal combustion engine. The automobile, in turn, altered the landscape of American society, creating a powerful, vibrant industry. New kinds of jobs sprouted up—for auto workers, mechanics, car dealers, service workers, and oil workers. The result of the transition to the automobile was so profound in its implications that it even changed our way of thinking, forging new attitudes and social trends based on the country's newfound mobility. It is now considered our birthright to be able to hop into a car and drive anywhere we please.

There were negative consequences as well—traffic jams, pollution, an exodus to the suburbs that emptied out the cities. Without a strong tax base, the inner city collapsed, helping to create huge slums. And the 40,000 people who die annually in car accidents in the United States (almost equal to all U.S. deaths during the Vietnam War) are taken for granted today, the price we pay for this birthright.

The point is that there are always trade-offs. The issue is not to debate the relative aesthetic merits of the horse and buggy industry versus the automobile industry. Both have their merits and drawbacks. The point is whether the jobs created by the new industry render the economy more efficient and make society as a whole more productive and prosperous.

### Jobs Which Will Flourish

So which jobs will flourish in 2020? Which will survive the onslaught of the computer and the information highway? Several classes of jobs, actually.

Once we understand the strengths and weaknesses of computers, we actually see that there will be many kinds of jobs that will not be replaced within the next fifty years. "There's no limit to the demand for human services," says Paul Krugman, an economist at MIT.

The following jobs will probably flourish even in the presence of the information highway. In addition to jobs which will survive by becoming more personalized and specialized, there are also the following:

**Entertainment.** Writers, entertainers, performers, actors and actresses who are engaged in the creative arts will flourish in the new era. The increasing abundance of leisure time in society will create an explosive demand for new forms of entertainment. For example, the recent proliferation of cable channels is creating a demand for new entertainment to fill the vacant hours of airtime. New types of entertainment which do not even exist today will generate entirely new industries.

**Software.** Although the price of a microchip may be less than a penny by 2020, jobs for software programmers will soar. Computer hardware will eventually become a commodity, like pork bellies. But software requires creative mathematical talent, which cannot be easily computerized. For example, the video game industry, which did not exist only a few years ago, is larger than the entire film industry today. Similarly, there is an insatiable demand for computer scientists who can design attractive Web pages for clients. Virtual reality, in particular, requires an inordinate amount of software. Ironically, the computer may put certain people out of work, but the software that the computer needs in order to operate cannot be computerized.

**Science and technology.** Computers cannot create new scientific theories. Although the job market for scientists and engineers will fluctuate with the economy, there will always be a demand for technical talent. Discoveries by scientists and engineers, in turn, will create entirely new industries.

**Service industry.** Chauffeurs, butlers, maids, personal trainers, bodyguards, doormen, police, lawyers, music teachers, private tutors, etc., are involved in intimate interactions with a variety of other people. Such jobs are almost impossible to replace with computers. For example, the travel industry, which is the fastest-growing industry in the world at present, requires tour guides, hotel managers, and service workers. Furthermore, computers are making the travel industry more efficient, flexible, and accessible via the Internet.

**Skilled and craft jobs.** Skilled workers, such as construction workers, repairmen, sanitation workers, highway work crews, park service personnel, forest rangers, teachers, etc., cannot easily be replaced by mass production. None of these jobs are repetitive; each new task requires an entirely different assessment of a problem. Some, like teaching, may be partly automated by being placed on the Internet, but ultimately students need human contact for specialized needs.

**Information services.** Workers in information services will service the information infrastructure, repairing and monitoring the cables, satellites, computers, relays, etc. The larger the information infrastructure becomes, the more workers will be required to build and maintain it. There will be a demand for maintenance workers who perform tasks requiring skills that cannot be automated, such as laying down wires and cables, replacing worn-out computer parts, and so on.

**Medical workers and biotechnicians.** As the country ages, there will be a growing demand for health-care workers to minister to the baby boomers. Robots, telemedicine, and so on may reduce the need for certain kinds of jobs, but they will never eliminate them totally. And the biotech revolution will open entirely new jobs which can only be imagined today.

## Industries Which Will Change—or Die

Even those industries that will suffer with the coming of the Internet may still survive and even flourish if they make the transition to offering *spe-*

*cialized, personalized services* that can't be replaced by machines. For example:

• Travel agents, who may lose the savvy business traveler to the Internet, might specialize in offering deluxe vacation packages providing many frills and can be responsive to personalized needs.

• While banks will lay off unskilled tellers, they will likely retain and promote skilled programmers and salespeople, and will likely move to sell specialized products, such as certificates of deposit, for which they can earn healthy commissions.

• While stockbrokerage houses may lose experienced speculators to the Internet, they may gain the less experienced customer who prizes individual attention and the in-house analyst's knowledge of market fluctuations.

• Real estate agents will lose the customer who likes to scan hundreds of listings in front of a keyboard, but will gain by catering to the specialized needs of the customer who may, for example, want to know where the good schools are located.

Another area that will undergo profound changes is the print industry. Instead of collapsing, as many have predicted, print media may undergo a metamorphosis to a new, higher form.

Columnist Charles Krauthammer paints a gloomy future for print media; he sees the inevitable demise of paper. "Imagine what the blacksmiths of 1896 felt when they looked up and saw their first automobile," he writes. "I know. I am a newspaper columnist in 1996, and for the last six months I've been trying out the Net. The future—mine, anyway, is bleak . . . the future is not print."

"Clay tablets," he continues, "gave way to papyrus, sheepskin scrolls to bound books, illuminated manuscripts to Gutenberg type. In the end, each revolution was for the better."

For those who say that computers are clunky, time-consuming, and full of jargon, he answers with the following analogy: "It was much easier to mount a horse than crank up the starter, release the hand brake, get in gear, and start yourself rolling down the road. Then came the ignition key."

It's debatable, however, whether computers will ever have the "ignition key" that will make them every bit as convenient as paper. People still like to browse the headlines before going to work and read paperback books at the beach, at home, or on the subway. Paper has become so convenient that computer screens may never approach its appeal.

The most likely future is that some of the functions of paper will indeed vanish. Newspapers are already shrinking because the younger generation was weaned on TV, rather than print. But in the future newspapers may survive by providing increasingly specialized services.

Because of the unfiltered cacophony of the Internet, newspapers must offer something that the Internet cannot, such as news analysis from authoritative sources and specialized services. The problem with the Internet is that everyone from cranks to seasoned experts offer their unsolicited advice, creating incessant noise. Editors will probably assume an increasingly important role by presenting something that the Internet cannot: wisdom. In an ocean of babble, those forms of media that can provide authoritative facts and penetrating analysis will serve an important function.

By 2020, people will probably receive personalized newspapers culled from the Internet, supplying the specific kinds of information each person needs from trusted sources.

## Winners and Losers

What about illiterate, unskilled laborers who fit none of the aforementioned categories? The fact is, each time society made an abrupt leap to a new level of production, there were losers and winners. It may well be that the computer revolution will exacerbate the existing fault lines of society, creating new "information ghettos."

Throughout history, there has been constant change in the social structure of society as a result of changes in the environment and in technology. When agriculture was introduced after the end of the last Ice Age 10,000 years ago, hunting and gathering, by and large, was replaced by farming. Farming was backbreaking work, but it was still preferable to leading a precarious nomadic life chasing after game.

With the coming of the industrial revolution, there was a mass exodus from the farms to urban factories. Today, only about 2 percent of the U.S. labor force is involved in farming. Modern industry gave unskilled labor the opportunity to enter the middle class within a generation or two.

At present, there are at least two revolutions going on in the world. In Asia, where two-thirds of humanity resides, world demand for products made cheaply by unskilled labor is creating an industrial revolution of unparalleled magnitude, lifting the lives of hundreds of millions of peasants into middle class status. These countries are making the transition from an agrarian society to an industrial one. These people, in turn, will hunger not for computer games and virtual reality, but for refrigerators, cars, TVs, dishwashers, and so on—merchandise created by conventional industry.

The trend in the West is toward an expanding service sector and a collapsing industrial sector. A typical industrialized country has a service sector which makes up 70 percent of the economy. In the United States, for example, industry makes up only 29.2 percent of the economy. In the U.K., it is 30 percent. In France, 28.7 percent.

Some economists have speculated that the industrial sector in the United States may eventually shrink as much as the agricultural sector—i.e., down to 2 percent. New jobs and industries will be created by the computer industry, but these jobs will demand increasing levels of education, which not everybody will have.

Michael Vlahos, a senior fellow at the conservative think tank Progress and Freedom Foundation, which is closely associated with Newt Gingrich, imagines that by 2020 America will have an information-stratified society, called "Byte City." At the top will be the Brain Lords (with techno-billionaires such as Bill Gates). Below them are the Upper Service workers (i.e., the cyber yuppies). And below them are menial workers (the cyber serfs). At the very bottom of society are the Lost People, those who are totally left out of the computer revolution. Frank Owen of *The Village Voice* sees this as a particularly bleak vision, what he describes as "Blade Runner Meets the Bell Curve."

In other words, it is possible that the information revolution may enrich the few at the expense of the many. As Barbara Ehrenreich observed: Gingrich "preaches third-wave futurism while representing the interests of second-wave plutocrats."

Several alternatives have been proposed to circumvent this potential polarization of society. One approach is a major effort to retrain workers, similar to the GI Bill initiated after World War II that catapulted so many young veterans into well-paying jobs. Retraining may well be a key part of any long-term solution to the problem, but one drawback is the cost of such a program, especially when many of the laid-off workers will be older and less resilient. A more radical proposal is to change the nature of work. Some have noted that the very idea of a "job" and "salary" evolved with the industrial revolution 300 years ago. Before that, people were locked into lifetime jobs with a guild or on a farm. One version of this would be to fund massive government service jobs, to clean up the environment and our cities or provide new forms of art and entertainment. If the economy is so productive that only a small percentage of the population can produce all the food and goods necessary to keep society going, why not use this bountiful wealth to put people to work to enrich society as a whole? Social critic Jeremy Rifkin, for example, sees a coming catastrophe for society unless workers share in the wealth created by the computer. He envisions vastly strengthening the "third sector" (i.e., civil society, in contrast to the public and private sectors), consisting of nonprofit organizations, civic groups, etc., to absorb the unemployed.

## Increase the Pie

Because of this economic dislocation, there is a debate in this country concerning how to divide up the pie, with one interest group often pitted against another, with one person's victory being another's defeat. Ultimately, it's a zero-sum game. A bigger slice for one interest group means a smaller slice for another, and the pie, in some quarters, is actually shrinking. The net result is that the natural fault lines of society around race and class are widening.

My own point of view is that we need to *increase the size of the pie.* Ultimately, science and technology have been a major source of our increased wealth in recent centuries. In the nineteenth century, for example, science and technology provided the underpinnings for mechanized industries based on railroads, steam engines, the telegraph, chemicals, utilities, textiles, and more. That propelled the United States to the forefront of the world's economy.

In the twenty-first century, because technological innovations today require much more scientific sophistication, we need to pump more resources into education and science, in order to reap great dividends in the future.

*The inevitable displacement that will result from the evolving computer revolution, therefore, is really a symptom, not the root cause of the problem.*

## What Creates Wealth in the Twenty-first Century?

Why has automation caused such anxiety among workers, especially middle-class managers? The roots of the current dislocations caused by the global marketplace run much deeper than simply the advent of the computer. According to Lester Thurow, former dean of MIT's Sloan School of Management, what is happening is nothing less than a seismic shift in the generation of wealth on this planet.

Beginning roughly 300 years ago, with the birth of capitalism, nations which accumulated vast wealth and exploited natural resources and capital rose to prominence, as chronicled by Adam Smith in *The Wealth of Nations.* But in the twenty-first century, Thurow writes, "brainpower and imagination, invention, and the organization of new technologies are the key strategic ingredients." Some of the countries likely to be economic giants of the twenty-first century, like Japan and China, are relatively poor in natural resources and arable land, but they have a trained, dedicated workforce and have placed a premium on science and technology.

"Today, knowledge and skills now stand alone as the source of comparative advantage," writes Thurow.

Many nations which are richly endowed with natural resources may slide into poverty, as the price of commodities continues to drop in the next century, until they grasp this fundamental fact.

Unfortunately, the United States has been slow to adjust to this new reality. As documented by David Halberstam in *The Next Century*, the U.S. economy was severely drained by the Cold War. Not only did the Cold War break the back of the Soviet Union, it also siphoned off trillions of dollars in precious resources from the U.S. economy.

Furthermore, today 50 percent fewer Ph.D.'s are being produced in science and engineering than just two decades ago. Around the country, science budgets are being downsized. The laying off of top scientists has become a symbol of the shortsighted search for short-term profits. (One reason the scientific establishment in the United States has remained as strong as it has is the large influx of educated immigrants, who, for example, account for a disproportionate share of the Ph.D.'s working in Silicon Valley. However, with the current vitriolic debate over immigration, even this resource may soon dry up, leaving the United States with a hollow scientific base.)

The sorry state of science education is another case in point. Students from the United States consistently score near the bottom in every international science test. This is unlikely to change. "Put bluntly," Thurow writes, "private capitalistic time horizons are simply too short to accommodate the time constraints of education."

In sum, the United States is eating up its own scientific seed corn. Only by changing national priorities will it be possible to reinvigorate the forces which made the United States a great power in the last century. But this requires not only making more investment in science but also changing our emphasis on short-term, instant gratification.

Thurow writes: "Technology and ideology are shaking the foundations of twenty-first century capitalism. Technology is making skills and knowledge the only sources of sustainable strategic advantage. Abetted by the electronic media, ideology is moving toward a radical form of short-run individual consumption maximization at precisely a time when economic success will depend upon the willingness and ability to make long-run social investments in skills, education, knowledge, and infrastructure. When technology and ideology start moving apart, the only question is when will the 'big one' (the earthquake that rocks the system) occur."

The real winners of the twenty-first century will be those nations which invest strategically in science and technology. After World War II, Germany and Japan, for example, were able to complete one of the most successful recoveries in history. One reason why this was possible was that their best and brightest were not working on hydrogen bombs, but on better cars and transistor radios. In the United States, the best scientific minds are often absorbed by the Pentagon, which uses their talents on new generations of weapons. Even redirecting a tiny percentage of the military budget to pure science could have a profound effect on the development of technology.

As with the job market, there will be winners and losers among nations in the next century. The winners will be those who see the computer not as the enemy, but as a tool by which to reinvigorate their scientific and technical base, to create entirely new industries which will absorb those who would ordinarily be left in the cold. The losers will be those who smash machines, squabble with each other over dwindling slices of the pie, and wallow in entitlements and short-term gains.

## Robotic Dangers: Self-aware Robots

By 2050, the nature of the debate may radically shift again as an entirely new class of machines enters the market: robots created with a limited amount of self-awareness. This may represent the fifth phase in the evolution of computing machines.

What happens when the interests of robots and humans diverge? Can robots harm us, even accidentally? Can they take over?

This is where AI collides with the realm of science fiction, since we are now dealing with machines with an independent will, armed with formidable mental and physical abilities which may easily surpass our own. The advantage here is that they can improve upon human commands, yielding strategies which are unforeseen by their human inventors. This could open up entirely new areas for science and industry. The problem, however, is that they can also contradict human orders, and hence pose a danger to humans. This is not a matter of idle speculation; AI researchers have devoted considerable thought to the question.

AI expert Daniel Crevier writes: "When machines acquire an intelligence superior to our own, they will be impossible to keep at bay. Episodes where a deputy rises and becomes the effective ruler of a nation have happened countless times in history. The evolution of life on earth is itself nothing but a four-billion-year-long tale of offspring superseding parents. The unrelenting progress of AI forces us to ask the inevitable question: Are we creating the next species of intelligent life on earth?" Or, as Arthur C. Clarke has said: "There's an element of fear involved because this challenges and threatens us, threatens our supremacy in one area in which we consider ourselves superior to all the other inhabitants of this planet."

Hans Moravec agrees: "Intelligent machines, however benevolent, threaten our existence because they are alternative inhabitants of our ecological niche. Machines merely as clever as human beings will have enormous advantages in competitive situations."

As robots become gradually more intelligent and humanlike in the next century, we can begin to quantify the dangers we can expect to face.

Scientists may gradually allow robots to assume control over our planet's vital functions. For example, in order to maintain a free flow of goods in the economy, as well as monitor and control power distribution, humans may give an extraordinary amount of control of the environment and the economy to computers.

Consider the simple question of "program trading" on Wall Street. Because humans are too slow to take advantage of tiny but rapid-fire movements of interest and monetary exchange rates, Wall Street firms have left hundreds of billions of dollars in the hands of computers. Since these computers are competing against each other, the smallest motion in interest rates may trigger an electronic stampede, like the one which precipitated the 1987 Wall Street crash. The problem is not that the computer is unsuccessful but that it is far too successful.

At present, this question is still simple enough so that it can be ameliorated by small changes in trading rules by the Securities and Exchange Commission. In the future, however, it is almost certain that some forms of artificial intelligence will be used to analyze trends in money, trade, and stocks.

It's conceivable that by the middle of the next century the sheer computer power necessary to run entire cities and nations, including electricity, banking and commerce, transportation, water, waste disposal, life support, etc., will become so great that society may leave this entirely to computers and robots. Only a handful of engineers may service the robots who, in turn, have the vast knowledge necessary to smoothly run the city. Any malfunction in the circuits of the system may cripple or paralyze an entire civilization. The more information is centralized, the easier it is to disrupt.

## Robots as Killing Machines

One reason robots may pose a threat is that their prime focus has been largely military—i.e., they have been specifically designed to kill other humans. The largest single benefactor by far has been the Pentagon's

DARPA, which has generously funded scores of AI projects, such as Shakey, for the single purpose of winning wars.

Perhaps the greatest threat posed by computers, according to AI investigators, is the control of our nuclear weapons by computer systems which contain autonomous AI capabilities. This was graphically explored in the 1970 movie *Colossus: The Forbin Project*, based on a novel by D. F. Jones, in which the United States gives control of its nuclear weapons to a supercomputer, the Colossus (named after Turing's historic machine).

Some critics have discounted the Colossus scenario, claiming that computers are just machines that do whatever we tell them to do, and hence pose no mortal danger. The problem is that the calculus of nuclear war is so swift that we may eventually carelessly cede control of our nuclear weapons to a computer with AI capability, the very scenario outlined in the movie *WarGames*, in which a computer, asked to initiate a game called Thermonuclear War, is unable to distinguish the game from an actual war, and as a result prepares to launch a first strike against the Russians.

MIT's Joseph Weizenbaum says, "To a certain extent we have crossed that threshold." During the Gulf War, he points out, "an American cruiser thought itself to be under attack by an airplane coming from Iran and shot it down. It turned out to be an Airbus with two hundred and thirty people on board. Had the ship's captain known the plane to be an airliner, he would, of course, never have ordered it to be fired at."

One way to partially solve this problem is to change the funding source for artificial intelligence. Since AI research is often costly, most AI researchers follow the source of their funding.

This problem should diminish with time as commercial enterprises begin to fund AI scientists. The goal would then be to satisfy the demands of consumers rather than to find ways to kill them off. The solution is to reduce the influence of the Pentagon in such research, rather than cut basic research itself.

A legitimate question is raised, however: can robots kill, even when we program them not to?

## Robots That Go Mad

In the movie *2001*, a ship's intelligent computer system, HAL 9000, malfunctions on a historic mission to Jupiter and systematically attempts to murder its crew. In a sequel to that movie, *2010*, an explanation is finally given why HAL became a serial killer. The problem began when it was given contradictory instructions. In order to carry out its mission, HAL was forced to lie to the crew. However, because HAL had no experience with lying, its circuits experienced an irreconcilable conflict. In order to stop lying to the humans, the solution it came up with was quite logical: destroy the humans. Then it wouldn't have to lie anymore.

This decision to become a mass murderer is actually quite understandable. The "H" in HAL stands for "heuristic," and "expert systems" like HAL suffer from what is called the "mesa effect." As long as an expert system stays within its comfort zone (i.e., its area of expertise), it performs admirably. However, once the system is forced to go even slightly outside this comfort zone (e.g., by lying to humans), it's like falling off a cliff or mesa. The system collapses.

Expert systems, confronted with a problem outside its normal range, will blindly continue to attempt to solve the problem, even if a solution is not possible, because the machine is unaware that it is outside its area of competence. Worse, as the machine falls off the mesa, it can be caught in a feedback loop which makes the system go berserk. The scenario in *2001* is certainly possible, in other words, precisely because of problems inherent in the mathematics of feedback—something that is sometimes called "the stability problem."

Although a computer may seem to perform flawlessly, tiny errors inherent in all feedback mechanisms can escalate until the system collapses. As Daniel Crevier has noted: "The results of these faultless steps will amount to irrational and imbalanced behavior: madness."

We humans, of course, have a wide array of feedback mechanisms that protect us from danger and help us adjust to the environment. That's why we have five senses and a brain to evaluate the messages from those senses. Nonetheless, feedback loops can also destroy humans. When one of us "cracks up," it's sometimes because of a feedback loop that escalated out of control.

Similarly, there is always a danger that AI systems trusted to control our nuclear weapons, our money supply, our life support systems, our cities' power supplies, etc., may experience a feedback loop with disastrous consequences to human life. Crevier claims that we will have to "take into account the possibility of madness and irrationality before handing over responsibilities to future intelligent machines."

There is no simple solution to the "madness" problem in feedback systems. Instead, scientists must design increasingly sophisticated mechanisms to shut down the system before it goes mad.

## Can Three Laws Protect Us?

Science fiction writers like Isaac Asimov have tried to eliminate the ability of robots to murder their human masters by encoding three "laws" of robotics directly into their programs. They are:

1) A robot may not injure a human being or, through inaction, allow a human being to come to harm.

2) A robot must obey the orders given it by human beings except where such orders would conflict with the First Law.

3) A robot must protect its own existence as long as such protection does not conflict with the First or Second Law.

But there is an element that is completely missed by the three laws—that robots may, in properly carrying out their orders, inadvertently threaten humanity.

Consider the laws of bureaucracy, which are similar to the laws within a robot's brain. A bureaucracy tends to expand, sometimes to the point that it destroys the economic base that made the bureaucracy possible in the first place. Several economists, for example, have written that the sudden collapse of the former Soviet Union was in part due to the bureaucracy's reaction to the arms race. The Soviet leadership gave its bureaucracy one mandate: to catch up to the West in the arms race. Given this single mission, the bureaucracy faithfully carried it out, even if it meant bleeding the economy dry building expensive nuclear weapons, until the system collapsed.

In a sense, the bureaucracy fell victim to the old right-wing strategy of "spending the Russians into a depression"—i.e., huge Pentagon expenditures forcing the Russians, with a limited economic base, to build similar weapons, which break their economy. The problem was not that the bureaucracy failed in its mission; the problem was that it was too successful, until the weight of its success crushed it into oblivion.

Likewise, a global economy controlled by AI systems could legitimately decide to accomplish its mission by expanding, like a bureaucracy. The three laws of robotics are useless against robots justifiably thinking they are carrying out their central mission. The problem is not that they have failed to carry out their individual orders; the problem is that their orders were inherently flawed in the first place. Nowhere in the three laws do we address the threat posed to humanity by well-intentioned robots.

The problem, in such an instance, is not with the computer; it is with humans, who may want to put electronic wonders on-line before they have electronic safeguards in place. Artificial intelligence, as the decades pass, must be kept on a tight leash. The more sophisticated the circuitry becomes, the more safeguards must be placed on them so that they do not have unintended consequences. There must be a feedback loop added to the design of AI systems so that these systems have fail-safe mechanisms and elaborate controls so they do not threaten human society. In fact, perhaps a new branch of AI will have to be created, specifically designed to keep AI systems under control.

At the very least, this means that robots will have to be hard-wired with a vast array of safety mechanisms so they do not overwhelm or replace their human masters. The three laws of robotics are not enough. There also have to be safeguards against well-intentioned robots as well.

Whether computers become our eternal helpmates or our masters, one thing is certain: they will not go away. Perhaps the thinking of most people working in artificial intelligence can be summarized by a statement made by Arthur C. Clarke:

"It is possible that we may become pets of the computers, leading pampered existences like lapdogs, but I hope that we will always retain the ability to pull the plug if we feel like it."

# Part Three

# The Biomolecular Revolution

# 7

# Personal DNA Codes

"We used to think our future was in the stars. Now we know it's in our genes."

—JAMES WATSON

THE NATIONAL INSTITUTES OF HEALTH (NIH), the world's premier medical complex, is the nucleus for revolutionary new research which will radically reshape our lives in the twenty-first century. It's a sprawling network of modern laboratories located in the leafy, tree-lined suburb of Bethesda, Maryland, just outside Washington, D.C. Begun as a humble one-room Laboratory of Hygiene in 1887 with a modest budget of just $300, it has since mushroomed to 70 buildings on 300 acres, with an annual budget soaring to $11 billion.

Perhaps the most pivotal and controversial of all the divisions of the NIH is the Human Genome Project (officially the National Center for Human Genome Research), one of the most ambitious projects in medical history, a $3 billion crash program to locate all the genes within the human body by 2005.

The man now in charge of the Human Genome Project is Francis Collins. On his shoulders rests much of the scientific, medical, and ethical responsibility for unraveling the secret of life.

Tall (he's six feet four inches), slim, well dressed with a dashing mustache, he reminds one of a distinguished version of Peter Sellers. But unlike Sellers, Collins rides to work at the NIH on a Honda Nighthawk 750 motorcycle wearing a black leather jacket. He is a far cry from a befuddled scientist or a gruff, uncaring bureaucrat. (He once posted a quote from Winston Churchill on his wall that read: "Success is nothing more than going from failure to failure with undiminished enthusiasm.")

Collins first won international acclaim for locating one of the most sought-after genes in the human genome, the gene for cystic fibrosis, the single most prevalent genetic disease among Caucasians in the United States. (It is so common that there is usually one child who is a carrier of this fatal, dreaded disease in most classrooms in the United States.) Accepting his position as director of the Human Genome Project meant tearing himself away from his beloved laboratory, but it has given him a chance to be part of scientific history.

"There is only one Human Genome Project. It will only happen once in human history, and this is that point in time. Without sounding corny, I do believe this is the most important scientific project mankind has ever mounted, this investigation into ourselves. . . . I feel I've been preparing for this job my whole life," he noted. The enormous power of our genes determines everything from the color of our hair, the shape of our nose, to the chemistry of our cells. But many people simplistically think that genes determine everything.

"It's almost comical these days to see people, sometimes tongue in cheek, sometimes not, talking about the gene for this or the gene for that," Collins says. "People are saying, 'Oh, I have the gene that makes me like sports cars.' *Time* magazine has a cover which says, 'Infidelity, it's caused by our genes.' I mean, come on! Behavior patterns, while they may be genetically influenced in modest ways, are never going to be understood by fleshing out all the DNA sequence of the human genome, at least in large part."

"We will not understand important things like 'love' by knowing the DNA sequence of *Homo sapiens*," he points out. "We have to be careful, in our enthusiasm for what we are doing, implying it's going to turn into more than what it is. That would be dangerous. If humanity begins to view itself as a machine, programmed by this DNA sequence, we've lost something really important."

### Mapping the Human Genome

The task Collins and his team have been given is the creation of a "map" of the 100,000 human genes hidden among 23 pairs of chromosomes in our cells—by the year 2005.

"What we have now is the road system of about 1850," he says. "You can get from one place to another but you may find it pretty hard slogging sometimes, and on occasion you may have to get out of the wagon and walk."

Eric Lander, the director of MIT's Whitehead Institute, adds: "What we will eventually get will approach the detail of something you'd get from the AAA."

The Human Genome Project is actually well ahead of schedule and under budget, as a result of the formidable firepower of the computer, biomolecular, and quantum revolutions being brought to bear on it. Many of the great advances in twentieth-century science are being focused on this major endeavor. Within a decade, gene hunting has accelerated by a factor of several thousand times with the introduction of computers, robotic laboratories, and neural networks. It is one of the most dramatic examples of the cross-pollination between the three revolutions which will set the pace for the twenty-first century.

The sequencing is so advanced that we can now give good estimates of the number of genes involved for each of the major body organs. For example, the human brain probably requires 3,195 genes, the heart 1,195, and the eye 547 genes.

The pace of DNA sequencing is breathtaking. Only a few years ago, scientists knew the location of only a handful of human genes. By mid-1994, the list had grown to 4,700 genes, or about 5 percent of the total. By late 1996, 16,354 human genes had been mapped, or about 16 percent of the total. Given the astonishing advances made in DNA sequencing, Collins says, "the sequencing part may be 99 percent done by 2002 or 2003, even though the annual budget has been 70 percent of what was originally proposed."

When completed, the impact of the Human Genome Project could be much greater than the discovery of Mendeleev's periodic chart of the elements in the nineteenth century, which finally brought order to the chaos of matter and gave birth to modern chemistry. By analyzing the periodic chart, new elements and their properties could be predicted from scratch. Modern civilization, with its dependence on metals, alloys, solvents, plastics, and high-tech substances, would not exist without the periodic chart. Similarly, biology and medicine of the twenty-first century may be unthinkable without the genetic map provided by the Human Genome Project.

## Predictions for the Future

I have noted that it is possible to make fairly reasonable estimates of computer technology into the next twenty-five years because of Moore's

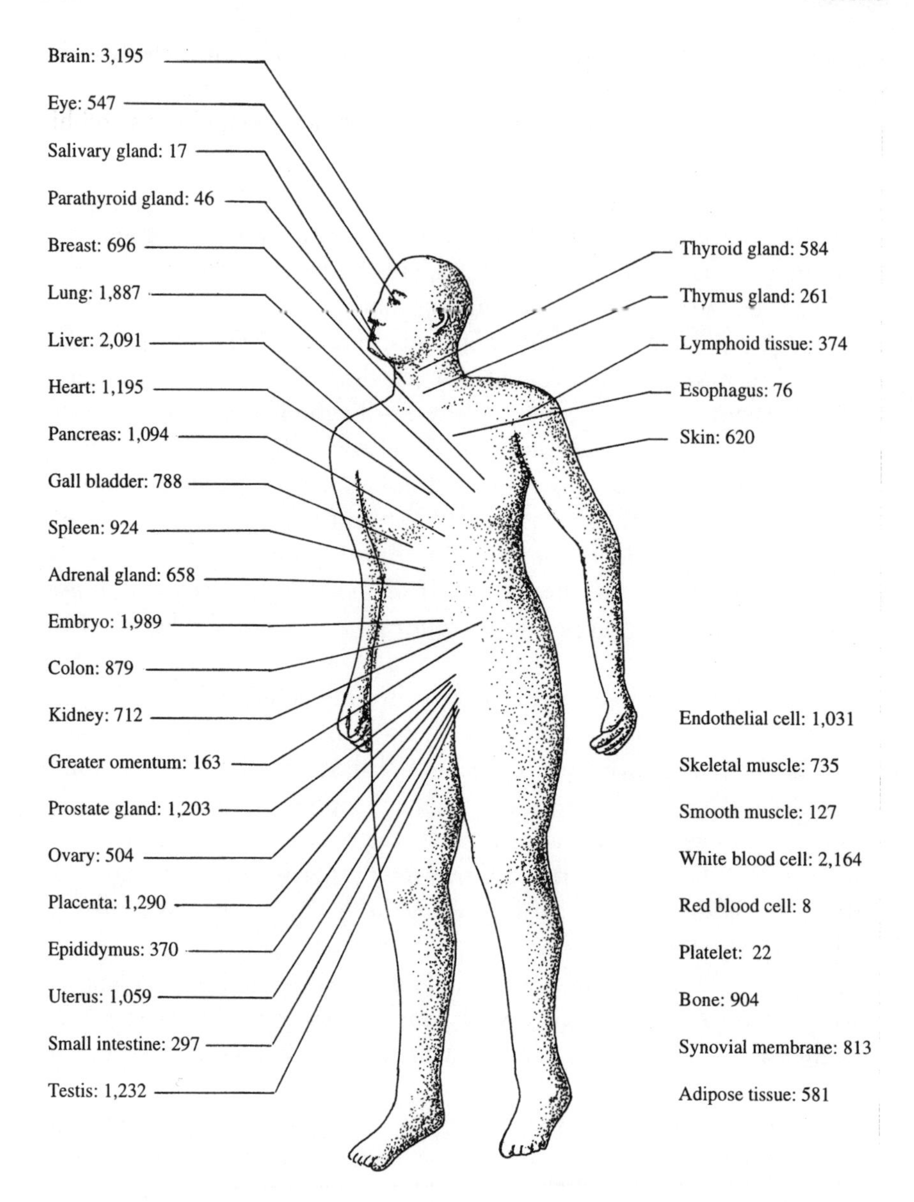


The number of genes for each human organ has been roughly determined. By 2005, the detailed molecular structure of each of 100,000 human genes should be completely determined. (Courtesy Robert O'Keefe)

law. Similarly, because DNA research is rapidly being computerized and roboticized, a new type of Moore's law has been taking hold recently in biology: *the number of DNA sequences that we can determine doubles roughly every two years.* As with computer technology, this predictive law, which has been so successful to date, makes it possible to peer into the future and makc reasonable estimates when certain medical milestones will be achieved.

Because the science of gene cloning is well understood, Collins and his colleague Walter Gilbert, a Nobel Laureate from Harvard, expect to see the following scenario from now through the year 2020.

• *By the year 2000,* Gilbert claims, scientists will have deciphered the genetic codes for twenty to fifty hereditary diseases which have caused untold suffering since the dawn of humanity, including cystic fibrosis, muscular dystrophy, sickle-cell anemia, Tay-Sachs disease, hemophilia, and Huntington's chorea.

• *No later than the year 2005,* the 100,000 or so genes that make up the human genome will have been deciphered by the Human Genome Project, which will open up the secrets locked for millions of years in our genes. For the first time, scientists will be able to view the complete genetic code of humanity.

• *By 2010,* the genetic profiles of hereditary diseases will balloon to approximately 2,000 to 5,000, giving us an almost complete understanding of the genetic basis of these ancient diseases. "It is reasonably likely that by the year 2010, when you reach your eighteenth birthday," Collins says, "you will be able to have your own report card printed out of your individual risks for future disease based on the genes you have inherited."

• *By the year 2020 or 2030,* all this will finally culminate in personalized DNA codes. Gilbert claims, "You'll be able to go to a drugstore and get your own DNA sequence on a CD, which you can then analyze at home on your Macintosh."

The next century, Gilbert predicts, will be a heady time when we "will be able to pull a CD out of one's pocket and say, 'Here's a human being; it's me!' "

This CD will be the crowning achievement of billions of dollars of research, the product of hundreds of dedicated scientists working to write the "encyclopedia of life," which will include everything necessary (in principle) to construct ourselves. Once it is completed, we will have an "owner's manual" for a human being.

The intense effort leading to personalized DNA codes is already reverberating throughout scientific laboratories around the world, giving us the promise of altering the course of medicine. By 2020, a map of the 100,000 genes in our human genome could revolutionize the way we treat disease, allowing us to create new classes of therapies and cure debilitating diseases once thought to be hopelessly incurable. Scientists will have a flood of new technologies, such as gene therapy and "smart molecules," to attack ancient diseases. Large classes of cancer should be curable by 2020, many scientists believe.

We can also reasonably predict what the "post-genome" world may look like from 2020 to 2050. Knowing the street addresses and telephone numbers of the people in the United States does not tell you how American society is constructed. It does not mean we know what people do for a living, or how business, government, schools, the arts and sciences, and other institutions are organized. In other words, possessing the human genome does not guarantee that we know how genes interact and how they function.

The explosive progress from now to 2020 is thus deceptive. From 2020 to 2050, scientists expect, progress will be much slower, because determining the function and interrelations of genes cannot be easily computerized. It may take many decades after 2020, but eventually we will understand the intricate web of interactions between genes, especially for polygenic diseases involving more than one gene, and how they are triggered by cues from the environment, including mental illness, Alzheimer's disease, arthritis, heart disease, and autoimmune diseases. Among the list of polygenic illnesses may appear aging. The "age genes," which some scientists believe might control the aging process, may offer the key to increasing our life span. Eventually doctors might treat aging as reversible phenomenon.

And beyond 2050, we may be able to manipulate life itself.

## Molecular Medicine

"The possession of a genetic map and the DNA sequence of a human being will transform medicine," Gilbert confidently predicts. This revolution is giving birth to a new form of medicine, sometimes called "theoretical medicine" or "molecular medicine," in which diseases can be battled at the molecular level. Computer simulations and virtual reality will enable us to attack viruses and bacteria at the precise genetic weak points in their molecular armor.

This does not mean, as molecular biologists are careful to point out, that medicine can be reduced to a set of molecules. That is a reductionist error. But the biomolecular revolution allows us to understand the complex interactions between genes, proteins, cells, and our environment and even psychology.

Today, having a physical exam is very much like going to an incompetent mechanic who diagnoses your car by listening to the engine. If the engine is purring smoothly, the mechanic says the car is perfectly fine. But internally the car could be on the verge of a major collapse. As you drive out the gas station, your brakes could very well fail or your steering wheel could fly off.

Similarly, a physical exam today usually consists of a few rudimentary tests on your body, such as taking a blood sample and determining your blood pressure. What's actually happening inside your body, especially at the genetic and molecular level, is completely unknown. Even after a thoroughgoing medical checkup with an electrocardiogram, you can still have a fatal heart attack even as you walk out the doctor's office. Ironically, the best technology available today can't predict with certainty whether you will drop dead on your doctor's floor. Furthermore, in the case of cancer, by the time the doctor spots a tumor, it may be too late: there may already be several hundred million cancer cells growing and spreading inside your body.

By contrast, imagine going to the doctor's office for a routine checkup in 2020, when personalized DNA sequences will be available. First, your doctor will take a blood sample, which will be sent to a genetics laboratory. Within perhaps a month, your complete DNA sequence will be provided.

Your doctor will be able to place your personalized DNA sequence into a computer, which will determine if you have any of the 5,000 known genetic diseases. Your doctor will also use your personalized DNA sequence to predict the mathematical chances of your getting any number of related diseases. He or she will then be able to recommend preventive measures years before any symptoms arise. Your personalized DNA sequence will therefore be the foundation on which your health can be analyzed. Gene therapy may then cure some of these previously incurable diseases.

"We are entering an era when disease will be predicted before it occurs," says William Haseltine of Human Genome Sciences. "Medicine is basically going to change from a treatment-based to a prevention-based discipline," he claims.

For good or evil, the biomolecular revolution promises an astounding array of applications, from bioengineered products which will flood the marketplace to the possibility of controlling life itself.

Whether we are mature enough to handle a technology this powerful and this volatile is another question. Some may welcome this revolution for the unquestioned benefits it will bring in relieving suffering and saving and prolonging the lives of millions. Others, for social or religious reasons, may oppose it for its excesses. But even its severest critics admit that all of us will be intimately touched by it.

## What Is Life?

To understand the fascinating science which lies behind the research that will make personalized DNA sequencing possible by 2020, it may be worthwhile to trace the curious twists in Francis Collins's career, which shed considerable light on the origins of molecular biology.

As a student, Collins was repelled by the dry memorization needed in biology, but was attracted to the rigor of the quantum theory and physical chemistry. In quantum chemistry, he could find elegant, precise mathematics governed by the Schrödinger wave equation in which one could calculate how electrons circle the nucleus, how atoms bond with each other, and how molecules create the complex chemical reactions which give life to our bodies. Quantum chemistry, he recalled fondly, "seemed very intellectually satisfying. The mathematical rigor, the sort of elegance of describing the universe with second-order differential equations—I liked that a lot. The ability to describe truth in that fashion appealed to me."

Unknown to him, however, a profound migration of quantum physicists and chemists was already taking place, initiated by the book *What Is Life?*, written in 1944 by Erwin Schrödinger himself, one of the founders of the quantum theory. Biologist Stephen Jay Gould calls *What Is Life?* "among the most important books in 20th century biology . . ."

Schrödinger, like Collins, was repelled by the sorry state of biology. In a time when many biologists were still influenced by "vitalism" (the belief that living things were animated by a mysterious and mystical "life force"), Schrödinger boldly asserted that living things could be understood by the quantum theory of atoms and that life was governed by a "genetic code" (a phrase he coined) locked in the arrangement of our molecules.

Molecules, instead of being merely idle building blocks of our bodies, now had a second function, to serve as repositories for the "code of life."

The problems identified in *What Is Life?* inspired a new generation of physicists to apply the quantum theory to help solve the secret of life, including George Gamow, Pascual Jordan, and Nobel Laureates Francis Crick, Linus Pauling, Walter Gilbert, and Max Delbrück.

*What Is Life?* also changed the life of a brash young student named James Watson. He recalled, "From the moment I read Schrödinger's *What Is Life?*, I became polarized towards finding out the secret of the gene." At Cambridge University, he teamed up with physicist Francis Crick, who was also deeply influenced by the book. Their work eventually identified the DNA molecule as the carrier of Schrödinger's genetic code.

## From Quantum Physics to DNA

Watson and Crick performed their historic work by using an important tool borrowed from quantum physics: X-ray crystallography, which fires a beam of X-rays through a crystallized sample.

To understand this process, think of the glittering crystal spheres that light up dance halls and discos around the country. The sphere is actually made of hundreds of tiny mirrors glued to a ball. When a beam of light is directed at the sphere as it rotates, the entire room is filled with a dazzling array of swirling dots. In principle, if you knew the location of all these dots, you could work backward and determine the precise location of all the mirrors glued on the ball.

Now replace these tiny mirrors with atoms in a crystal, and the light beam with a powerful beam of X-rays. As the X-ray beam bounces off the individual atoms, it creates thousands of tiny scattered waves which interfere with each other and spread out in space. (This expanding wave front of X-rays produces a pattern of bright and dark spots which can be captured on special photographic film. Encoded in this pattern of seemingly random dots is all the information necessary to locate the atoms within the crystal. By using the quantum physics of X-rays, one can then determine the precise atomic structure of the crystal.)

Rosalind Franklin employed this technique to obtain X-ray photographs of crystallized DNA. Using this result, Watson and Crick proved that DNA contained Schrödinger's "genetic code."

DNA, they showed, consisted of two tightly coiled strands arranged like a double helix, making up the celebrated double-stranded "molecule of life." Like brilliant pearls on a string, the genes which make up our body lie along these strands of DNA, which form the 23 pairs of chromosomes locked inside the cell nucleus. These chromosomes pack so much information that if all the DNA in just one microscopic cell were fully stretched out, it would be six feet long!

Along this six feet of DNA lie all of our 100,000 genes; what makes the difference between a virus, a fish, an insect, a mouse, and a human is encoded in this sequence of genes. The DNA, in turn, consists of even smaller units, called nucleic acids, of which there are four, labeled A, T, C, and G. Like rungs on a spiraling staircase, the nucleic acids along the double helix are paired off. (Each pair of nucleic acids is called a "base pair.") The precise sequence of A, T, C, and G strung along the DNA make up Schrödinger's "genetic code."

A single gene may be made up of thousands of base pairs. Each performs its magic by creating a template of itself made of RNA, which in

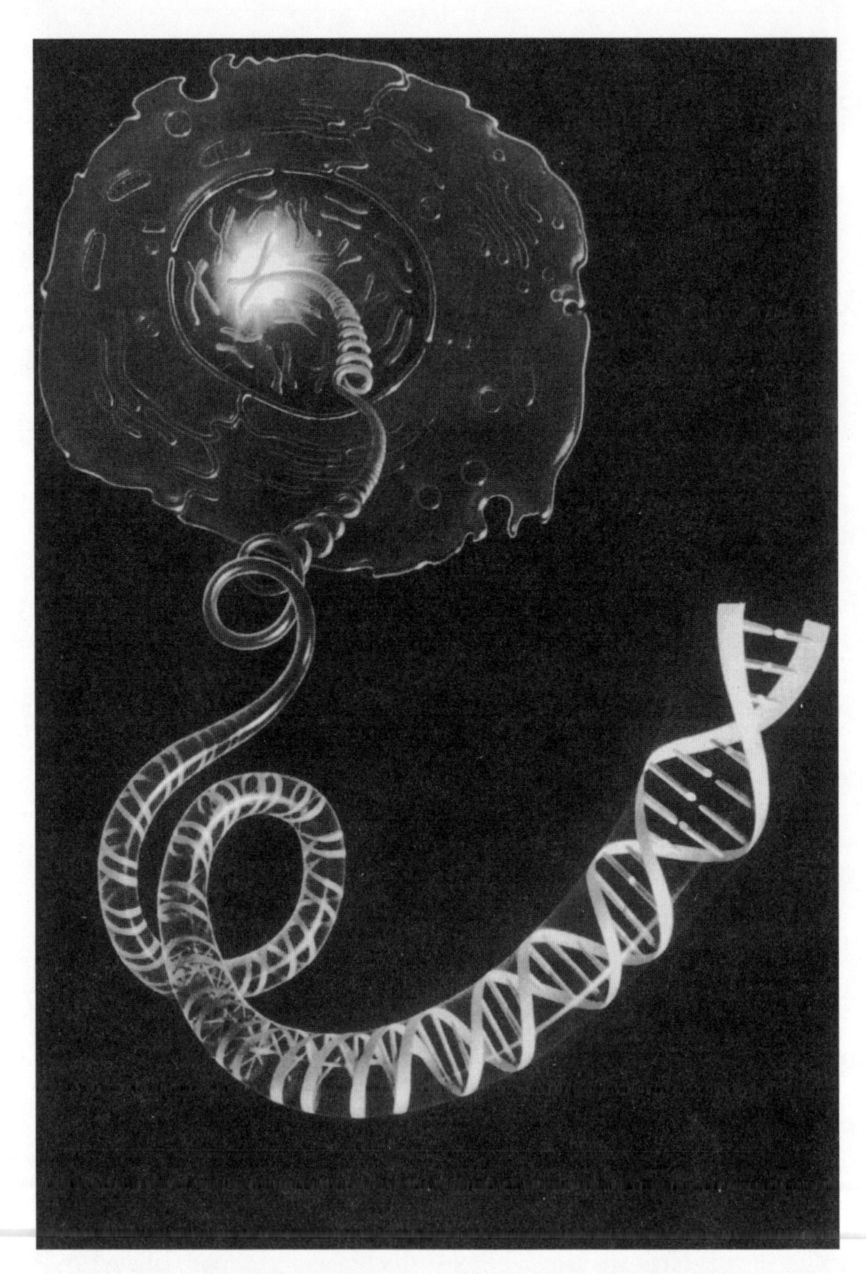

In the nucleus of our cells, 23 pairs of chromosomes are made of DNA molecules. Along the double helix of DNA lie our genes like pearls on a string. These genes, in turn, consist of thousands of "base pairs" consisting of pairs of nucleic acids A, T, C, and G. (Courtesy National Institutes of Health)

turn contains the code necessary to manufacture a single protein molecule. (More precisely, it takes three base pairs, or a codon, to code for a single amino acid, which in turn are the building blocks for proteins. Since there are four types of base pairs, there are thus $4 \times 4 \times 4 = 64$ possible amino acids that can be created by the DNA molecule. Since this is larger than the number of amino acids found in nature, more than one codon can code for the same amino acid.) Our bodies, in turn, are made of these proteins. That is the key: each gene produces one protein, which in turn may circulate in the body to perform a specific function, such as triggering chemical reactions as an enzyme or serving as a building block for tissue.

By the time Collins was a graduate student at Yale, he could feel the revolutionary winds unleashed by the decoding of the DNA molecule. Biology, he realized belatedly, was no longer memorizing the parts of a flower; biology was undergoing a profound change, similar to the epic birth of quantum mechanics itself back in 1925. "I realized, 'Oh my gosh, this is where the real golden era is happening.' I was worried that I would be teaching thermodynamics to a bunch of students who absolutely hated the subject. Whereas what was going on in biology seemed like quantum mechanics in the 1920s . . . I was completely blown away."

At this point, Collins took the biggest gamble of his career, switching fields. Like a long string of other quantum scientists before him, he held his breath, took the plunge, and never looked back.

## Reading the Code of Life

Just as the engine driving the explosive growth in computer technology for the next twenty-five years is the use of photolithography to exploit smaller and smaller wavelengths of light, the engine driving the explosive growth in DNA sequencing for the next twenty-five years is the automation of the techniques originally pioneered by Frederick Sanger, Walter Gilbert, and Allen Maxam.

To understand how DNA sequencing is done and how easy it is to computerize, imagine finding a treasure map which is written with impossibly dense cryptic symbols. To decipher the treasure map, you might proceed in three steps.

First, with a pair of scissors you snip off key segments of the code for detailed analysis. Second, you enlarge these segments with a magnifying lens. Third, you peer into the lens and read the coded letters on each segment.

Each of these three steps has a counterpart in DNA sequencing. For example, snipping off segments of the code corresponds to using organic chemicals, called "restriction enzymes," which can slice up DNA at specific points. (Fortunately, these remarkable restriction enzymes are produced naturally by certain bacteria when they combat viruses, slicing up the attacking viral DNA into ribbons). About 400 of these restriction enzymes have now been identified, each one able to slice up a DNA segment at specific points.

In the next step, scientists have to magnify these microscopic segments. The counterpart of a magnifying lens is to insert these DNA segments into bacteria (such as *Escherichia coli)* which then make millions of copies of the fragments, somewhat similar to the way that fermenting creates alcohol.

Finally, the counterpart of reading the code on these segments is to use a device called "gel electrophoresis" to separate out these microscopic segments.

To see how this separation works, imagine a group of children engaged in a race. In general, the heavier children will be slower than the lighter ones, so eventually the children begin to separate out. At the finish line, the lighter children will be first, followed by the heavier ones, who will be lagging at the back.

Similarly, heavier (or longer) gene fragments placed in a viscous gel move slower than lighter (or shorter) ones. By watching these fragments move sluggishly in a gel inside an electric field, you can distinguish the heavy ones from the lighter ones. At the "finish line" we see a series of bands, each band representing a distinct gene sequence of a certain weight or length. The distance between the bands tells us the relative weight of these bands, with the heaviest on one side and the lightest on the other. (For example, if we use a restriction enzyme which cuts DNA only whenever G appears, then the distance separating each of these bands tells us where along the DNA sequence the various G's lie. We can repeat this process with different restriction enzymes, and hence locate the position of all the A's, T's, and C's in the same way. After several iterations of this technique, we can read off the sequence of A, T, C, and G.)

Previously, biologists could only guess at the riches that lay hidden within the gene. But with this pioneering work, biologists began to read the ancient code of life for the first time in its three-billion-year history.

These DNA sequencing methods are so easily automated that, in fact, we expect the tree of life for thousands of life forms to be sequenced by 2020. Thus we will know the genetic relationship between many organisms on earth—and when they separated from each other. The details of the evolution of life on earth, once a matter of endless speculation, will be reduced to mathematics. This will give new meaning to the expression "the web of life."

One factor accelerating this process is the simple economics of gene sequencing. In 1986, Gilbert shocked audiences at the Cold Spring Harbor conference when he estimated it would cost one dollar to sequence one base pair or $3 billion for the entire Human Genome Project.

"The audience was stunned," recalled Robert Cook-Deegan of the National Academy of Sciences, by Gilbert's projection, which many felt was too low. "Gilbert's cost projections provoked an uproar." As outrageous as the claim sounded in 1986, today Gilbert's estimates sound quaint and even conservative; by 1990, the cost per base pair fell to ten dollars. Now it costs less than fifty cents to sequence a single base pair, and that price continues to drop like a rock. In fact, by 2020, many scientists expect that the cost per base pair could be an infinitesimal fraction of a penny, making personalized DNA sequences economically feasible.

## Of Microbes, Mice, and Men

Not surprisingly, the preliminary information scientists have obtained about these genes is profoundly altering our understanding of our bodies and the origin of our species, as well as our relationship to the rest of the animal kingdom.

For example, below is a table showing the number of base pairs in various forms of life on earth:

| Organism | Base Pairs (in millions) |
|---|---|
| Viruses | .01 |
| *E. coli* | 5 |
| Yeast | 12 |
| Nematode (worm) | 100 |
| Drosophila (fruit fly) | 180 |
| Tomato | 700 |
| Mouse | 3,000 |
| Human | 3,000 |

Scientists are completing the sequencing of these organisms in roughly this order.

Viruses, the simplest of all organisms, were the first to be completely sequenced since they consist of short DNA or RNA strands surrounded by protein coats. In 1977, Frederick Sanger and his colleagues found the complete DNA sequence for the first virus, phi-X174. This virus was chosen because of its relative simplicity: it has only nine genes, arranged on a single chromosome 5,375 base pairs long. If spelled out in terms of

A, T, C, and G, it would fill up only one page of this book with dense script. The human genome, by contrast, would take up 500,000 pages.

Some of the greatest killers in history, like smallpox, are now yielding to the gene hunters. Smallpox is known to have 186,000 base pairs, polio 7,700 base pairs, rabies 13,000, measles 18,000, influenza A 18,000, and the common cold 7,500. One of the longest viruses to be sequenced is the human cytomegalovirus, which has 230,000 base pairs and causes flu-like symptoms.

The pace of sequencing other forms of life is accelerating. The next milestone was reached in late 1995, when the first gene map of an entire cell was deciphered. The cell, *Hemophilus influenzae*, contains 1,743 genes, located on a single, circular chromosome made of 1,830,137 base pairs.

In early 1996, even this herculean feat was surpassed when the genome for ordinary baker's yeast was decoded. Yeast contains 12,057,000 million base pairs, divided into 6,000 genes, arranged on 16 chromosomes. Yeast is of particular importance because it shares so many of its genes with humans.

In 1997, scientists at the University of Wisconsin–Madison announced they had unraveled the genome of the bacterium *E. coli*, which contains 4,638,858 base pairs and 4,300 genes. Fifteen percent of human gene sequences contain parts of the *E. coli* genome.

Presently, scientists are working on many fronts, performing DNA sequences on a variety of organisms simultaneously. We expect to see scientists in the next few years announcing the DNA sequence of increasingly complex organisms, including nematode worms, fruit flies, mice, and eventually humans (in roughly this order). The culmination of this long process will be personalized DNA sequencing.

## The Human Family Tree

Human sequencing has resulted in many surprises. One profound discovery is how close we are linked genetically to other animals in the web of life. Since all life on earth probably originated from a single DNA or RNA molecule, the genetic overlap between any two life forms gives us a numerical way of calculating how close they are in terms of their evolution. The greater the overlap, the closer they are on the evolutionary tree.

Mapping the DNA tree of life also has enormous implications for human medicine. We often find genes in animals which are similar in function to genes in human beings; these are called "homologous genes." (Homologous genes are related genes found in entirely different species, having a common ancestor with that gene, and often, but not always, performing the same function. Wings and hands are homologous organs.)

Finding homologues to human genes in the animal kingdom, which are often less complex, can save scientists thousands of hours of rummaging through the human genome.

For example, about a billion years of evolution separate humans from yeast. This is reflected in the fact that a third of the genes in yeast are also found in humans. About 40 percent of the genes in roundworms are also found in humans. The genetic overlap between mice and humans is about 75 percent.

Molecular biology indicates that our ancestors (i.e., hominids) separated from the ape family roughly 5 million years ago. We find that we share fully *98.4 percent* of our DNA with our closest genetic relative, the chimpanzee.

By determining how close any two humans are genetically, we can also determine how close they are by blood. For example, two identical twins have the same genetic code, and therefore the "genetic distance" between them is zero. (Actually, even identical twins have a few dozen genetic differences between them caused by random mutations.) If we compare parents and children, or two siblings, we find that, on average, they differ by about 0.05 percent of their genetic code. (This means that close relatives differ by about 1.5 million nucleotides.) And if we take two humans purely at random, then we find that, on the average, their genetic code differs by about double that amount, or 0.1 percent. This is summarized in the following table:

| ORGANISM | % GENETIC OVERLAP WITH A HUMAN |
|---|---|
| *E. coli* | 15 |
| Yeast | 30 |
| Worm (Nematode) | 40 |
| Mouse | 75 |
| Cow | 90 |
| Chimpanzee | 98.4 |
| Another human | 99.9 |
| Sibling | 99.95 |

By computing the "genetic distance" between any two humans, we can also reconstruct the outline of the human evolutionary family tree. For example, we can calculate that the genes in the human body diverge at the rate of 2 to 4 percent per million years. From this, some scientists have asserted that humans probably branched from a common ancestor sometime between 140,000 to 290,000 years ago.

Using this technique, we can now construct the complete "family tree" of the human race, putting in all the details that were lost thousands of years ago in prehistory. Analyzing just a few proteins and genes already has given us startling insight into the origin of all the races and peoples of the world. By 2020, when personalized DNA sequencing is possible, our ancestral family tree should be nearly filled in, including all the branches which have been forgotten for tens of thousands of years.

Not only does the map fill in gaps in the linguistic and archaeological theories about the origin of humanity, it even gives the dates at which missing branches in our family tree diverged from other branches thousands of years before the first written records.

### DNA Testing

Although finding all the genes within our DNA will take many years, there are several easy spin-offs which are already rippling through society. For example, it is very easy to use DNA sequencing techniques to identify a handful of "markers" along the genome which are unique to every individual. This makes possible DNA testing, much like fingerprinting in the last century, an indispensable part of criminology.

In the next century, DNA testing will continue to have important applications in many areas, including:

**Paternity and immigration suits.** About 285,000 paternity suits are filed nationwide every year in the United States, of which 60,000 are disputed and require testing. In the future, not only will all paternity suits be settled definitively; it should be possible to establish the precise genetic relationship between any two individuals.

**Solving historical mysteries and exposing frauds.** In 1997, DNA evidence apparently vindicated Sam Sheppard, the doctor accused of killing his wife in 1954 (whose story became the basis of the movie and TV series *The Fugitive*), and identified the likely murderer.

**Analyzing DNA from ancient people.** The DNA of the "Man in the Ice," frozen in a glacier thousands of years ago, is now being analyzed, as well as the DNA from scores of ancient Egyptian mummies. This is already yielding new information about the history of disease and the way ancient people lived.

**Analyzing DNA in amber.** DNA samples from insects preserved in amber can date back to before the time of the dinosaurs, which died out 65 million years ago. George Poinar of Oregon State University, for example, has even extracted muscle tissue from a 125-million-year-old Lebanese weevil. This is, says Poinar, "the best preserved protein on the face of the earth." So far, DNA has been successfully extracted from about a half dozen ancient samples in amber.

The poet Alexander Pope wrote in his poem "Hesperides":

> I saw a fly within a bead
> Of amber clearly buried;
> The urn was little, but the room
> More rich than Cleopatra's tomb

We now realize that the DNA in amber is indeed much richer than Cleopatra's tomb.

**Predicting medical disorders.** The skating world was shocked when two-time Olympic gold medalist Sergei Grinkov suddenly collapsed and died of a heart attack in 1995 when he was only twenty-eight years old. His blood was subsequently analyzed by DNA sequencing. As expected, a genetic defect was found. The gene *PLA2*, which he inherited from his father (who also died young), was responsible for clogging his heart prematurely.

For the public, however, the most sensationalized impact of DNA testing has been in the area of DNA fingerprinting and criminology. Being able to read just a handful of markers on DNA samples has completely overturned the study of criminology.

### Crime, Punishment, and DNA

In 1983, DNA testing burst on the international scene with a sensational rape/murder case that began in the village of Narborough, England. After many surprising twists and turns in this historic case, DNA evidence was crucial in both exonerating the chief suspect and convicting the true killer.

Since then, DNA tests have had a profound effect on crime and punishment. They have reversed the verdict in fully 25 percent of the sexual assault cases referred to the FBI since 1989. President Bill Clinton's 1994 Crime Control Act contained a little-noticed provision which called for the establishment of a national DNA data bank.

Since then, forty-two states have passed laws to require prison inmates to give blood or saliva samples for DNA analysis. Of those, twenty-six states have started setting up DNA data banks of their own, which will be eventually linked into the national data bank.

By 2020, however, there will be a significant change that will make DNA fingerprinting obsolete. The handful of markers used in DNA fingerprinting will give way to the infinitely more sophisticated personalized DNA sequences. Fingerprints may give us a match, but DNA may tell us what the person looked like and give us his or her medical history.

For example, by examining a *single cell* from a person's dandruff, it is in principle possible (using the polymerase chain reaction process) to reconstruct the entire genome of an individual. Possessing the personalized

DNA sequence, one can reconstruct important details of the person, including blood type, eye and hair color, sex, genetic diseases, general body shape, medical status, disposition to baldness, approximate height and weight, even body chemistry. (Certain features whose genetic origins are still unclear, such as the details of the face, probably will not be available even in 2020.)

## Synergy Between Computers and DNA Research

Personalized DNA sequencing by 2020 is no longer a farfetched idea because of the intense synergy between the computer and the biomolecular revolution. What is forcing this cross-pollination is the sheer volume of work involved in sequencing and then analyzing three billion base pairs, which has overwhelmed the molecular biology community. Inevitably, they are turning to computer scientists.

"We've always known that the day would come when engineering would play a critical role," said David Botstein. "That day has arrived." The rapid progress since the introduction of computer sequencing is breathtaking. In the 1980s, it took a biologist an entire year to sequence 10,000 base pairs. By 1992, a single machine could analyze that many base pairs in a single day. Leroy Hood of the University of Washington predicts that by 2002 scientists will be able to sequence one to ten million base pairs per day per technician! This is an advance by a factor of a third of a million in just a decade.

Today, the world's largest repository for gene sequencing for all life forms is GenBank, located at the Los Alamos National Laboratory. (GenBank was started in 1982 by mathematician Stanislaw Ulam, who earlier won notoriety for helping to create the hydrogen bomb along with Edward Teller. Ulam, like Schrödinger, Delbrück, Crick, Pauling, Gamow, Jordan, and Gilbert before him, was fascinated by the ability of quantum physics to decode the secret of life. From all over the world, scientists send their DNA sequences by e-mail to the Los Alamos computer, which acts as a huge clearinghouse for genetic information.

By 1990, 60 million base pairs had been sequenced, and 50 million base pairs (a quarter of which are human) were stored in GenBank. By 1997, GenBank contained over *843 million* base pairs.

To fully understand why computers and robots will eventually take over the sequencing of DNA, imagine that DNA is a long ribbon stretching off into the horizon. This ribbon has tiny stripes on it which are only one millimeter wide, about the thickness of a pencil line. In this analogy, each stripe represents a single base pair. On a scale of this magnitude, the ribbon representing the DNA of a worm would extend 120 miles. The ribbon representing the human genome would extend *1,600 miles*, or roughly halfway across the United States.

Robert Waterson, a mathematician who is now director of the sequencing effort at the largest DNA center in the United States, at the Washington University in St. Louis, notes, "In the last six years, with all this sequencing, we are not yet even halfway to Columbia yet [160 miles away from St. Louis]. And now we have the temerity to suggest that it is time to set out for L.A. [1,600 miles away]."

Yet, according to Waterson, every week his group generates 27,000 DNA segments, each made of 500 nucleotides. He hopes to get up to 40,000 a week within a year. "We figure that for us to do one-third of the human genome in five to six years, we'd have to get reads of 80,000 to 90,000 a week."

## The Birth of a New Science: Computational Biology

Computer scientists engaged in DNA sequencing will not simply pack up their bags and go back home in 2005 when the project is done. This is because the Human Genome Project is just the beginning of an entirely new science.

"It's turning biology into an information science. Many biologists consider the acquisition of sequencing to be boring. But from a computer science point of view, these are first-rate and challenging algorithmic questions," says Richard Karp of the University of Washington, one of the leading computer scientists in the country.

Computer science first invaded the world of biology in 1983 when Russell Doolittle and his colleagues rocked the closed world of molecular biology by making a major biological discovery by simply reading computer printouts. Without performing a single experiment, Doolittle was able to find a similarity between two dissimilar proteins involving different areas of biology: the *sis* cancer gene and a cellular growth factor. He and his colleagues noticed that the DNA sequence found in this particular type of cancer was also the same DNA sequence involved in cellular growth, thus showing that cancer genes created abnormal growth in cells. Biology was not supposed to be done this way.

Robert Cook-Deegan of the National Academy of Sciences asked the rhetorical question: "Why should he be able to publish a major discovery that came from just sitting at a computer terminal? That wasn't biology, was it?"

This dramatic discovery heralded the beginning of using computers to spot patterns in DNA sequences rather than getting one's hands dirty with test tubes of proteins. "The intrusion of computers into molecular biology shifted power into the hands of those with mathematical aptitudes and computer savvy," Cook-Deegan notes. "A new breed of scientist began to rise through the ranks, with expertise in molecular biology, computers, and mathematical analysis."

In the past, biologists learned about life by analyzing the interior of living specimens (i.e., *in vivo).* In the last century, they learned to study life in glass (i.e., *in vitro).* In the future, they will study life via computers (i.e., *in silico).*

## DNA on a Chip

What will the sequencing process look like in 2020? Will we have thousands of acres devoted to housing monstrous computers and robot factories that sequence people's DNA?

Probably not. Just as the future of computer technology lies in miniaturization via the microchip, many scientists feel the future of DNA sequencing will be the "bio chip" and the "DNA chip," in a grand merger of the computer and biomolecular revolutions.

The bio chip is a microchip designed specifically to perform "homology" searches between similar human and animal genes. This bio chip is enormously useful for biologists, because if a certain genetic sequence in an animal is already known to control a certain protein, searching for its counterpart, or homologue, in humans reduces the guesswork involved in identifying unknown human genes. In the future, in line with Moore's law, the bio chip will eventually take over the business of DNA analysis.

A primitive bio chip already exists—it is a quarter of an inch across, contains 400,000 transistors, and "is the most complex chip that the Jet Propulsion Laboratory at Caltech has ever designed," according to Leroy Hood. It is about 5,000 times faster than a Sun Sparcstation 1. When instructed to identify a 500 base sequence among 40 million bases, the Sun Sparcstation computer took five hours, while the chip took only *3.5 seconds.*

Scientists are now perfecting a DNA chip as well, a microchip that can almost instantly screen a person's DNA for selected genes. DNA chips, which will soon begin to enter the marketplace, can test for HIV, cancer, and thousands of genetic diseases within a matter of hours. This new diagnostic tool may revolutionize the $17.5 billion diagnostics industry.

With the advent of the DNA chip, Gilbert's dream of "personalized" DNA records that contain all our genes is no longer pie in the sky. Already, several start-up biotech companies are racing to read our DNA by scanning them onto microchips. The merger of computers and molecular biology on a DNA chip may signal a new era in cheap, rapid genetic screening.

To the naked eye, the DNA chip seems to be rather unremarkable. Only the size of a fingernail, it looks very much like the microchip that is used in most PCs. But under a microscope, you would see a most unusual pattern. With the same photolithography techniques used to etch microscopic grooves in tiny transistors, scientists use a template to etch the outline of DNA strands corresponding to a particular set of base pairs. By washing a solution of DNA over these templates, those sequences which remain are those which fit precisely into each probe.

The trick is that the only DNA strands which stick to the microchip are those which precisely fit the template pattern etched onto the chip. All other strands are washed away. A laser then makes some of these sequences fluorescent and a computer makes the final identification.

Already, the Affeymetrix corporation is marketing a chip with 65,536 probes etched onto it, each probe being the template for eight base pairs. "We have actually produced a prototype chip containing a million probes," claims Robert J. Lipshutz, the company's director of advanced technology. Affeymetrix has already succeeded in placing all the genes for HIV on a DNA chip, which can accelerate AIDS screening dramatically.

The potential for the DNA chip, which squeezes an entire DNA laboratory onto a single chip, is enormous. Already, one can use it to screen for the notorious *p-53* mutation, which is implicated in over half of all cancers. Cystic fibrosis, which comes in any of 450 different mutations, can be screened by the DNA chip in a few hours at a cost of only a few dollars. (The traditional process of identifying these cystic fibrosis genes is quite expensive and takes at least a week.)

## Post-Genome Era: From 2020 to 2050

The rapid progress toward personalized DNA sequencing should continue steadily and unabated for the next twenty-five years. It is largely a by-product of the fact that DNA sequencing techniques are easily automated and computerized. By 2020, we should have a nearly complete Encyclopedia of Life.

After 2020, when everyone has their own personalized DNA code, the problem will increasingly shift to understanding how the genes perform their magic in our bodies. As Walter Gilbert says: "Science will have moved on to the problem of what a sequence *means*, what the gene actually does."

After 2020, molecular biologists will be flooded with millions upon millions of genes from different organisms whose function must be tediously determined. Just possessing a personalized CD with 3 billion impenetrably dense symbols on it does not mean that we know how our genes function.

For example, now that yeast has been decoded, it will become a laboratory for human gene research to determine how genes function. Usually, scientists determine what a specific gene does by deleting or mutating it in an organism and watching what happens to the organism. (This painfully crude method can be compared to trying to figure out how a supercomputer works by smashing individual components to see what happens.) Because yeast cells reproduce in a few days, they have an enormous advantage over mice, for example. Already, scientists have found that the human cancer gene *ras* is also found in yeast and, as in humans, makes yeast organisms lose control over their reproductive processes. Yeast has proven to be a gold mine for other human genes as well, including those for neurological and skeletal disorders (even though yeast has no nervous system or bone!).

One roadblock which will impede progress in the "post-genome era" is the infamous "protein folding problem." In biology, molecular structure is destiny. Knowing the shape of an organic molecule often helps to determine its function. For example, many organic molecules which interact with each other resemble a "lock and key," with one molecule shaped like a key entering a keyhole in the other molecule. Unfortunately, X-ray crystallography, the workhorse of molecular biology, depends on being able to crystallize the sample. If a particular protein cannot be crystallized, then it cannot be probed by X-rays.

Using standard chemical methods, one can determine the atoms within a protein molecule (which may number in the thousands) and the sequence of its amino acids. But this tells us nothing about how the amino acids are physically arranged in three dimensions. In general, protein molecules look like a series of amino acids arranged in many ribbons and helixes glued together in a bizarre fashion. At first glance, attempting to determine the shape of a complex protein molecule without X-ray crystallography seems hopeless.

But here is where the quantum theory comes in. Quantum mechanics gives us the bonding angles between each atom, allowing us to determine how these ribbons and helixes rotate with respect to each other. But to determine the precise shape of these ribbons and helixes, one must use a powerful supercomputer.

In general, all physical systems tend toward a state of minimum energy. Imagine a protein molecule that consists of a large collection of Slinky coils bound together by string. Now shake this strange contraption. At first, it seems as if the resulting motions are completely random and impossible to predict. But actually the final configuration of these coils, no matter how complicated, is nothing but the state of minimum energy.

Using a supercomputer, one can calculate the energy of millions of possible arrangements of these ribbons and helixes. By selecting the configuration of ribbons and helixes with the lowest energy, one can determine how proteins "fold" into the proper shape.

Not surprisingly, the protein folding problem is a stubborn one requiring a lot of computer time and ingenuity. Since by 2005 we will in principle have all 100,000 protein molecules required to build a human being, scientists will be relying on supercomputers for many decades to tell us how these thousands of proteins fold in three dimensions.

As you can imagine, progress beyond 2020 will likely slow down significantly as the difficult problems of gene function, polygenic diseases, and protein folding begin to dominate research. It will be a slow, labor-intensive process. However, the dividend will be enormous, unraveling a host of genetic diseases which have plagued humanity since we first walked the earth. It may also give us the molecular tools necessary to conquer one of the greatest killers of modern times, cancer.

# 8

# Conquering Cancer—Fixing Our Genes

"How cancer develops is no longer a mystery."
—ROBERT A. WEINBERG, MIT

". . . the time is coming when there will be magic bullets to treat cancer the way we now treat many infectious diseases with vaccines and antibiotics."
—FRANCIS COLLINS, NIH

REBECCA LILLY is a typical bubbly sixteen-year-old suburban high school girl. Smart, athletic, and vivacious, she frets about the things that worry most teenagers, such as school, grades, and being accepted by her friends. Her softball coach, Tom Mayers, says proudly, "Becca's got a heart as big as this ball field. She's gutsy, she never complains, she never gives up. She keeps us all in the game."

One of the highlights of her life was a surprise sweet-sixteen birthday party. She mingled easily on the dance floor, dancing to the Macarena and laughing with her friends. Like most teenage girls, she dreams of boys, parties, and her future. Unfortunately, all these dreams are on hold, for an unspoken chasm separates her from her friends. She suffers from an incurable brain tumor. On November 1995, she became the first person in history to be treated for brain cancer using gene therapy.

Since age ten, she has suffered from the knowledge that she may die of a brain tumor, a high-grade malignant glioma growing relentlessly inside her skull. Untreated, the tumor will expand without mercy until it slowly crushes her brain. She has been in and out of hospitals ever since. She is more familiar with brain scans, MRI, radiation treatments, and brain surgery than she is with the SAT.

She has had open-brain surgery four times to remove the tumor, but each time it has grown back. Chemotherapy is out, since the toxic drugs designed to kill cancer cells cannot penetrate the blood-brain barrier and hence attack the brain tumor.

When, one by one, all the alternatives failed, her parents finally opted for a radical, last-ditch experimental treatment: gene therapy.

During a grueling nine-hour operation to remove most of her tumor, doctors injected a harmless virus into her brain. Scientists had altered the genetic code of the virus so that it was no longer harmful, and inserted a gene into the virus designed to infect the cancer cells and make them self-destruct. The virus was like a Trojan horse, designed to trick the cancer cells into dying.

For a while, this new treatment of tomorrow seemed to work. Rebecca recovered her usual sense of humor, her memory improved, and she seemed to return to normal. "She had six good months," says her doctor.

Unfortunately, in May 1996, the latest MRI scan showed that the brain tumor was growing back. "This is a desperate field," admits Roger Packer, a neurologist at her hospital in Bethesda, Maryland.

But 1996 also saw new hope for gene therapy. By replacing the mutated gene *p-53*, which is found in over 50 percent of all common cancers (which codes for a protein that weighs 53,000 atomic units), doctors at the University of Texas were able to shrink lung tumors in two cases, stop lung cancer from growing in three others, and even wipe it out completely in another.

Although no one is claiming that this is a cure for cancer, this kind of gene therapy may one day revolutionize the way doctors treat cancer and genetic diseases. Gene therapy may eventually help to combat HIV and even chronic diseases like Alzheimer's, mental illness, arthritis, and aging.

By 2020, doctors may well regard chemotherapy, radiation, and surgery to treat cancer with the same dismay we now feel for the use of arsenic, bloodletting, and leeches to treat diseases years ago. By 2020, entire classes of genetic diseases, including many forms of cancer, may be viewed the same way we view smallpox today.

## The Father of Gene Therapy

"The floodgates are wide open," declares W. French Anderson, sometimes called "the father of gene therapy," who heads his own institute at the University of Southern California. "We've got the green light. We're going to have to proceed a step at a time. But we are at the beginning of what promises to be the most exciting time in the history of medicine. What an incredible time to be alive!"

Anderson predicts that, by 2020, not only will we have personalized DNA sequencing but "virtually every disease will have gene therapy as one of its treatments." His enthusiasm is shared by many of his colleagues. Leroy Hood of the University of Washington has confidently predicted, "Over the next twenty to forty years, we will have the potential for eradicating the major diseases that plague the American population."

Behind these auspicious predictions is the realization that the more we study diseases, the more we understand and appreciate their genetic and even molecular origin. In fact, Nobel Laureate Paul Berg of Stanford University even believes that *all* diseases are, in the final analysis, genetic in nature. He says, "You can sit here for an hour, and you can't get me to conclude that any disease that you can think of is not genetic."

Anderson, who dabbles in Formula One auto racing, archaeology, sports medicine, and his specialty, Tae Kwon Do, the Korean martial arts form, is a pioneer in this burgeoning field. He holds a fourth-degree black belt, and for relaxation he occasionally slams his right foot into a stack of wooden boards, cleanly severing five at a time. He was even the chief physician to the U.S. Olympic Tae Kwon Do team in Seoul in 1988.

Anderson likes to compare martial arts to doing research on the basic genetic mechanisms of the cell. Science "is something best done without thinking, something transcendent, and intuitive," he claims. Unlike other sciences, whose basic laws are well established, gene therapy is a new, by-the-seat-of-your-pants field that, like Tae Kwon Do, requires bold innovation and creativity as well as hard work.

In 1990, Anderson's team was the first in the world to get permission to perform an experiment that might revolutionize science well into the next century: to fix the defective genes of a human being. Within a few short years, a formidable army of doctors were following Anderson's lead and performing gene therapy experiments on a variety of diseases. By 1993, there were 40 gene therapy trials. By 1996, this number grew to 200, involving 1,500 patients. About 30 diseases are being studied, approximately half of them involving cancer. Gene therapy experiments were absorbing $200 million of the NIH budget.

On these clinical trials ride the hopes and prayers of youngsters like Rebecca.

## Three Stages in Medicine

Like computers, medicine is being thrust into its third stage by the biomolecular revolution, Anderson claims. During the first stage of medicine, shamans and mystics painfully scoured the plant kingdom for thousands of years looking for herbs that might scare dreaded spirits away, at times stumbling upon valuable remedies that are used even today. Some of our common drugs have their origin during this primitive but important stage. But for every herb that was, by trial and error, found to be effective against certain ailments, there were thousands more which did not work, some of which even injured the patients.

For example, a country doctor who became one of the founders of the famed Mayo Clinic in Rochester, Minnesota, recorded with rare candor that most of his potions were worthless, but there were two things in his black bag which were guaranteed to work every time: morphine and his saw, which were used in amputations.

In the second stage of medicine, which began after World War II, the mass distribution of vaccines and antibiotics temporarily vanquished whole classes of diseases. Abigail Salyers and Dixie Whitt, authors of *Bacterial Pathogenesis*, write: "One of the main reasons for the elevation of physicians to their current status as respected professionals was that antibiotics actually enabled them to cure diseases for which in the past they had only been able to provide ameliorative (and largely ineffective) therapy."

Fortunately, we are now entering the third stage of medicine, "molecular medicine," perhaps the most exciting and profound of all. For the first time in history, each level of pathogenesis, protein for protein, molecule for molecule, even atom for atom, is now being revealed. Like a general eagerly reading the map of the enemy's defenses, scientists today can read a germ's complete genome and identify the molecular weak spots in its armor.

As Sherwin B. Nuland of the Yale University School of Medicine says: "In a 20-year period, the ancient art of healing passed from the relatively simple and restricted optimism of the antibiotic era to the seemingly endless vistas of the molecular age."

## The Scourge of Cancer

The one disease that has frustrated the most intensive crash program in history is at last yielding its secrets to molecular medicine. Cancer, one of the most dreaded of all diseases, is the second leading cause of death in the United States (after heart disease), killing half a million Americans every year. It is also one of the most pervasive. Altogether, there are 200 forms of cancer (affecting virtually every type of cell in the human body). Unlike ordinary cells, cancer cells have lost their ability to stop dividing. They are immortal—they proliferate without limit until they choke off normal bodily functions and kill the victim. (This does not mean that each cancer cell is immortal. Cancer cells can die, just like ordinary cells. The difference is that cancer cells proliferate indefinitely, so the cell line is immortal.)

Scientists are now on the threshold of a complete understanding of how cancer develops at the molecular level. In the main, *the mystery of cancer has been solved.* Cancer has now been revealed to be a genetic disease, and the precise sequence of four to six mutations necessary to create a cancer cell for many common cancers is now known. Not only have the main genes involved been identified; scientists also know the basic molecular steps through which a normal cell suddenly becomes cancerous.

"The pieces of the puzzle have finally fallen into place," claims Robert A. Weinberg of MIT. Cancer research centers are now bursting with activity as they close in on the fine details of how cancers form and grow. As Dennis Salmon, a cancer specialist at UCLA, says: "This is the most exciting time imaginable!"

Molecular medicine has already given us the answer to one of the central mysteries of cancer—i.e., why it has such a bewildering variety of causes, from lifestyle, environment, viruses, toxins, diet, radiation, tobacco smoke, animal fat, sex hormones like estrogen, etc. About 30 percent of all cancers, in fact, can be traced to tobacco smoking alone. If we include the contribution from diet, we can establish a link to roughly 60 percent of all cancers. And by comparing ethnic groups that mature in different regions (e.g., Africans and Japanese growing up in the United States), epidemiologists have determined that a vast majority, perhaps as many as *70 to 90 percent,* of all cancers can be correlated to the environment and lifestyle.

## A Unified Theory of Cancer

There are two major kinds of genes involved with cancer: *oncogenes* and *tumor suppressors.* To understand how they work, think of a speeding car which has both an accelerator (oncogene) and a brake (tumor suppressor). One speeds up the car, the other stops it. The car can go out of control in two ways: either the accelerator can be stuck (an activated oncogene) or the brake can be defective (an inactivated tumor suppressor). In other words, a cell can go berserk either if it divides uncontrollably or if it loses its ability to stop dividing.

Scientists have found over 50 types of oncogenes for cancer of the breast, colon, bladder, and lungs. These oncogenes include the gene which codes for the protein p-21 (which derives its name from the fact that it weighs 21,000 atomic units, or as much as 21,000 hydrogen atoms), as well as p-60.

The second class of genes that can cause cancer, the tumor suppressors, include mutated versions of the genes *DCC* and especially *p-53*, which scientists are now realizing are found in the majority of common cancers. Unlike the oncogenes, these defects occur in genes which normally shut off the reproductive process; with mutations in these genes, the cells reproduce out of control, almost forever.

Doctors expect to have by 2020 almost a complete encyclopedia of perhaps hundreds of oncogenes and tumor suppressor genes, giving us an understanding of the molecular basis for cancer and opening up scores of new ways of attacking it.

## P-53: The Key to Most Cancers

One reason why scientists feel confident in predicting that whole classes of cancers may be curable by 2020 is that most cancers are caused by mutations in just a handful of genes, the most significant being *p-53*. Although hundreds of genes involving cancer may exist, the key to curing most cancers may be to focus on the common ones implicated in the vast majority of cancers and neutralize them via gene therapy or "smart molecules."

Every year, we find that mutated versions of *p-53* are implicated in more and more cancers, from cancer of the lung, colon, breast, esophagus, liver, brain, and skin to leukemia. It has been found in 52 common forms of cancer, and the percentage of cancers that have faulty *p-53* is staggering: 90 percent of all cervical cancers, 80 percent of all colon cancers, 40 to 60 percent of all ovarian cancers, 35 to 60 percent of all bladder cancers, and 50 percent of all brain cancers. "This quite clearly is the most commonly mutated gene we've yet found in human cancers," notes Bert Vogelstein of the Johns Hopkins School of Medicine. It is so important that scientists have dubbed *p-53*, when it functions normally, the "guardian of the genome." *P-53* is so essential for cancer formation that in 1994 *Science* magazine named it "Molecule of the Year."

Understanding *p-53* has also solved some long-standing mysteries which have dogged the field for decades.

*P-53* normally prevents reproduction in a damaged or mutated cell and promotes cell suicide (called apoptosis). When *p-53* is mutated or neutralized, deranged cells can continue to proliferate within the body, thereby creating tumors.

As we now understand, the reason for its appearance in a wide variety of cancers lies in its molecular structure; it is extremely long and delicate (consisting of 2,362 base pairs). Mutations in *p-53*, which is located on the short arm of chromosome 17, can occur at over 100 sites along the gene. Hence, *p-53* is riddled with potential sites for mutations. (By contrast, other commonly found genes involving cancer usually have harmful mutations occurring at only a half dozen sites.)

The gene is actually an aggregate, consisting of four or more identical copies of a smaller subunit. All four subunits must act correctly in order for *p-53* to properly control cell multiplication. The fact that *p-53* is such an unwieldy molecule makes it particularly vulnerable to mutations. For example, colon cancer results from the mutation of perhaps four to six genes. A typical cancer of the colon may proceed in the following fashion: the loss of function of the *APC* gene, the activation of the *K-ras* gene, and the loss of the *DCC* and *p-53* genes.

This, in turn, solves one of the central riddles of cancer, why it often takes twenty to forty years for a cancer to develop after the first exposure to radiation, asbestos, and other carcinogenic materials. The reason it takes so long is that a series of multiple mutations must occur before the growth mechanism of a cell is finally disrupted. This successive disabling of the cell's reproductive mechanism usually takes time, often decades, to occur.

All this has tremendous practical implications. Blood tests are becoming available to find out if people have a mutated version of *p-53*. Although it takes three to five more mutations to trigger a cancer, a mutation in *p-53* may be the most important of them all. By 2020, tests for defective *p-53* and hundreds of other genes implicated in cancer will be commonplace.

Second, gene therapy will target defective *p-53* genes to see if they can be replaced by a normal version of the gene.

Third, *p-53* will give us an understanding of why certain classes of chemicals and agents in the environment cause cancer. *P-53* has several "hot spots" where chemical toxins can bring about mutations. For example, aflatoxin, a potent cancer-causing chemical found in moldy food, which can lead to liver cancer, is known to cause a mutation in *p-53* by changing G to T. By analyzing the ways in which certain chemicals cause mutations in *p-53*, one may be able to understand why environmental factors and toxins can cause cancer.

Such discoveries could significantly affect the fortunes of multibillion-dollar industries. The tobacco industry, for example, has been able to defeat lawsuits brought by the families of smokers who died of lung cancer by claiming that no one can definitely prove that tobacco smoke causes cancer. Since the link between tobacco smoke and lung cancer is indirectly established through epidemiology and statistics rather than biochemistry, the tobacco industry has always claimed in court there is no "smoking gun" which implicates tobacco smoke.

All this changed in 1996, when scientists proved that the chemical benzoapyrene diol epoxide (BPDE), which is commonly found in tobacco smoke, causes a characteristic set of mutations in *p-53* at three specific sites. These three mutations are the "fingerprint" of BPDE and are easily detected in *p-53* mutated by tobacco smoke. These are precisely the mutations implicated in lung cancer.

Since over 400,000 Americans die of lung cancer each year (80 to 90 percent linked to smoking, according to the American Cancer Society), this could have enormous political and economic repercussions. In the future, lawsuits may be decided on the basis of whether cancers can be traced to specific molecular "fingerprints" along key genes such as *p-53*, *p-16*, *ras*, and so on.

By 2020, scientists will have found the genetic fingerprints of hundreds of different kinds of chemical pollutants in our environment. By matching a person's cancer with the genetic fingerprint left by a carcinogen, scientists in many cases will be able to tell precisely what gave this person cancer. This could have a profound effect on how pollutants are regulated and who pays for the damage. It may also help solve the mystery of why breast cancer is on the rise in the West, which has stumped epidemiologists around the country.

But perhaps one of the most intriguing discoveries in recent years involves something called telomeres, which are now recognized as a kind of biological "clock." By resetting the clock, one may be able to order cancer cells to die.

## Telomeres: The "Fuse" for Cancer

Since the beginning of cell research, scientists have dreamed of being able to understand the mysterious biological clock that determines when normal cells die and explains why cancer cells are immortal. Within the last few years, this clock has been discovered, opening up an entirely new field for the twenty-first century.

Since the 1960s, scientists have known that cultured cells of newborns will divide 80 to 90 times, whereas cells of seventy-year-olds will divide only 20 to 30 times. But if a cell contains a time bomb, then what is the fuse? We now know.

In the 1970s, it was noticed that the ends of our chromosomes have a "cap" on them called telomeres, much like the plastic tips on the ends of shoelaces that prevent them from getting frayed. If these telomeres are lost, then the chromosomes stick to each other and the cell eventually dies. In a normal cell, the telomere fuse gradually becomes shorter and shorter, until the cell commits suicide. When the fuse is gone, that is the end of the cell. But certain abnormal cells, we now understand, have the remarkable ability to keep the telomere fuse perpetually long. They become immortal. Such cells are called cancer cells.

Close examination of the telomeres showed they consisted of the genetic sequence: TTAGGG . . . repeated over and over again, up to 2,000 times. It was found that the older the cell, the shorter the telomere. The cell loses about 10 to 20 of these segments every time it divides—the fuse gets shorter after each division. It was theorized, therefore, that when the fuse (telomere) gets too short after too many cell divisions, the telomere disappears and the cell dies.

In 1984, the enzyme "telomerase" was discovered; this enzyme could reverse the process and lengthen the telomeres, thereby preventing cell suicide. Telomerase, however, is absent in most cells of the body.

In 1994, Christopher M. Counter, Silvia Bacchetti, and their colleagues at McMaster University made a crucial discovery. They showed that telomerase is found in a wide variety of cancers, which have a genetic mutation allowing them to manufacture telomerase. This, in turn, prevented their telomeres from disappearing and thereby made them immortal.

With these discoveries, we now have a working hypothesis about cell aging, cell death, and cancers. The telomere acts like a clock which measures the process of cell aging and death. The shorter the telomere, the older the cell. Cancer cells, because they can manufacture telomerase, which freezes the contraction of the telomeres, "have forgotten how to die," as Samuel Broder, director of the National Cancer Institute, puts it.

This discovery opens up new avenues for cancer detection and therapies in the twenty-first century. One method would be to detect telomerase in the body. Since normal cells lack telomerase, the presence of this key enzyme would signal the presence of growing cancer cells. Another possibility is to neutralize the telomerase, so that cancer cells begin to age normally. Since telomerase is not found in normal cells, this therapy would target only cancer cells. (Chemotherapy, by contrast, acts more like a blunderbuss, striking normal and cancer cells alike.)

## Cancer in 2020

Cancer, because it is a crazy quilt of at least 200 different kinds of diseases, one for every kind of human tissue, will not be cured in its entirety by 2020. As Richard Klausner of the National Cancer Institute says: "There will never be a single cure for cancer."

However, by 2020 scientists should have an almost complete catalog of the mutations involved in these 200 cancers, which will trigger an explosive growth in radically new cancer therapies and detections, including a variety of startling new strategies for attacking cancer's molecular weak spots and vulnerabilities.

There are several new avenues generating intense interest, many of which should reach fruition by 2020.

The first has to do with *cancer detection.* Imagine being able to detect a tiny colony of cancer cells a decade before a visible tumor forms. Extremely sensitive tests are now being devised (and will soon hit the market) which can detect infinitesimal amounts of proteins that are emitted by only a few hundred cancer cells as they grow and eventually create blood vessels. These proteins can be detected by analyzing one's urine and blood. Similarly, doctors will be able to test directly for the presence of cancer genes in our genetic makeup. About half of all cancers are found in our hollow organs (lung, colon, bladder), which often have a mutant *ras* gene. By devising simple tests in our urine and blood for the *ras* gene (which in the future will be performed in our own home), we will be able to detect a majority of all cancers years before they form tumors or spread.

The second approach has to do with the development of *natural cancer fighters.* Science is beginning to understand at the molecular level why certain natural products and vitamins help guard against cancer. Genistein, which is found in soybeans and cabbage, is found in high concentrations in the Japanese diet and is known to suppress the formation of blood vessels in cancer tumors. (The Japanese, in fact, have concentrations of genistein in their urine 30 times that of Westerners.) Antioxidants in foods (like vitamin C and E, and lycopene in tomatoes, catechins in berries, and carotenoids in carrots) are known to reduce the mutation rate in cells by suppressing free radicals. Other vegetables contain chemicals that create enzymes which protect against cancer (such as indoles in cabbage, limonoids in citrus fruits, isothlocyanates in mustard).

The third approach is *enhancing the immune system.* Normally, the antibodies created by the immune system are not sufficiently powerful to target a cancer cell. One can, however, create "monoclonal antibodies," or chemicals which specifically target the proteins found on the surface of the cancer cell. After an initial wave of enthusiasm for such antibodies, the scientific community experienced intense disappointment. But Lloyd Old, formerly of the Memorial Sloan-Kettering Institute for Cancer Research in New York, says, "The concept remains sound, and slow, steady progress is being made in developing antibody therapies."

A fourth approach has to do with *targeting cancer genes.* Gene therapy can inject the correct gene to replace the defective ones causing the cancer. Scientists have successfully injected the correct *p-53* gene into cancer cells in cell cultures, thereby stopping their reproduction, and are performing human experiments as well. Alternatively, scientists could develop inhibiters to block the defective protein created by the cancer gene. For example, the protein produced by the *ras* oncogene can be stopped by farnesyl transferase inhibitors.

A fifth approach centers on *cancer vaccines.* Although this approach was one of the first to be tried and was later abandoned, new interest in cancer vaccines has been stimulated by the biomolecular revolution. With modern techniques, one can accurately monitor the effectiveness of certain vaccines, which was almost impossible before.

A sixth approach doctors can take is to *shut off the cancer's blood supply.* In order for a cancer to grow beyond the size of a pea, it has to stimulate the growth of blood vessels and capillaries to supply nourishment for the tumor. This process of growing blood vessels is called "angiogenesis." The strategy to block blood vessel growth is to develop angiogenesis blockers. Already, thirty biotech firms around the world are creating such angiogenesis blockers, such as TNP-470, some of which are now in clinical trials.

Yet another approach targets *telomerase.* If we can neutralize telomerase, we can make the cells mortal again, just like other cells.

No one knows precisely which therapy will be most effective against cancer. But the point is that the biomolecular revolution has now cracked the mystery of cancer and has given us a wealth of extremely promising new avenues for attacking cancer which will eventually replace the primitive tools of chemotherapy, surgery, and radiation available today.

Many scientists believe that by 2020 entire classes of cancers may be curable.

## Hereditary Diseases: Ancient Scourge

By 2020, the biomolecular revolution may also bring another class of ancient diseases under control: hereditary diseases.

Stephen Hawking, one of the world's great cosmologists, suffers from

ALS (amyotrophic lateral sclerosis), the same hereditary disease which took lives of baseball player Lou Gehrig, Senator Jacob Javits, and actor David Niven. Although Hawking's thinking is as sharp and penetrating as ever, he has totally lost control of his hands, arms, legs, tongue, even his vocal cords, and communicates with the world via a voice synthesizer while sitting completely helpless in a wheelchair. He performs all his complex mathematical manipulations entirely in his head.

Throughout history, horrible genetic defects such as ALS have tormented the human race. Frédéric Chopin may have suffered from cystic fibrosis, Henri de Toulouse-Lautrec from pycnodysostosis, Vincent van Gogh and King George III from acute intermittent porphyria (causing intermittent bouts of insanity), songwriter Woody Guthrie from Huntington's disease, and Niccolò Paganini from Ehlers-Danlos syndrome.

There are about 5,000 human genetic diseases, including muscular dystrophy, hemophilia, cystic fibrosis, sickle-cell anemia, and Tay-Sachs disease. Genetic diseases take a particularly heavy toll among the young, resulting in one-fifth of all infant mortalities, half of all miscarriages, and 80 percent of all cases of mental retardation. Genetic diseases afflict perhaps 15 percent of the general population, but if one considers diseases which are polygenic or have a strong genetic component (such as cancer, Alzheimer's disease, diabetes, and cardiovascular disease), then they account for fully 75 percent of all deaths in the United States.

Although medicine was helpless for thousands of years against these ancient diseases, molecular medicine promises us new therapies and strategies in the battle against them, and possibly even cures.

However, it is a battle that must be waged indefinitely, because there is a never-ending struggle between evolution (which gradually eliminates these harmful genes by natural selection) and mutations (which are constantly being replenished by random errors, cosmic rays, toxins, environmental contaminations, etc.). In each generation, a few hundred mutations occur in the DNA of each one of us. If we assume that a small percentage of these are harmful, then perhaps two or three harmful genes creep into our bodies by mutation. Thus perhaps 10 billion new harmful genes enter the human gene pool every generation. As a result, the battle against genetic disease will never end.

## How Hereditary Diseases Have Altered History

Only in the last ten years or so, with the coming of biotechnology, have these genetic diseases finally been understood at the molecular level. But some genetic diseases have been recognized for millennia. Hemophilia, a rare blood disease which prevents the blood from clotting normally, was known as far back as biblical times. The Talmud excused male babies from circumcision if the child had siblings who bled uncontrollably; it also recognized that the disease was hereditary, passed from the mother to the son. These diseases have even altered the destiny of nations, often because of intense inbreeding within the ruling monarchies of Europe.

In the eighteenth century, King George III of England suffered from periodic bouts of madness brought on by acute intermittent porphyria. It was, apparently, during one of these episodes of dementia that his Prime Minister, Lord North, mismanaged his American colonies, thereby triggering the American Revolution and the birth of the United States. In the nineteenth century, one of his successors, Queen Victoria of England, was a carrier of hemophilia; when many of her nine children married, they spread the hemophilia gene into the royal courts of Europe, creating havoc. (Three of her daughters were carriers, like herself, and her son Leopold was a hemophiliac.) "Our poor family seems persecuted by this disease, the worst I know," she lamented. In Russia, Victoria's gene was passed on to her granddaughter Alexandra, who married Czar Nicholas II. Their son Alexis had hemophilia, and the unscrupulous but charismatic monk Rasputin used his hypnotic powers to control his bleeding and thereby exert enormous power over the royal family. Some historians claim Rasputin paralyzed the Russian royal court, delayed badly needed reforms, and helped set the stage for the Bolshevik Revolution of 1917. As geneticist Steve Jones of University College, London, writes: "It is odd to reflect that both the Russian and the American Revolutions may have resulted from accidents to royal DNA."

Many of these diseases have heartbreaking symptoms which lead to a slow, painful death. Some of them are truly bizarre, like Lesch-Nyhan syndrome, which affects 2,000 people in the United States, in which young patients literally chew off their fingers uncontrollably in fits of self-mutilation. Other genetic diseases can be horribly disfiguring, such as neurofibromatosis, affecting one person in 4,000, in which the victim's skin is covered with scores of tiny brown tumors. (The most famous victim of this disease may have been John Merrick, the celebrated "elephant man" of the late nineteenth century.)

Historically, the most feared hereditary disease is Huntington's disease, which has long been associated with witchcraft and devil worship (including the famous Groton witch of 1671). Families of victims were mercilessly harassed and exiled into camps as if they were lepers. Patients with Huntington's gradually lose control of their muscles and their mind. The body often undergoes violent convulsions and bizarre dancing movements until it is covered with black-and-blue spots. Many die of respiratory problems or starvation, because their violent flailing is so fierce they cannot be physically fed. In the United States, it affects some 30,000 people and 150,000 more are at risk.

Some genetic diseases, such as muscular dystrophy, have become well known because of televised fund-raisers.

Many genetic diseases affect specific races and ethnic groups:

**Cystic fibrosis.** This is the most common genetic ailment affecting Caucasians. CF is potentially a widespread problem, because as many as one in every 25 Caucasians is a carrier. In the white population, it affects one in 1,800 babies and strikes 35,000 young people in the United States and Canada. About 1,000 new cases of this disease are recorded each year in the United States.

CF is a parent's nightmare: the disease thickens the mucus in children's lungs, which weakens the lungs and clogs the pancreatic ducts, so the body cannot digest nutrients well. One of the earliest records of this disease dates from the Middle Ages, when there was a common adage among the people of Northern Europe: "Woe to that child which when kissed on the forehead tastes salty. He is bewitched and soon must die."

**Tay-Sachs.** Fortunately, screening for certain diseases has brought some genetic defects under control, even without gene therapy. Tay-Sachs is one example. It affects one in 3,600 Jewish children of mainly Eastern European descent. Within this population, as many as one in 30 is a carrier of this disease. Tay-Sachs attacks the nervous system; children appear normal at birth, but then suffer progressive mental retardation, blindness, loss of muscle control, and usually death before age four.

**Sickle-cell anemia.** This disease strikes 4,000 children per year in the United States, mainly African-Americans. Roughly one in 500 African-Americans has this disease, but as much as 10 percent of the entire African-American population are carriers. It is a chronic disease in Africa, where 120,000 children are born with it each year. In South Africa, as many as 40 percent of the people have the gene.

## From Now to 2010: Gene Hunting

By 2005, when the first human DNA is completely decoded, scientists will have a broad map on which to locate the genes of our bodies. By 2010, we should have a genetic listing of almost all 5,000 genetic diseases.

At times, searching for these defects can be slow. As Francis Collins has said, locating a particular gene from scratch, without any guideposts, is like "trying to find a burned-out lightbulb in a house located somewhere between the East and West coasts without knowing the state, much less the town or street the house is on." Imagine, for the moment, assembling all the telephone books listing the people in the United States. Let us say that we are looking for just one misspelled name out of 3 billion letters. Possessing the complete collection of phone books for the country does little to identify the single misspelled letter.

So far, in decoding the mystery of hereditary diseases, the biomolecular revolution has already revealed some surprises. In general, the defective genes found are extraordinarily long, which vastly increases the probability of error. In many cases, a hereditary disease is caused by a single error. Other times, it is caused by strange repetitions of certain genetic fragments.

Here is a brief listing of some of the genetic errors that have so far been isolated, which reveal how even the tiniest of misspellings within the human genome can cause unending suffering.

**Huntington's disease.** Located on the short arm of chromosome 4, the Huntington's gene *IT-15* is 200,000 base pairs long. This gene is involved in producing two brain neurotransmitters, acetylcholine and gamma aminobutyric acid. In normal people, there is a repetition of the triplet CAG, which may repeat up to 11 to 34 times. In a diseased patient, the CAG repeats far more than that, sometimes more than 80 times, and causes dramatically reduced production of these two chemicals. The longer the triplet repeats beyond a total of 40, the more severe the disease.

**Cystic fibrosis.** In 1989, the gene was finally identified by Francis Collins and Lap-Chee Tsui on chromosome 7. It is 250,000 base pairs long. CF can be caused by the omission of as few as three base pairs, an infinitesimal fraction of the total. The mutation is caused by the following deletion of nucleotides:

ATCTTT → ATT

This, in turn, triggers cystic fibrosis by deleting just one amino acid (phenylalanine) out of the 1,480 amino acids for which the gene encodes.

**Lesch-Nyhan syndrome.** This disease is caused by a single mutation which renders a key gene on the X chromosome, which stretches for 50,000 base pairs, incapable of producing the enzyme HGPRT (hypoxanthine guanine phosphoribosyl transferase).

**Duchenne's muscular dystrophy.** In 1986, scientists finally isolated the gene for Duchenne's, which produces a protein called dystrophin. It's one of the longest genes so far isolated, stretching across 2.5 million base pairs. In fact, its extraordinary length explains its high rate of mutation.

Unfortunately, although we should have a good listing of the precise mutations causing thousands of genetic diseases by 2010, it may take until 2020 or beyond before we have cures for many of them.

"The gap between the ability to diagnose and the ability to treat genetic diseases could well be five to twenty years or more," states Leroy Hood of the University of Washington.

In this never-never land before gene therapy becomes a reality, how can the information we have learned be applied? Nancy Wexler, who helped to track down Huntington's disease, notes that some people, upon being told that they have an incurable genetic disease, "end up hospitalized—not for the disease, but for depression."

Ultimately, the most promising strategy to combat genetic diseases is to directly intervene via gene therapy.

### The Final Frontier: SCIDS

Gene therapy is the path advocated by W. French Anderson of the University of Southern California, the leading pioneer in the field. Anderson is tackling a rare hereditary disease called SCIDS (severe combined immunodeficiency diseases), which the press has dubbed the "bubble boy syndrome." The most famous case was "David," a child born without a normal immune system, who could be killed by the common cold. David spent his life imprisoned in a sterilized plastic bubble; even his mother could hug him only through special plastic gloves. Without healthy white blood cells capable of warding off diseases, children with SCIDS usually succumb to disease in childhood. Before he died in 1984, David became a symbol of the terrible hereditary diseases which have stalked the human race.

Curing defective genes is no trivial task; the body contains 100 trillion cells. However, millions of years of evolution have created perhaps the most efficient "vector" for altering these cells: the virus. By first neutralizing a virus (so it cannot make the patient sick), scientists can insert the correct gene into the virus, and then insert the virus into the patient.

Anderson's experiments may prove to be the prototype for gene therapy in 2020. He first extracted blood from his young patients, and then infected the blood with the modified virus. After the virus infected the blood and inserted the correct gene into the blood cells, the blood was injected back into the patient. The world's first such gene therapy patient was a four-year-old girl, Ashanthi DeSilva. In 1995, Anderson's team claimed that 50 percent of the child's white blood cells had their genetic mechanisms corrected.

After seven intensive years of gene therapy experiments, however, many of the results are still disappointing. One frustrating problem is that the body's immune system sometimes attacks the virus and the modified cells, thereby preventing the corrected genes from proliferating in the body. The entire field was reeling from the impact of a scathing 1995 report to the NIH, which stated that gene therapy was "oversold" to the American people and that most of the experiments had failed to show any significant medical progress.

David Rimoin of the Cedars-Sinai Medical Center, echoing the skepticism of the report, has said, "You need a smart bomb to get the DNA to the right place, and a smart detonator to set it off at the right time, and for the most part, those mechanisms are not yet available."

People with cystic fibrosis in the trials, for example, suffered complications when their immune system reacted negatively with the injected virus, the "smart bomb" which was carrying the correct gene.

The report was a sober reality check to gene therapy, but certainly not a deathblow. Yes, experiments were overblown. Yes, the experiments have in the main failed to show much progress. But that does not dampen the optimism of scientists and victims of genetic diseases.

As Francis Collins says: "This is a new field. Would you criticize a baby while still in its cradle for not getting up and quoting Shakespeare? Come on, give us a break here!"

Since that 1995 report, there have been a number of partial successes. As noted earlier, the University of Texas group found tumors reduced in size or even eradicated totally when gene therapy with *p-53* was carried out.

The NIH's Michael Blaese, a collaborator of Anderson's, has stressed that the progress of gene therapy can be compared to the initial flights of the Wright brothers. Although there were those who snickered at the strange experiments conducted by these bicycle makers, their logic and science was sound, and within a few decades the air was filled with flying machines.

## From 2020 to 2050: Polygenic Diseases

In a sense, the intensive progress in genetic diseases we will make from now to 2020 is deceiving. The exponential growth in our understanding of our genome is due to the computerization and automation of DNA sequencing.

Progress beyond 2020 will become increasingly difficult as we confront the next class of genetic diseases: *polygenic diseases*, diseases caused by more than one gene. A cure for polygenic diseases may still be elusive far into the foreseeable future because they are caused by the interaction of an unknown multiplicity of genes. Thus, the techniques for isolating the genes for polygenic diseases are not easily computerized. Furthermore, they may need to be triggered by some unknown change in the environment.

One such disease is schizophrenia, which slowly destroys the mind and spirit of a human being, leaving him or her at the mercy of disembodied voices. *Nature* called it "arguably the worst disease affecting mankind." This disease, which strikes 1 percent of the human race, uses up 30 percent of all hospital beds in the country, more than any other disease.

There is a definite genetic link to schizophrenia. However, the link is weak: for twins, there is a 50 percent chance that one twin will be schizophrenic if the other twin is, which means that there is a definite genetic component to the disease. But the fact that this is not a 100 percent correlation indicates that many genes are involved, some of which may be triggered by cues from the environment.

There is tantalizing evidence that at least one of these many genes for schizophrenia lies on chromosome 5. In 1988, Canadian scientists found that, in one clan with 104 family members, 39 were schizophrenic and 15 had other mental disorders. The chances of this being purely random is one in 50 million. However, the hope that this was the only gene for schizophrenia was dashed when other studies found no link to chromosome 5. In 1995, another series of studies showed promising clues in chromosome 6, in the area known as *6p21* through *6p24*.

Walter Gilbert believes that by 2010 many of the genes for schizophrenia may be found. By 2020 we will probably have a good understanding of how these genes interact with each other and with the environment, but it is likely that a cure will still be elusive.

## From 2020 to 2050: Germ-Line Therapy?

So far, the excitement about fixing our genes has focused on *somatic* cell gene therapy—i.e., cells in our body which are not involved with reproduction. When the individual dies, the corrected genes die with that person. More controversial is *germ-line* gene therapy, which involves manipulating the DNA of our sex cells. In principle, germ-line therapy can banish genetic diseases in future generations. If successful, descendants would never again have to fear a particular hereditary disease. But such therapy also raises grave moral and ethical questions, which I'll address in Chapter 12, since it involves tinkering with the DNA of the human race.

Scientists expect someday to make eye-opening discoveries which will make germ-line therapy a realistic possibility for humans. Scientists can already perform simple germ-line manipulation in animals, and there is no foreseeable barrier to extending this technology to humans. It is clearly a technology that has the potential to be used in disturbing as well as beneficial ways.

# 9

# Molecular Medicine and the Mind/Body Link

"I start with the premise that all human disease is genetic."
—PAUL BERG, Nobel Laureate

"My crystal ball tells me that an increased understanding of the immune system and an increased ability to manipulate it genetically will have a major impact in the next ten to twenty years."
—STEVEN ROSENBERG, chief of surgery, NIH

IN 1994, THE BRITISH TABLOID PRESS screamed the headline: "Flesh Bug Ate My Brother in 18 Hours!" Gruesome pictures of people's faces being eaten by killer microbes dominated the news. The next year, the headlines blared a grim tale about an Ebola outbreak in Zaire. Emergency teams from around the world quickly converged on rural villages in Zaire to halt the spread of this mysterious, incurable disease, which kills over 90 percent of its victims. The fact that Ebola does not spread very rapidly (because it kills its victims so swiftly they have no time to infect many others) did not stop a jittery public from snapping up any book on the subject, sending several soaring onto the best-seller lists.

Twentieth-century medical science, it seems, was caught by surprise with the horrid tales of "flesh-eating" bacteria, the Ebola outbreak, the unrelenting spread of AIDS, the "Mad Cow" disease, the deaths of schoolchildren due to *E. coli* 0157:H7, and the arrival of waves of bacteria resistant to all known antibiotics.

For the past fifty years, the successes of twentieth-century medical science have been so dramatic that doctors were lulled into thinking that many infectious diseases had been conquered forever, only to find new lethal strains of such diseases endangering society.

Back in 1969, the Surgeon General of the United States, William H. Stewart, solemnly announced that it was "time to close the book on infectious disease." Many futurists, echoing his comments, predicted that the world would be free of infectious diseases in the twenty-first century. In fact, as we're discovering, the reverse is true, as the equivalent of a medieval bestiary of microbes makes an unpleasant comeback.

Doctors did not fully appreciate the fact that bacteria and viruses are constantly mutating and evolving, sometimes millions of times faster than humans, to evade and overcome our best defenses. Despite the efforts by modern medicine, infectious diseases, which have been on the earth for billions of years before humans, will probably be around for several billion more.

But even as the news headlines announce resistant and incurable diseases which are now breaching our medical defenses, the real story is being missed: we have a new weapon in our ancient, perpetual war against disease. The powerful convergence of the quantum, computer, and DNA revolutions is giving birth to a new science, "molecular medicine," which promises to offer new ways of combating the challenges posed by these virulent infectious diseases in the twenty-first century.

Already, new medicines are being created by analyzing the molecular weak spots of diseases on the computer, in virtual reality. HIV is the first virus to be attacked by the full force of molecular medicine. It has been systematically taken apart, protein for protein, almost atom for atom, until all its molecular weak spots have been exposed. As a result, for the first time, scientists have new hope for a possible cure. The concentrated attack on HIV will set the pace for molecular medicine in the twenty-first century.

By 2020, doctors will possess voluminous catalogs containing the complete genomes of hundreds of viruses and bacteria, as well as our personalized DNA sequences, giving us an unprecedented view into the inner machinery of how diseases enter our body, reproduce, and cause their debilitating effects.

## Molecular Medicine Through 2020: Eliminating Killer Viruses

One of the missions in the twenty-first century of the Centers for Disease Control and Prevention (CDC) in Atlanta and the heavily guarded facilities at Fort Detrick at the U.S. Army Medical Research Institute for Infectious Diseases in Frederick, Maryland, is to control the outbreak of viruses, "the greatest threat to the survival of our species." The outbreak of a "Doomsday Virus," such as an airborne AIDS or Ebola virus, could threaten the very existence of human life.

One of the greatest killers in human history has been the virus for smallpox. This disease, which probably crossed over from animals to humans about 10,000 years ago, has been a deadly killer of humans ever since. It laid waste to Alexander the Great's army in the fourth century B.C. and killed Roman emperor Marcus Aurelius. It has destroyed entire cultures and has torn apart great empires. As late as the 1960s, it afflicted 10 million people worldwide and killed more than 2 million people every year.

But back in 1966 the UN's World Health Organization (WHO) began a massive antismallpox vaccination program targeting thirty-one countries. As the number of smallpox cases plunged rapidly to zero, the life cycle of smallpox (which infects only human hosts) was finally broken. When no new infections were reported, the chain of infections was finally shattered. By May 8, 1980, the World Health Assembly officially declared smallpox eradicated.

Today, there are only two vials of the deadly smallpox virus left on earth: one in the maximum-security wing of the CDC in Room 318B and one 5,000 miles away in Russia, at the Research Center for Virology and Biotechnology in Koltsovo, Novosibirsk. In June 1999, scientists at the two centers are scheduled to simultaneously heat up their samples to 250 degrees Fahrenheit and eradicate smallpox from the face of the earth.

This spectacularly successful campaign, in fact, sets the tone for public health campaigns well into the twenty-first century, when other diseases are expected to be systematically hunted down and eliminated forever. By 2020, other diseases will join smallpox in the maximum-security wing of the CDC, including polio and leprosy, which the WHO expects will be eliminated around 2000. Scientists expect measles to be eliminated soon after that. Other diseases that may join the list include neonatal tetanus, dracunculiasis, Chagas' disease, and onchocerciasis. More difficult but also possible are tuberculosis and malaria.

By 2020, Building 15, not far from Room 318B, will contain a rogues' gallery of the greatest killers in human history. It will also be a death row for these diseases, housing the last vials of these diseases before they are heated up. To prevent any escapes, these rooms are kept continually at the highest level of security, called biosafety level 4 (BSL4), which is sometimes dubbed "the hot zone" or "blue suit lab," and are kept at negative pressure, so air will flow inward, rather than outward releasing the virus. These viruses are so dangerous that each worker dresses as if for a trip to outer space. One doctor described the feeling of being on the BSL4 floor: "Once inside the lab in the suit, you're pretty much isolated. It's just you and your air hose. It's a bit like scuba diving."

But although smallpox and certain other diseases may face extinction early in the next century, outbreaks of diseases like Ebola are expected to increase. Ebola is one of several "emergent" diseases, probably ancient pathogens which crossed over from animals and infected small, isolated villages for centuries. It is likely that modern technology and development released them to the general population.

By 2020, with scientists possessing a vast encyclopedia containing the DNA and RNA genomes of hundreds of animal and human viruses, they will be able to reconstruct their broad evolutionary history and their family tree relatively quickly. Although viruses mutate extremely rapidly, even an entirely new virus will have some DNA that resembles the DNA of some known virus.

To visualize the virus itself at the atomic level, scientists will crystallize the viral sample and place it in the beam of an X-ray machine. Often, the shape of the virus yields important clues as to how it latches on to a human cell, penetrates the cell membrane, and hijacks its reproductive machinery to produce copies of itself. At present, only few viruses have been decoded at the atomic level. By 2020, however, the atomic structure of hundreds of viruses will be known.

This technique was first demonstrated in 1985 when scientists were able to obtain a complete three-dimensional molecular picture of a virus, rhinovirus 14, one of 200 to 300 viruses that cause the common cold. Physicists placed the crystallized rhinovirus in an atom smasher, which hit the crystal with an intense beam of X-rays. The deflected X-ray beam produced 6 million bits of data, which were then fed into a Cyber supercomputer. Because of the complexity of the information, it took about a month of computer time to finally reassemble the virus in its memory and give a three-dimensional representation of the virus.

The computer printout revealed a surprise: the virus looked very much like a soccer ball, with twenty triangles that fit together making a rough sphere. The skin of the soccer ball consisted of a protein coat; inside were the nucleic acids. By analyzing the three-dimensional structure of the soccer ball, it was easy to see how the virus evaded the human body's defenses. These triangles fit together very tightly, making it difficult for antibodies to penetrate the defenses of the virus. We can also see why it is so easy to contract colds.

Analyses of other viruses by X-ray diffraction show that some resemble space capsules with specialized mechanisms to grab on to cells and penetrate them. One virus resembles a spaceship with landing gear, by which it docks with the "mother ship" (i.e., the cell). Similarly, these techniques have given us new insight into how rabies and polio operate.

Discovering the three-dimensional shapes of hundreds of these viruses should give us the ability to create new ways to penetrate their defenses.

## The Origins of Viruses

For most of human history, the origin of viral plagues was a complete mystery, making them extremely difficult to prevent. But by 2020, scientists will know the molecular origin of entire classes of viruses, which will give us important clues on how to contain and fight them.

It was long thought, for example, that most viruses originate as crossovers from animals. But this was just speculation until the coming of molecular medicine. The flu virus, for instance, is one of the first viruses to have its genome traced back into the past, to reveal its intriguing and curious origin in the animal kingdom.

The influenza virus has been one of the great scourges of humanity. The worldwide pandemic of 1918, for example, killed over 20 million people, more than the number killed during World War I. Fully half the people on the entire planet were affected by the disease. It killed half a million people in the United States alone, making it the most deadly demographic catastrophe in the century. It was so virulent that it actually depressed the life expectancy in the United States from fifty-two to thirty-nine years.

According to one theory, the origin of Asian influenza goes back to a form of Chinese farming, called polyculture, an age-old custom in which peasants often engage in a unique form of pig and duck farming, living in close proximity to these animals. It is likely that duck viruses are passed on to the pigs when they eat the droppings of the ducks. Similarly, fish and duck ponds are fertilized with pig manure. Pigs apparently act as a "mixing bowl" for the viral genes from the ducks. Pigs get infected by both duck and human viruses, mix them up genetically, and then pass them along. The Nobel Laureate geneticist Joshua Lederberg warns that "new varieties of flu are formed every few years through natural crossbreeding of those prevalent in birds and swine."

One frustrating fact was that the original 1918 flu virus disappeared without a trace, leaving molecular biologists unable to determine precisely why it killed so many millions of people. But in 1997 scientists announced the major discovery that rare samples of the 1918 flu virus were preserved in old tissue samples left over from the pandemic. Scientists, who are now analyzing the flu virus's genetic material, believe that this may one day save the lives of tens of millions if they can prevent a disastrous recurrence of this deadly disease.

In fact, genetic analysis of the flu is so accurate that scientists can identify various strains of the flu, calculate the "genetic distance" between these strains, and even predict when the next epidemic may occur.

Molecular medicine gives us, for the first time, the ability to draw the family tree of viruses, track their origin, and perhaps recommend ways to control them at their source, as in polyculture agriculture. By 2020, we should have an almost complete understanding of how viruses evolve and spread, which, in turn, may help us to defeat one of the greatest challenges of the twenty-first century: AIDS.

### HIV: Paradigm for the Twenty-first Century

Some scientists believe that today's concentrated attack on the HIV may be a new paradigm for the future, using molecular medicine to spot the genetic molecular weak spots in a virus or bacterium and then fashion new therapies. The HIV is one of the first viruses to be attacked primarily at the molecular level, creating new therapies designed by computer simulations of chemical processes.

For the first time, there is guarded optimism that progress is being made—that "cocktails" of anti-HIV drugs can lower HIV levels to the point where the virus count can no longer be measured. Although these drugs are extremely expensive (costing about $15,000 per year per patient) and resistant strains may eventually develop, this is the first good news in the battle against the HIV in many years. Unfortunately, the HIV has built up such momentum over the last thirty years that it will be difficult for any therapy to reverse its advance completely until well into the twenty-first century.

Although it was reported in 1997 that deaths due to AIDS dropped 50 percent in New York City and 12 percent throughout the United States for the previous year, the AIDS epidemic is spreading almost unopposed throughout most of the world; it will not reach its peak until early in the twenty-first century. Some epidemiologists have estimated that by 2000 there may be 100 million people infected by AIDS, far more than all the people killed by the world wars in the twentieth century.

In 1996, the United Nations Joint Program on HIV-AIDS released their latest figures, which starkly laid out the grim spread of this disease. Worldwide, 1.3 million people have developed full-blown AIDS symptoms, a rise of 25 percent within just one year. The number of people who died of AIDS-related diseases in 1995 was 900,000. More significant is the number of people in the world infected with AIDS: 21 million people, 42 percent of whom are women.

Every day, 8,500 more people are infected with the HIV. Of these, two-thirds are in sub-Saharan African countries. This translates into a world tragedy for decades to come.

## Computers and the AIDS Family Tree

By analyzing the HIV at the molecular level, scientists are now explaining the mysteries surrounding AIDS, such as why it takes ten years to kill its victims and why it is so difficult to cure.

The HIV, far from being dormant for a decade, wages a fierce, continuous battle with the body's immune system from the moment of infection. The body's immune system destroys the virus at the rate of about a billion particles per day (about a third of the total). The virus, in turn, destroys about a billion CD4 helper T cells per day, which the body tries desperately to replenish. This ferocious struggle, with literally billions of HIV particles and immune system helper cells dying each day, goes on for several years, until the number of helper cells slowly drops, from 1,000 cells per microliter of blood down to 200, at which point the symptoms for AIDS begin. Death usually follows within two years.

Molecular biologists now see why a cure for AIDS is so frustrating. The HIV lacks the usual repair mechanisms which correct for genetic errors each time the virus reproduces. As a result, it mutates extremely rapidly each time it reproduces, at the rate of one error for every 2,000 nucleotides, a fantastic rate. In ten years, the HIV can undergo the equivalent of a million years of human genetic mutations. The HIV which finally kills the host may be several thousand generations removed from the HIV which originally infected the patient.

From a molecular point of view, the HIV is surprisingly simple, consisting of only nine genes; it bears a striking resemblance to its simian cousin, SIV.

Gerald Meyers of the Los Alamos National Laboratory in New Mexico, who has analyzed hundreds of HIV sequences from around the world, has determined that the HIV-1 genome seems to mutate at the phenomenal rate of about 1 percent per year. (It took humans about 5 million years to diverge by 1.6 percent of their genome from chimpanzees.) In this way, one can use the HIV as a "molecular clock" to detect when different varieties branched off from others.

There are six subclasses of HIV, which differ by 30 percent in their genes. We know that each mutates at 1 percent per year; thus perhaps thirty years ago there was a major branching or bursting of the disease, which Meyers calls the "Big Bang." "We seem to be in possession of a robust molecular clock," says Meyers. "There is no way to pinpoint where the burst occurred, but it appears to have been in the early 1970s. And the various strains have evolved in parallel with the pandemic."

In the United States, the main variety of HIV is Type B, which can be transmitted homosexually, usually through the transmission of bodily fluids through small tears in the skin. The rate of new HIV cases is reaching a plateau in the homosexual population, but health workers are bracing for another invasion if new strains of HIV which are mainly spread through heterosexual contact arrive on our shores. Scientists fear this could create a "breakout" into the larger U.S. population, as it already has in many parts of the globe.

In countries like Thailand, 90 percent of the HIV is Type E, which is spread heterosexually. Studies done by Max Essex at Harvard explain the differences in who contracts the disease; it turns out that Type E infects cells of the vaginal walls much more readily than Type B. Given the large number of GIs in Thailand, its flourishing sex industry, and the ease of international travel, it is only a matter of time before these other varieties of HIV spread here.

## Unraveling the AIDS Genome

Knowing the genetic makeup of the HIV is the key to ultimately devising a cure. When scientists first unraveled its genetic makeup, they were surprised to discover that this retrovirus, which contains RNA rather than DNA, was the most complicated they had ever seen. Most retroviruses have only three genes, called *gag*, *env*, and *pol*. The HIV, however, has as many as nine, with 9,200 base pairs.

The most important genes code for three enzymes: HIV protease, reverse transcriptase and HIV integrase. By 1994, the three dimensional molecular structure of each of these three enzymes was determined. Scientists' strategy to attack the virus is based on creating drugs which interfere with these critical enzymes. By preparing a cocktail consisting of drugs which inhibit these enzymes, doctors hope to contain the disease.

The AIDS virus attacks cells in at least four major steps, each of which provides an opportunity for scientists to devise new cures. (In step one of this simplified picture, the virus attaches itself to receptors located on the surface of the host cell, such as a CD4 cell. It injects its RNA into the cell, and an enzyme converts it to DNA. In the second step, the alien viral DNA then penetrates into the cell's nucleus and hijacks its machinery to produce long strands of RNA and protein needed to make copies of the virus. In the third step, the protease enzyme slices up these viral proteins into shorter strips, suitable for making new viruses. And last, thousands of newly formed HIV capsules form inside the cell and finally burst from the cell membrane, flooding the body with a new generation of the deadly virus.)

Each step reveals the HIV's vulnerable points. AZT (azidothymidine), for example, is a drug that strikes the AIDS virus in step one, by preventing the conversion of RNA into DNA. Because the AZT molecule looks very much like the thymidine molecule (except for a missing hydroxyl group) the virus is fooled into incorporating AZT into its replication instead of thymidine. With AZT substituting for thymidine, the DNA synthesis comes to a halt, since the missing hydroxyl group is essential for creating the backbone of the DNA molecule.

Initially, AZT gave almost immediate relief from AIDS symptoms, raising false hopes. However, the euphoria soon evaporated. In all tests, AZT-treated victims were eventually overrun by mutant AIDS viruses within one to two years.

Yet a new series of drugs were developed in 1996 which attacked the HIV at a different point, at step three, where the protease enzyme cuts up viral proteins into the shorter pieces necessary to make new HIV viruses. These protease blockers prevent the reassembly of HIV particles at a late stage in their reproduction, just before they burst out of the cell.

Initial studies with protease inhibitors have been startling. After four months of treatment with one such drug, Indinavir, scientists at New York University found that 13 of 26 patients had no detectable virus in their blood at all. And 24 out of 26 patients taking a combination of Indinavir, AZT, and 3TC also had no trace of the AIDS virus. This makes it the most powerful form of therapy ever developed against AIDS. Another protease inhibitor, Ritonavir, also has had promising results, reducing the mortality rate by half among 1,100 patients. "We have shown that we can suppress viral replication and keep it there," said Julio S. G. Montaner of the University of British Columbia.

At present, AIDS researchers are cautious. As Harvey Kakadon of Beth Israel Hospital in Boston said: "Don't believe anyone who tells you we are not flying by the seat of our pants." Furthermore, the drugs are very costly, and it is not known if a resistant strain will eventually crop up.

But every day it seems that new developments surface concerning the genetic nature of the HIV. In late 1996, a gene was discovered, called

*CKR5* (found in 1 percent of the Caucasian population), whose mutation renders a person completely immune to the HIV. These individuals' immune cells lack the "docking sites" necessary for the HIV to latch on to them. This in turn could spawn an entirely different approach to the HIV, using gene therapy to alter T cells so that the HIV cannot attach to them.

Although no one is claiming to have turned the corner on the HIV yet, the point is that molecular biology has opened up several promising new avenues for cures, raising optimism to new levels. And the methods scientists have used—molecular medicine—will be the way much of the research into disease will be conducted by 2020. The tedious, painful, and often dangerous trial-and-error method pioneered in the distant past will soon give way to molecular biologists using DNA research, computer modeling, and virtual reality to find a molecular cure.

## New Microbes

Ironically, bacteria are easy targets for molecular science. Viruses, which are tiny strands of DNA or RNA which reproduce by hijacking our cell's metabolic system, are difficult to attack (without destroying our own cells). Unfortunately, resistant bacteria were allowed to flourish due to foolish, shortsighted health policies. The flesh-eating germ (necrotizing fasciitis), for example, is one of a series of virulent mutant strains of streptococcus, which causes strep throat. The most active mutant strain can destroy up to one square inch of tissue per hour, a phenomenal rate. Although it is still treatable with antibiotics, its related strains have already developed immunity to the antibiotic erythromycin. Jim Henson, founder of the famed Muppets and beloved by millions of children for his creative work with puppets, died of a resistant form of strep in 1990.

As the pace of industrialization speeds up, we expect more emergent and resistant diseases to appear. In the jet age, only a few hours of air travel separates most parts of the world from the United States. As James Hughes of the Centers for Disease Control and Prevention has remarked: "A disease that's in a faraway place today may be in our own backyard tomorrow. We're certainly not immune." For example, Legionnaires' disease, toxic shock syndrome, and Lyme disease are examples of diseases which have spread as a consequence of modernization within the United States.

The lesson here is that we are swimming in an ocean of disease. Sitting in our chairs eating dinner, or idly walking through a park, we are blissfully unaware that there are millions of germs covering almost every square inch of our surroundings. Inside our own body there are more germs than the total number of humans who have ever walked the earth.

We forget that for billions of years on this planet, our ancestors' immune systems waged a silent but relentless war against disease, developing millions of different molecular possibilities to destroy these unwelcome invaders. As our DNA evolved to create new defenses against these diseases, germs evolved ingenious mechanisms to penetrate these defenses, in a never-ending dance of life and death. As one author puts it: "Although man can build a better mousetrap, nature always seems to build a better mouse."

Unfortunately, bacteria have the upper hand, since they can evolve as much in a day as we evolve in a thousand years, which gives them a decided advantage in evolving new mechanisms to evade our defenses.

With the harnessing of antibiotics after World War II, however, killer diseases like pneumonia, tuberculosis, cholera, malaria, syphilis, meningitis, etc., were temporarily brought under control for the first time. In fact, it is estimated that our life expectancy was extended about ten years because of antibiotics alone. Today, more than 8,000 antibiotics are known, and about 100 are widely prescribed, which are effective against a wide variety of bacteria. Worldwide sales of antibiotics currently top $23 billion per year.

But as the death rate from these bacterial infections plunged, the public and the pharmaceuticals gradually lost interest in these ancient diseases. Not surprisingly, the inevitable loss of this bulwark is having grave implications for the future of public health.

## Too Much of a Good Thing

When doctors of 2020 look back at twentieth-century science, they will be astonished at the shortsighted, foolish policies of the past, marveling that doctors in the twentieth century thought that winning a minor skirmish against bacteria was tantamount to winning the war.

The careless and rampant overuse of antibiotics today has killed off all but the strongest and most resistant bacteria. Our own bodies have become a Darwinian battleground where only the nastiest mutant strains of bacteria survive and thrive.

About one in every 10 million bacteria will be resistant to a particular antibiotic. The handful of bacteria that are resistant eventually emerge and proliferate when antibiotics are overprescribed. Using two antibiotics increases the potency of the medicine, since on average only one in 100 trillion (= 10 million times 10 million) bacteria will be resistant to both antibiotics. However, if we wait long enough, sooner or later mutant bacteria will emerge which are resistant to both.

In 1977, for example, the first strains of *Streptococcus pneumoniae* (which causes pneumonia) resistant to penicillin were discovered. Today, there are mutant strains of the pneumonia bacteria which are resistant to penicillin, cephalosporin, and other antibiotics.

In 1992, 19,000 patients in the United States were killed by resistant infections. These resistant strains also contributed indirectly to the deaths of 58,000 more.

As Robert E. Shope, professor of epidemiology at Yale Medical School, has said: "If we don't gear up to bring matters under control, we could face new crises similar to the AIDS epidemic or the influenza epidemic."

"There are now organisms, still fortunately rare, resistant to every antibiotic known," warns Fred Tenover of the Centers for Disease Control. "There are only so many ways you can attack a bacterium biochemically, and we've exhausted the majority of the simple targets. For some organisms, in fact, we're at the end of the road."

Worse, the drug companies have been lax in searching for new antibiotics. "We're running out of drugs. We're approaching an era of killer bugs, and it could be a disaster," says Mitchell L. Cohen, a specialist in infectious diseases at the CDC.

The problem is that the drug companies have to spend a large amount of resources on new medicines. Because it often takes ten to fifteen years and $300 million to bring a new drug onto the market, we may be defenseless against certain resistant diseases early in the next century.

The drug companies have also sold enormous quantities of antibiotics to farmers to treat their livestock as "growth promoters." In one study, scientists fed farm chickens food that was supplemented with tetracycline, and within six months, seven of eleven chickens carried large numbers of bacteria resistant to tetracycline in their intestines. Some bacteria even developed resistance to four other antibiotics.

This raises serious questions about large-scale practices carried out by agribusiness, which routinely uses huge quantities of antibiotics to produce greater profits. Twenty-five million pounds of antibiotics are routinely fed each year to farm animals such as pigs and chickens. That is a staggering 50 percent of all the antibiotics consumed in the United States. Not surprisingly, this practice is banned in many parts of Europe.

## Designer Molecules

Once we know how various organic chemicals work at the molecular level, scientists will be able to create new molecular variations, often by computer simulation and virtual reality, allowing us to develop drugs with no side effects and new antibiotics for resistant diseases.

Side effects, some of which can be fatal or debilitating, are among the most frustrating features of many new drugs, often nullifying promising new therapies. Most side effects occur because a molecule does more than perform its desired function. In the molecular world, structure is decisive. Using a drug with side effects is like using a worn-out key that opens not just one lock but several. This can cause unintended effects.

In the future, computers may be able to construct molecular "keys" that fit only one molecular "lock." Molecules will be tested in virtual reality so that they don't trigger undesired chemical reactions.

Virtual reality can also be used to design new classes of antibiotics. The penicillin molecule, for example, contains what is called a "beta-lactam ring," which is responsible for its potent powers because it helps to destroy the cell walls of many bacteria. The beta-lactam ring of penicillin does its work by removing the bacteria's ability to control endogenous enzymes, which are used to dissolve the bacteria's cell walls. Unable to control this volatile enzyme, the bacteria literally disintegrate.

"It's like throwing a monkey wrench into an essential machine that assures the stability of the cell wall," says George Jacoby at the Lahey Clinic in Burlington, Massachusetts.

This, in turn, gives us new avenues to create "designer drugs." By analyzing the molecular machinery of bacteria, scientists by 2020 will be able to systematically pinpoint other sites in the bacteria which are particularly vulnerable to antibiotics, such as the ribosomes (tiny protein factories inside the cell) and the pathway for folic acid production. Then they can design new antibiotics via three-dimensional computer simulation of proteins to attack these sites. Streptomycin, gentamicin, and tetracycline, for example, target the ribosomes of the bacteria, and sulfa drugs and trimethoprim act by blocking the folic acid pathway.

Using molecular biology, we can also come to see how bacteria evolved sophisticated ways to neutralize antibiotics. Resistant bacteria have evolved a way to shatter the beta-lactam ring of penicillin by producing beta-lactamase, which breaks the beta-lactam ring at the carbon-nitrogen joint, rendering it inactive.

"It's a straightforward counterattack by the bacteria to break up the antibiotic before it can reach its target," comments Jacoby. Many classes of bacteria can now manufacture enzymes which, as in the case of the beta-lactam ring, break down or neutralize the active part of the drug. Some bacteria, by contrast, have evolved a novel system of literally pumping the drug out of their cells.

Scientists also know now, at the molecular level, how bacteria spread this resistance to other bacteria. Inside the bacterial cell there are tiny circular pieces of DNA strands, called plasmids. Within the plasmid's DNA code, one can identify the precise sequences which produce beta-lactamase, which confers resistance to penicillin. These plasmids are often freely exchanged between bacteria, even bacteria of different species. As a result, if one bacterium develops an immunity to antibiotics, it can spread this plasmid to other bacteria. If a bacterium formed a resistance to all antibiotics, this resistance could, in principle, cause a major collapse of our medical system, a nightmare scenario. In 1976, for example, it was discovered that gonorrhea was able to resist penicillin by exchanging plasmids with a resistant form of *E. coli* bacteria. Today, 90 percent of gonorrhea bacteria in Thailand and the Philippines have become resistant.

## Of Molecules and Mystics

How will doctors in 2020 and beyond deal with these problems? One novel possibility is to rediscover the ancient wisdom of the shamans.

The doctors of 2020 and beyond will scour the world looking for new sources of antibiotics found in nature. Common drugs such as aspirin, codeine, quinine, reserpine (for high blood pressure), vinblastine (combats Hodgkin's disease), and ipecac (induces vomiting) all have their origins in ancient folklore. New anesthetics, for example, have been found by analyzing the contents of blowgun poisons, which contain curare. The skin of an African frog, which contains a chemical that may be the first of a new line of antibiotics, is used to treat the skin infection impetigo.

Paul Cox, a former Mormon missionary, exemplifies the kind of pioneer who has attracted the keen interest of the medical community. He has been particularly successful in searching for new medicines from Samoa. (Speaking the language fluently, recently he was even made a full-fledged Samoan chief.) Cox heard of shamans on the island of Upolu near Samoa telling of a plant which was effective against yellow fever. He sent the plant to the National Cancer Institute, which isolated a powerful antiviral agent called prostratin, which is now one of the NCI's candidates for HIV therapy.

Normally, the success rate for finding promising drugs among plant life is low, less than 1 percent. Cox, by patiently listening to the lore and wisdom of the indigenous people, has been able to achieve a success rate of 7 percent. In fact, 86 percent of the plants recently analyzed by Cox and by scientists from the University of Uppsala have been found to display significant biological activity against disease.

After analyzing thousands of promising plants and animals, there is the laborious process of extracting the active ingredient of each one. This is where the computer revolution comes in. Previously, searching for new medicines was always a hit-or-miss process, usually miss. For every 10,000 substances analyzed, a hundred show promise, ten may be tried on humans, and one might eventually be found effective and placed on the market.

This will change dramatically in the twenty-first century. Already, the advent of robotic laboratories is accelerating this process several thousand times. While it takes years to screen a few thousand chemicals by hand with laborious animal trials, the new automated "combinatorial chemistry" laboratories can screen millions of compounds within a few months without the use of any animals whatsoever. This will not only increase the efficiency of our search for new antibiotics; it will also drive down the costs significantly, which has been a sensitive topic as the costs of exotic new drugs have soared, putting a strain on people's paychecks and the U.S. health-care system. (Presently, a single chemist may create 50 new compounds per year, at a cost of $5,000 to $7,000 each. New computerized methods can drive down that cost so that each chemist can create 100,000 new compounds for a few dollars each.)

How does this new robotic approach work? In a single test involving 100,000 chemical compounds, promising compounds are placed in long arrays of tubes, which contain a protein or substance that causes a specific disease. Optical scanners then look for unusual activity within the tubes, such as increased production of ultraviolet radiation, signaling that a reaction has taken place. The compounds that reacted are then selected out, and new chemicals, which are variations of the original compounds, are inserted into the tubes. (The process begins again, each time homing in on the specific chemical in the compound which is causing the reaction. Edward Hurwitz, a biotechnology analyst at Robertson, Stephens and Co., calls it "a fundamental paradigm shift in drug development." Once the active ingredient is isolated, biochemists can scan the molecule to find out precisely how the antibiotic works its magic.)

Already, scientists are isolating new antibiotics that target weak spots in the bacterial cells' machinery. For example, one strategy is to interfere with the bacteria's way of harvesting amino acids, the building blocks of proteins. Once the machinery for producing amino acids is blocked, the bacteria will not be able to reproduce.

By the twenty-first century, this computerized approach to searching for antibiotics should yield hundreds of new biomolecular ways in which to attack the bacteria's cell walls, ribosomes, and other key cell structures.

## Mind/Body Link

Numerous experiments have revealed that our moods, including stress, and our social contacts have an immediate impact on the level of activity of our immune system, and hence our ability to counteract germs.

In the third stage of medicine, one of the frontiers of research will be to explore this link using the tools of molecular biology. The mind/body link, sometimes viewed by traditional medicine as bordering on quackery, will soon reveal its secrets to molecular medicine, which will be able to explore *how* the mind affects the immune system, and vice versa, at the cellular and molecular level. In a sense, we are going full circle on the mind/body link, except our understanding is at a much higher level.

Historically, one frustrating problem with exploring the mind/body link has been the reliance on anecdotal data, which is notoriously subject to extraneous effects, such as the placebo effect, the power of suggestion, and subjective judgments. Without careful experiments, control groups, and meticulous records, it becomes nearly impossible to verify the first-person accounts of remarkable cures and remissions.

Within the last few years, however, there has been a flood of solid new experiments and analyses which point to the existence of this mind/body link. In 1996, a definitive study done at the Johns Hopkins School of Hygiene and Public Health showed a link between heart attacks and depression. Doctors followed 1,551 people for over thirteen years and found that those who were depressed were four times as likely to have a heart attack. In 1993, a landmark study of 752 men analyzed over a seven-year period in Göteborg, Sweden, showed that men who exhibited unusual amounts of stress in their lives died at a rate three times greater than those who were calm, showing a direct link between one's longevity and one's emotional state. High levels of stress were, in fact, better predictors of one's death rate than high blood pressure, cholesterol, or triglyceride levels.

But perhaps more interesting was the finding that for people who led a full social life, with rich interactions with friends, wives, and family, there was no relation between one's life expectancy and one's level of stress. This indicated that social contact helped to assuage the effects of stress on the body. Social isolation, in fact, has been shown to result in alarmingly high death rates.

In 1991, scientists at Carnegie-Mellon University demonstrated how stress can suppress the immune system's response to colds. By deliberately exposing students to cold viruses, they found that among students with stress, 47 percent came down with colds, compared with only 27 percent of those without stress.

By examining a subject's blood several times a day, one can in fact find a direct correlation between white blood cell activity and levels of stress. Our immune system was shown to be, in a sense, a barometer to our emotional state.

In an important paper in 1993, scientists at Yale University compiled an extensive list of mind/body research, including the harmful effects of stress on diabetes, heart disease, metastasis of cancer, asthma attacks, and bowel disease. Stress even adversely affected the nervous system itself, causing damage to the hippocampus and hence to our memory.

Other studies done recently strengthen the link between stress and other diseases:

- flare-ups in herpes due to stress
- incidence of colon cancer and stress
- incidence of heart disease and hopelessness
- surviving bypass heart surgery and optimism
- surviving second heart attacks and anger
- heart attack rate and depression
- survival rates from breast cancer and participation in support groups

The list of experimental and epidemiological results is quite extensive and has survived peer review in established medical journals.

One of the tasks of twenty-first-century medical science will be to flesh out precisely how this mind/body link operates at the molecular level. On the one hand, there is the well-established relationship between our emotions and our endocrine system. When faced with a life-threatening emergency, our brain sends electrical signals to our glands to emit adrenaline, noradrenaline, and cortisol, which then circulate in our blood and prepare the body for the "flight or fight" response. The brain also signals the glands to produce natural opiates like beta-endorphin and enkephalin to prepare for possible pain. The flooding of our body with these powerful hormones suppresses our immune system (perhaps an ancient evolutionary response to conserve our resources in an emergency).

In 1996, scientists at the National Institute of Mental Health did a careful study of the effect of depression in women (the average age of the women was forty-one). They found that depressed women suffered from 6.5 to 14 percent lower bone density. They also found that these women had higher levels of the hormone cortisol, which can cause bone loss. In a third of the women studied, the loss of bone was so severe that it matched the level of bone loss usually seen after menopause. One theory is that depression triggers the release of cortisol, which in turn accelerates the loss of bone.

Others believe that there may be a three-way linkage among our immune system, our endocrine system, and our nervous system, which communicate with each other via peptides that travel through the blood, providing a constant feedback among all three, using the blood as its communication system.

These novel discoveries, most of them made within the past five years, may affect how medicine is practiced in the next century. In the future, doctors may take a more comprehensive look at our lifestyles and emotional states, analyzing whether we have social support networks, engage in regular exercise and relaxation (e.g., yoga, meditation, vacations), and have ways to vent our anger and stress. Molecular medicine will force doctors to view the body as a complex web of interacting systems.

## Imaging Devices in the Twenty-first Century

Molecular medicine will also be aided by new advances in quantum physics, paving the way for a new generation of imaging devices, including new types of MRI, CAT, and PET scans. Already, these devices have opened up entirely new areas of medicine, allowing us for the first time to view the living brain as it thinks and the inside of the body as it functions. In the twenty-first century, a new generation of these imaging devices will give us the unprecedented ability to see fine details of the living body, such as clogged arteries, microscopic tumors, etc., that up to now have eluded scientists.

Each of these devices originates in a principle in quantum physics. (*CAT scans* use multiple X-ray photographs to create cross sections of the living body. These X-rays are shot through the body at different angles. Computers are then used to reassemble these multiple photographs to produce cross-sectional pictures of the body. *PET scans* use radioactive glucose to detect neural activity within the brain. Since brain activity increases the consumption of glucose, the energy source of the brain, scientists can assess brain activity by measuring concentrations of radioactive glucose, which emits an antielectron [a positron] that is easily detected. *MRI machines* make use of the fact that the nucleus of the atom is spinning like a top. When placed in a powerful magnetic field, these spinning nuclei are all aligned with respect to the field. By applying an external high-frequency signal, one can actually flip these nuclei upside down. When the nuclei revert to their original configuration, they emit a small burst of energy, which can then be detected. Since different nuclei emit different signals, one can differentiate between the various atoms found in the body.)

At present, the resolution of these devices is not very great. X-rays are difficult to focus, and the resolution of PET scans is not very good. In the twenty-first century, however, a new variation of MRI imaging, called echoplanar imaging, will provide imaging speeds which are 1,000 times faster than those presently available. These high-resolution machines will be able to take images at 30 frames per second, which is the rate at which television images appear on the screen. The advantage of this speed is that it will enable doctors to freeze images of the body which are blurred by body fluids or by motion. MRI images, for example, presently cannot take accurate picture of fatty deposits in the heart because the deposits are tiny and the heart is constantly in motion and filled with fluid. This new generation of echoplanar imaging will eventually make it possible to take rapid still pictures of the heart in action, enabling doctors to peer into the various arteries and veins to determine the degree of blockage. This, in turn, could help to control the greatest single health hazard in the Western world: heart disease.

X-ray photographs are foggy because X-ray beams are difficult to focus and manipulate. But in 1996 scientists were able to focus an X-ray beam by shooting it through a block of aluminum. X-rays will travel right through aluminum, but will bend slightly in the process. This small deflection can be exploited by having rows of thin holes drilled into the aluminum block. Each hole will bend the original beam a bit, until the entire beam can be focused down to a tiny spot a few millionths of an inch in diameter. Not only is this cheaper and more reliable than previous techniques; it may have widespread application in etching silicon wafers and in improving imaging equipment using X-rays.

Presently, these imaging machines are primarily used once a problem has already occurred, to check for and measure the amount of damage that has been done. In the future, the quantum theory will make possible a new generation of imaging machines which will detect potential problems years to decades before they actually become problems.

But perhaps the most interesting aspect of the future of molecular medicine is that aging itself might prove to be a treatable disease.

# 10

# To Live Forever?

"The LORD God . . . said, 'The man has become one of us, knowing good and evil; what if he now reaches out and takes fruit from the tree of life also, and eats it and lives forever?' So the LORD God banished him from the garden of Eden to till the ground . . . and He stationed the cherubim and a sword whirling and flashing to guard the way to the tree of life."

—Genesis, 3:22–24

"By design, the body should go on forever."

—ELLIOT CROOKE, Stanford University biochemist

"I don't want to live forever through my works. I want to live forever by not dying!"

—WOODY ALLEN

THE SEARCH for eternal youth has fired the imagination of aging kings, emperors, and ordinary people for countless millennia. Since antiquity, rulers, in their relentless quest for eternal life, have dispatched teams of explorers to track down the fabled fountain of youth, accidentally altering the course of history on several occasions.

This quest is with us even today. The baby-boom generation, particularly with its emphasis on youth, seems determined to resist surrendering to Father Time, and has poured $40 billion into fueling the current exercise and diet fads.

Anyone who has ever stared in a mirror and watched the inexorable spread of wrinkles, sagging features, and graying hair has yearned for perpetual youth at some point. Aging is no fun: it involves a profound loss in muscle mass, increase in body fat (especially around the waist in men and in the buttocks in women), weakening of our bones, decline of our immune system, and loss of vigor.

No matter how rich, powerful, glamorous, or influential you might be, to confront aging is to confront the reality of your mortality. Or as Butch Cassidy said to the Sundance Kid: "Every day you get older. It's a law." Unfortunately, the secret of aging and eternal youth has always been shrouded in mystery, if not quackery and outright fraud.

By rights, however, the body should live forever. Surprisingly, certain organisms, in fact, live indefinitely. Certain cells, and even animals, routinely defy the laws of aging and have no measurable life span. So if living forever does not violate any known law of cell biology, then why can't we stay eternally young?

A number of tantalizing and remarkable discoveries indicate that the genetic and molecular origin of aging may be within sight. Wild speculations and ancient folklore are, for the first time in human history, being replaced by hard data and concrete, reproducible results. The excitement is palpable among researchers. Leonard Hayflick of the University of California at San Francisco, sometimes called the "dean of biogerontology," states, "Gerontology is now at a stage where several of the theories are being collapsed into each other, and, although much important information is not yet included in the merger, we are making good progress toward the biogerontological counterpart of the physicists' Grand Unified Theory."

Some biogerontologists have made some cautious but reasonable predictions for the future. From now to 2020, perhaps the best bet in terms of delaying or maybe reversing some of the diseases and symptoms of aging will be carefully monitored hormone treatments. There are severe drawbacks to this volatile but promising technique. But if its side effects can be contained, then a combination of antioxidant/hormone treatments may reverse some of the ravages of aging (although they will probably not extend the human life span).

After 2020, however, when we have personalized DNA sequencing, an entirely new avenue will open up—i.e., identifying the fabled "age genes," if in fact they exist. It should be stressed that not all scientists believe that there are such things as age genes. And even if they do exist, the task of sifting through thousands of genes to locate the age genes will be a tedious one, but some biogerontologists claim to have found some age genes in animals, and they may have homologues in humans. One promising avenue would be to study the personalized DNA sequences of people who live exceptionally healthy and long lives, and correlate them by computer to see if they share key genetic factors.

From 2020 to 2050, yet another promising approach will open up: growing new organs. It is of no use to have a long life span if we are stuck with bodies that are crumbling with decay. Already, skin and other tissues can be grown in the laboratory, and plans exist to grow entire organs, including kidneys, hearts, and even possibly hands. Eventually, growing new organs may become as common as heart and kidney transplants today.

## The Search for the Fountain of Youth

Almost every society has its mythical tales of immortality. The Hindus, the Romans, the Chinese, all have their mythology of the fountain of youth, which on several occasions even changed the course of modern history. Greek legend gives the saddest warning to those seeking to outwit the natural order of things. Eos, the beautiful goddess of the dawn, fell in love with and married a mere mortal, Tithonus. But while the gods stayed eternally young, Tithonus began to age, so Eos begged Zeus to make her lover immortal like the gods. Zeus granted her wish, but Eos made a fatal mistake: she forgot to ask for eternal youth for Tithonus. Eventually Tithonus became a shriveled cripple who incessantly babbled to himself. This so irritated the gods that they changed him into the cicada.

The tale of Tithonus is a clear challenge facing modern science. Not only must science and medicine extend the human life span; it must also reinvigorate and revitalize our bodies, so we don't become a nation of nursing home patients.

## Animals That Are "Immortal"

Before the biomolecular revolution, scientists were forced to speculate about human aging from indirect clues. Perhaps the simplest clues to aging came from the animal kingdom and evolutionary biology.

All mammals eventually reach a fixed body size as they age; however, certain animals which do not have a fixed body size (such as some lobsters, flounders, sturgeons, sharks, and alligators) simply increase in body size with time but show no noticeable sign of aging. *These animals are "immortal"* in the sense that their aging process is so slow that it is either nonexistent or too slow to be measured reliably in the laboratory. Many textbooks incorrectly state that these animals have a finite life span like other animals. These texts confuse "life expectancy" with "life span." Life expectancy refers to the average age an organism lives until it dies of disease, predators, or starvation, whereas life span refers to the maximum age an organism can live if these external causes of death are removed. This is the reason why we do not see 500-year-old alligators the size of houses prowling about—because they have succumbed to the perils of living in the wild.

However, when these animals are kept in zoos, they are largely immune from these external factors and simply grow indefinitely, with almost no diminution of their physical functions after reaching sexual maturity. The classic example of this is the flounder. The male flounder reaches a fixed size and ages normally. However, the female flounder grows indefinitely and shows no signs of aging or loss of function with time.

The existence of animals with no fixed life span seems to indicate that "age genes" do exist. Apparently, the cells of these animals never lose their vigor or their ability to reproduce.

From a strictly evolutionary point of view, however, aging may serve a purpose. Nature has little use for an aging animal well past his or her vital, childbearing age; such an animal is a drain on the rest of the herd or pack. Perhaps nature planned for organisms to gracefully age and die, leaving precious resources for the next generation to perpetuate the species.

Generalizing from lower animals to humans is always dangerous, but aging in humans also seems to obey an evolutionary path. Paleontologists, analyzing the remains of our ancestor *Australopithecus*, are now convinced that our lineage separated from other primates about 5 million years ago. Within that 5-million-year period, our life span has more than doubled compared with our primate cousins. In evolutionary terms, in almost a blink of an eye our brain size, body weight, and life span ballooned, which is extraordinary for any species in the animal kingdom. The relative brevity of this remarkable expansion indicates, but does not prove, that our life span is basically controlled by a handful of age genes.

Since we share 98.4 percent of our genes with the chimpanzees, by systematically focusing on the genes that separate us from our primate cousins, perhaps we can locate the age genes among them.

## How Old Was Juliet?

Generations of high school children gasp when they read Shakespeare's *Romeo and Juliet*, for they are amazed to discover that Juliet was only thirteen years old.

We sometimes forget that, for most of human existence, our lives were short, miserable, and brutish. Sadly, for most of human history, we repeated the same wretched cycle: as soon as we reached puberty, we were expected to toil or hunt with our elders, find a mate and produce children. We would then have a large number of them, with most of them dying in childbirth.

As Leonard Hayflick says: "It is astonishing to realize that the human species survived hundreds of thousands of years, more than 99 percent of its time on this planet, with a life expectancy of only eighteen years."

Since the industrial revolution, thanks to increased sanitation, sewage systems, better food supplies, labor-saving machines, the germ theory, and modern medicine, our life expectancy has risen dramatically. At the turn of the century, the average life expectancy in the United States was forty-nine. Now, it is around seventy-six, a 55 percent increase in a century. As Joshua Lederberg notes: "In the U.S., greater life expectancy . . . can be attributed almost entirely to this mastery of infection, this annihilation of the bugs." And today, the fastest-growing segment of our population is the group that is over a hundred years old. (Recently, a curious new phenomenon has been observed, the "robust elderly," which may lessen the burden of the elderly on society a bit.)

## The Physics of Aging

Central to a "unified theory of aging" are physics, information theory, and genetics. There is, first of all, the Second Law of Thermodynamics, which states that disorder (or entropy) must increase in any closed system. In short, things run down. In the words of George Harrison: "All things must pass." Our bodies, our machines, our creations, *even the universe*, must eventually wear out.

Applied to the universe, it means the stars will eventually exhaust their nuclear fuel, plunging temperatures down to nearly absolute zero, creating a dismal universe consisting of dead stars, black holes, and cold formless gas. The destiny of the universe is to reach a state of maximum chaos.

In our bodies, this increase in entropy is manifested by the *loss of information*. Each time our cells reproduce or are battered by toxic chemicals, tiny errors in the information of our DNA begin to accumulate, until our cells can no longer repair themselves and function normally. Eventually, the Second Law of Thermodynamics catches up with our cells, and aging becomes irreversible. As entropy increases, our cells no longer have their original resilience and vitality because of accumulated information loss. Hayflick calls this "molecular mischief"—i.e., the idea that aging is caused by the gradual buildup of errors in our molecular code, which slowly reduces the efficiency and vigor of our cells. Aging may be caused by the loss of our ability to repair this molecular damage.

If the Second Law is an ironclad law of physics, at first it seems a hopeless task to try to reverse aging. But there is a loophole in the Second Law: it refers only to a "closed system." This means we can have a tradeoff: we can reduce entropy in one area (and hence reverse aging) as long as we increase it in others, so the total amount of entropy still increases.

For example, the creation of a baby represents a massive decrease in entropy. But this is compensated for by the chaos that the baby produces elsewhere (in the stress on the mother's body, in the increased consumption of food, and in the vast resources needed to create the baby). In other words, the loophole in the Second Law may be exploited by the age genes, whose purpose is to repair the molecular damage caused by aging.

## Aging: You're Just Getting Rusty

When middle-aged people, bemoaning creaky joints and aching muscles, claim they are "getting rusty," they may be closer to the truth than they realize.

One of the most seminal ideas about aging is the oxidation theory, which states that aging is driven by the same process that makes iron rust, silver tarnish, and fires burn—namely, oxidation, a volatile, corrosive process caused by unleashing the chemical force locked within the oxygen that we breathe in the atmosphere. Oxidation is one important way in which the Second Law is manifested in our bodies.

On the one hand, oxidation is the energy source that fuels our body. When we take a deep breath, the oxygen filling our lungs filters down into our cells, which uses the chemical ATP (adenosine triphosphate) to carefully release the energy necessary to flex our muscles and move our bodies.

However, there is a dark side to this process. Unchecked, oxidation by itself also wreaks havoc in our system, creating "free radicals" in our body which, like a monkey wrench thrown into a finely tuned machine, disrupts cell functions. These volatile free radicals, because of their electrical nature, can rip apart proteins and nucleic acids, disrupting the delicately balanced machinery of the cell.

That aging could be linked to the damage caused by oxidation was first proposed by R. Gerschman in 1954, and further advanced by Denham Harman of the University of Nebraska. They reasoned that if aging was caused by oxidation brought on by free radicals, then aging might be slowed by the neutralizing action of antioxidants. The most common antioxidants include vitamins E, C, and A, as well as beta-carotene, superoxide dismutase (SOD), catalase, and glutathione peroxidase.

Antioxidants are commonly found within the body as well as in our food. (Antioxidants are often added to cereals and baked goods; they are used to slow down the oxidation process, which causes the food to become rancid and stale.)

In controlled experiments, the life spans of certain animals (mice, fruit flies, rats, nematodes, rotifers, and the mold called neurospora) were shown to be lengthened with antioxidants. In fact, the life span of mice can be increased by 30 percent. These animals did not become decrepit like poor Tithonus. Studies have shown that antioxidants postpone the appearance of cancer, cardiovascular disease, and diseases of the nervous system and the immune system.

One testable prediction of the oxidation theory is that animals with short life spans should have higher levels of free radicals. Laboratory studies have borne this out.

The oxidation or free radical theory of aging may give us a crucial clue to how, at the molecular level, damage accumulates in our bodies. But still the question remains: how do we slow or even prevent the damage?

## From the Present to 2020: Hormones, the Elixir of Life?

By far the most painless and medically proven way to increase our life expectancy (and prevent the country from going bankrupt from skyrocketing medical expenses) is to lead a healthy life—i.e., quit smoking, exercise regularly, and eat a low-fat, low-cholesterol, high-fiber diet. Study after study has shown that the American people are grossly indulgent in their lifestyle, leading to a host of chronic illnesses.

However, medical research is gradually changing its opinion about one of the more unsavory areas of biogerontology research, hormone therapy. Traditionally, hormone therapy has had the reputation of being a haven for charlatans, faddists, and outright crooks. The field of hormone treatments has a colorful history, with scores of scandalous and even hilarious encounters with hormone quacks making preposterous claims.

In the 1920s, a colorful fundamentalist preacher named John "Doc" Brinkley claimed that transplanting the testicles of goats and other animals could reverse aging. Thousands of elderly people heard his claims via the radio station he founded and made the pilgrimage to his clinic in Kansas. He became so wealthy and powerful that he even ran for the governorship of the state of Kansas (he lost).

However, hormone therapy is rapidly shedding its snake-oil image and entering the ranks of rigorous science with a series of new studies. In fact, with a boost of $2 million from the National Institute on Aging of the NIH, nine research teams are currently conducting studies on "tropic factors" like hormones which promote growth and maintenance of tissue.

From now to 2020, hormone therapy may blossom into an important way to control some of the ravages of aging and protect against disease (although it will probably *not* extend our maximum life span).

It's well known that women are protected from many of the diseases and symptoms of aging via the sex hormone estrogen during their reproductive years. However, when a woman undergoes menopause, the levels of estrogen drop and bone loss and heart disease increases. Evolutionary biologists have concluded that women were not meant to live very long after menopause. Charles Hammond of the Duke University Medical Center put it bluntly when he said, "At the turn of the century, women died soon after their ovaries quit."

Already, estrogen is the number one prescription drug in the United States and also one of the most exhaustively studied. The famous Nurses Health Study, which followed 120,000 nurses for more than ten years, found that postmenopausal women who received estrogen had half the incidence of heart disease. Other studies have shown that in older women estrogen reduces hip fractures by 50 percent, improves memory, lowers the incidence of colon cancer by up to 55 percent, and maintains collagen that keeps the skin supple and moist.

Furthermore, the biomolecular revolution has deciphered how hormones like estrogen work. They perform their magic by stimulating the genes of the target cells to produce certain proteins (like prolactin) that carry out specific functions in the body. In other words, hormones act to "turn on" certain genes within the cell.

## Cancer and Aging

There is a dark side to estrogen treatments, however, and that is the increased possibility of breast cancer. A study of 240,000 women by the American Cancer Society showed that those taking estrogen for at least six years had a 40 percent increase of fatal ovarian cancer. For those taking estrogen for eleven or more years, the risk increased by 70 percent.

In general, there is a trade-off between slowing the diseases of the aging process with hormone treatments and increasing the risk of cancer. The origin of this trade-off comes directly from physics and molecular biology. Hormones like estrogen act to speed up the cells' metabolic and reproductive abilities, thereby accelerating the rate at which it performs complex genetic functions. But this increases the probability that errors will be introduced into cell reproduction and function.

Think of running an engine at maximum performance. The greater the activity, the greater the wear and tear on the engine. Likewise, the increase in vigor brought about by hormone therapy inevitably increases oxidation, releasing free radicals, and causes mutations, which lead to information loss and cancer.

In other words, *aging may be the price we pay to protect ourselves from cancer.* As V. K. Cristofalo of the Center for Gerontological Research at the Medical College of Pennsylvania says, "Every cell in your body is a stick of dynamite. If it turns neoplastic, you're a goner. In order to survive as a species, we had to evolve mechanisms that would allow us to control cell division long enough to reproduce."

Mice, for example, age thirty times as fast as humans, but they also have thirty times the rate of cancer.

There are, however, some ways of reducing the risk from cancer. Cutting down the amount of estrogen and introducing another hormone, progesterone, may lower the incidence of breast cancer, according to some studies. Isaac Schiff of the Massachusetts General Hospital puts it bluntly: "Basically, you're presenting women with the possibility of increasing the risk of getting breast cancer at age sixty in order to prevent a heart attack at age seventy and a hip fracture at age eighty. How can you make that decision for a patient?"

With 19 million male baby boomers reaching their fifties over the next decade, there will be a similar explosive interest in reversing male aging. Already, men account for about a quarter of all cosmetic surgery procedures (mainly hair treatments and liposuction). However, what is causing the most interest is hormone therapy, in the form of testosterone, the male sex hormone.

The male menopause is more subtle than the female's; instead of the sharp decline in a woman's health in her fifties, the level of testosterone in men drops about 1 percent a year after the age of forty. That testosterone can increase an aging male's vigor is also well known. Men with unusually low levels of testosterone (called hypogonadism) suffer from deterioration of their bones, muscles, energy, and sex drive. In 1992, Joyce Tenover of Emory University Medical School showed that thirteen elderly men placed on testosterone gained in muscle mass and general vigor and excreted less bone material.

Unfortunately, most of what is known about the side effects of testosterone therapy comes from an unusual source: muscle builders, who are notorious for injecting themselves with large amounts of powerful but faddish chemicals. The side effects of taking large quantities of testosterone are fairly well established: enlarged breasts, sterility (large doses of testosterone have been analyzed as a potential contraceptive), cancer (in the form of prostate tumors), and thickening of the blood (which increases the chances of a stroke).

From now to 2020, scientists will be working on ways to control the side effects of this powerful treatment. Fortunately, there are a number of ways this can be done. It may, for example, be possible to use gene therapy to control cancer. A more practical method is to use virtual reality to model the proteins controlled by these hormones which trigger only the desired effect, without the side effects. This means creating a protein shaped like a "key" which fits into only one molecular "lock."

## Long Shots

In 1997, the National Institute on Aging, alarmed that unproven "antiaging" hormones were becoming widespread, even issued a press advisory warning about them. One controversial experimental therapy uses the human growth hormone (HGH). Historically, HGH was available only in minute quantities via the pituitary glands of cadavers. Since 1985, however, biologists have used bioengineered bacteria to artificially generate a large number of human chemicals, including insulin and HGH. In 1990, Daniel Rudman, professor of medicine at the Medical College of Wisconsin in Milwaukee, conducted controlled injections of this hormone on twelve healthy but aging men for six months, and claimed to see almost instant changes, replacing slowly decaying lives with young, vibrant ones.

One of the first recipients of Rudman's experiments was retired auto worker Fred McCullough. Although sixty-five at the time of the study, he said, "I felt like a teenager again. I mean, man, I never felt so strong in my life!" His flabby skin became smooth and youthful. His fat disappeared. His flabby muscles became harder. Shrunken internal organs were restored in size and vigor. His remarkable story was typical of the people used in this study. Rudman's report soon sparked a black market in growth hormone, especially among athletes and bodybuilders and others seeking reinvigorated, rejuvenated bodies.

Attempts to reproduce Rudman's pioneering results were only partly successful, however. In 1996, Maxine Papadakis of the University of California at San Francisco studied fifty-two men, aged seventy or older, and verified that there was a 4 percent increase in lean body mass and a 13 percent decrease in fat, as expected. However, what counts more than muscle mass is strength, endurance, and mental ability. On this level, they found no improvement at all, and in fact documented a series of unpleasant side effects, including swelling ankles, aching joints, and stiff hands. "It's not the fountain of youth," says Papadakis. "We cannot recommend it," her group concluded.

Another popular hormone is DHEA (dehydroepiandrosterone), a steroid secreted by the adrenal gland which increases in levels during puberty and drops after the age of twenty-five to thirty. Although it was discovered back in 1934, only recently has its role as an anticancer and antiaging drug in animals been recognized. When given to mice, DHEA reduces the rate of breast cancer, lengthens their life span, and increases their vigor. Skeptics point out that when given DHEA mice also eat less, so perhaps it is caloric reduction, rather than DHEA itself, which causes cancer rates to fall and life span to rise. More research will clarify this.

Microbiologist Arthur Schwartz has been investigating DHEA for over fifteen years and sees its potential as a drug against colon cancer. He says, "In animals, the evidence against cancer is indisputable. If the same thing happens in humans, we really will have something."

In 1995, a study done at the University of California in San Diego treated sixteen elderly people with DHEA and found a 75 percent increase in their overall health and well-being. By raising their DHEA level to that of a thirty-year-old, scientists were able to reduce joint pain, enhance the quality of sleep, improve mobility, and (in men but not women) increase lean muscle mass.

Since DHEA raises the level of sex hormones in the body, the leading theory is that it works its magic via stimulating the production of these hormones. (If true, then DHEA may eventually cause the same serious side effects as sex hormone therapy.)

One hormone that seems to be less promising is melatonin, a natural hormone secreted by the pineal gland, which apparently helps to control the rhythms of our sleep cycle. Most of the clinical studies done on melatonin have concentrated on jet lag, insomnia, and other aspects of sleep. But because melatonin levels drop in middle age, some have prematurely claimed (in several bestsellers) that melatonin actually reduces the effects of aging.

There may be less here than meets the eye. At a conference held in 1996 by the National Institutes of Health, doctors blasted the widespread publicity surrounding this faddish hormone (the only one available without a prescription or approval by the Food and Drug Administration). Richard J. Wurtman of MIT, whose 1994 study on melatonin and sleep unexpectedly set off the fad, criticized the lack of oversight into the use of the hormone, saying, "There is nobody minding the store."

So far, speculation has far outraced the facts. The melatonin fad has been fueled mainly by anecdotes, for double-blind clinical trials on melatonin are virtually nonexistent. The baby boomers are essentially human guinea pigs testing these unproven therapies.

In the future, there may be many other sensational claims of this or that hormone being the elixir of youth because it diminishes in the body with age. However, as Hayflick stresses, *all* hormones decrease with age; hence this does not prove that they are the source of eternal youth. It is not clear whether reduced levels of hormones are the *cause* of aging or the *result* of aging.

## Beyond 2020: It's All in Our Genes

Beyond 2020, scientists will be searching for the "age genes" which may retard or repair molecular damage due to aging and the Second Law of Thermodynamics. Assuming that such genes exist and can be isolated, then perhaps through gene therapy the process of aging may be arrested and one's maximum life span can be extended.

One intriguing clue to the puzzle was provided by Michael Rose of the University of California at Irvine. By selective breeding, he was able to increase the life span of fruit flies by 70 percent. "That's what makes the field very exciting now—we are doing things that work," he says. His "superflies" were also much more physically robust than ordinary fruit flies. Significantly, he has also found that his long-lived flies produced more quantities of the antioxidant superoxide dismutase (SOD), which helps to neutralize the effects of the dangerous free radical superoxide. Do the flies have an extended life span because they are able to resist the degrading effects of oxidation?

In 1991, biologist Thomas Johnson of the Institute for Behavior Genetics at the University of Colorado stunned the scientific world with his announcement that, for the first time in history, he was able to genetically change the life span of another organism. He isolated within the nematode, a tiny worm, a new gene which he auspiciously called *age-1*. By manipulating this gene, he could increase the three-week life span of the nematode by 110 percent, a dazzling achievement which seemed to prove, once and for all, that, at least for certain organisms, there is an age gene and that it can be systematically manipulated.

As Johnson says: "If something like *age-1* exists in humans, we might really be able to do something spectacular." His next goal is to see if any counterpart to *age-1* lies within the human genome.

Others scientists have produced encouraging results as well. Cynthia Kenyon of the University of California at San Francisco has shown that worms *(Caenorhabditis elegans)* with a mutation in the gene *daf-2* live more than twice as long as normal, 42 days to the usual 18 days. She noted that the mutated worms "looked pretty happy and healthy" even as their normal counterparts were dying off of old age.

Siegfried Hekimi of McGill University in Montreal produced mutant worms that lived five times the normal life span, a record for any animal. "These animals are as close to immortality as worms can get," he said. He isolated four genes which slowed down not only aging but everything else in the worm, including eating, cell division, and swimming. He called them "clock genes."

## Reversing Aging: From Animals to Humans

Extrapolating from nematodes to humans, however, is a daunting challenge. The nematode consists of only 959 somatic cells. But as Tom Johnson of the University of Colorado points out, "of the 8,000 genes already found in this worm, 40 percent have mammalian homologues." In fact, human and worm genes are so close that human genes have been shown to restore normal function in mutant worm genes.

Siegfried Hekimi believes these clock genes work because they slow down the metabolic rate of the worm, thereby reducing damage to tissue. The work of S. Michael Jazwinski of Louisiana State University Medical Center on baker's yeast strengthens these arguments. He has identified several genes which seem to influence the yeast's life span. The best-studied yeast gene is called *LAG1* (for longevity assurance gene 1), which, when introduced into older yeast cells, extends their life span by about a third. Furthermore, he has found what seems to be the counterpart of this gene within the human genome. He speculates that the human counterpart may be useful in expanding the life span of human cells.

Another gene has been discovered which controls SOD, a powerful antioxidant discussed earlier. A gene about 4 million base pairs from the outermost marker on chromosome 21, numbered *D21S58*, controls the production of this enzyme. In the body, the superoxide radical combines naturally with hydrogen peroxide to create the toxic hydroxyl molecule, which has been known to shatter genes and destroy entire cells. There may be a link between the free radical theory and the genetic theory: genes control the production of antioxidants which reduce the damage to the DNA caused by oxidation. In support of this, James Fleming, of the Linus Pauling Institute in Palo Alto, California, has been able to lengthen the life span of fruit flies by giving them an extra copy of the SOD gene.

All this forms an important link in a "unified theory of aging" which brings together DNA, information loss, oxidation, and genes.

## How Long Can We Live?

Perhaps the simplest way to determine whether longevity in humans is genetic is to find out if longevity is inherited. The first of many exhaustive studies on the inheritance of longevity was done in 1934 by Raymond Ruth Pearl. Scientists have found that 87 percent of the people in their nineties and hundreds had at least one parent who lived beyond the age of seventy.

One convincing test of the inheritance of human "age genes" is to analyze identical twins. Studies have shown that identical twins usually die within three years of each other. (By contrast, fraternal twins of the same sex were found to differ from each other in life span by six or more years.) Most biogerontologists would conclude that there is a weak, but measurable, link between longevity and heredity.

Similarly, there are bizarre inherited diseases, like progeria and Werner's syndrome, which seem to wildly accelerate the aging process, transforming cuddly, cherubic infants into decrepit, aged individuals within a span of a few years. Research isolating these strange aging diseases have shown abnormal amounts of the enzyme helicase, which is vital for DNA repair. Again, aging seems to be directly related to failure in DNA repair mechanisms.

So far, no one has come close to isolating the age genes within humans, if they exist at all. However, Michael West, a molecular biologist at the University of Texas Southwestern Medical Center in Dallas, claims to have made a promising first step by isolating the "mortality genes" in human cells which control the aging process in cells in the skin, lungs, and blood vessels. These genes have such a dramatic effect that he has dubbed them *M-1* and *M-2*.

He predicts: "In the next few years, we will fully characterize the genes that regulate the aging of cells. Then you'll see an aggressive application of that understanding to age-related diseases like atherosclerosis, coronary artery disease, aging of the brain, and other maladies like osteoarthritis and skin aging."

To back up this claim, he recites the almost magical power locked within *M-1* and *M-2*. By turning these genes on and off, he has demonstrated the ability to turn the aging process on and off, demonstrating even to his critics the tight cause-and-effect relationship between genes and aging in humans.

Normally, *M-1* and *M-2* are both switched on in aging cells. But West has demonstrated that, by chemically switching off the *M-1* gene, he can restore youthfulness, doubling the number of times they divide. By chemically turning off the *M-2* gene, he can produce an even more dramatic effect: these altered cells divide indefinitely. Then, by turning on these genes, he can make the cells start to age once again. He claims that "by switching these genes on and off, we can cause the cells to become younger or grow older at will."

One gene whose effect on aging has been well documented is the *apo-E* gene, which codes for the protein apolipoprotein. The *apo-E* gene comes in three varieties, E2, E3, and E4, and is closely related to Alzheimer's disease. People with two copies of E2 have eight times the normal risk of getting Alzheimer's. People with two copies of E3 usually come down with Alzheimer's by age seventy-five.

But what is interesting in terms of aging is that there seems to be a relationship between the E4 gene and how long you will live. Studies done on people up to the age of 103 have shown that the smaller the incidence of the E4 gene, the longer-lived people seem to be. For people under sixty-five, the incidence of E4 was 25 percent. But for the age group ninety to 103, this number dropped to 14 percent. One possibility is that having the E4 gene lowers your life expectancy by increasing your chances of coming down with Alzheimer's.

## Aging Research from 2020 to 2050

How will all this progress develop in the future?

The mounting evidence for age genes which influence the aging process is by no means conclusive, but it is quite impressive, coming from a variety of independent research, from aging in worms and fruit flies to antioxidants and gene repair mechanisms, and human mutations. Still, the connections are circumstantial.

Christopher Wills, professor of biology at the University of California in San Diego, thinks that by 2025 science will likely isolate the mammalian age genes in mice. We share 75 percent of our genes with mice and have much of the same body chemistry; this is a strong reason to believe that an age gene found in mice could also be at work in humans. If such genes are located, the next step would be to find out if these age genes have their counterparts in humans. Wills believes that if they are found in humans, they may extend the human life span perhaps to 150 years.

But by 2020, when personalized DNA sequencing becomes widespread, a second tactic may prove fruitful as well. By analyzing populations of healthy individuals in their nineties and beyond, it will be possible to use computers to compare their genetic backgrounds and cross-check for similarities in key genes which are suspected of influencing aging. A combination of studies on the DNA of long-lived animals and the personalized DNA sequences on elderly individuals may considerably narrow down the search for the age gene.

### You Aren't What You Eat: The Caloric Theory

As yet, none of these methods can prove that we can increase the human life span. Perhaps the only theory with a proven track record of extending the life span of animals is the caloric restriction theory, which states that animals which consume calories just above starvation levels live significantly longer than the average. Although this offbeat theory flies in the face of common sense (because a well-fed animal is well nourished and healthy, and should have greater resistance to disease and aging), it has held up under repeated testing among a wide range of animals. Scientists have consistently increased the life span of mice and rats in the lab by 50 to 100 percent. It is the *only* laboratory-tested theory of age extension for animals which has held up under decades of careful scrutiny. Why?

Across the animal kingdom, the life span of animals is roughly inversely correlated to the metabolism rate. The slower the metabolism rate, the longer the life span. When the metabolism rate is artificially reduced by restricting calories, the life span is also lengthened.

The caloric theory is the inverse of the folk wisdom of "Live fast and die young." It seems to say, "Live slowly and live longer."

This effect was first noticed near the turn of the century, firmly established in 1934 by Cornell University researcher Clive MacKay, and further studied by pathologist Roy L. Walford of the University of California at Los Angeles (who believes he may live to be 140 by going on a near-starvation diet himself). Today, the most rigorous test of this theory is being carried out by the FDA's National Center for Toxicological Research, in Jefferson, Arkansas.

In 1996, a study done at the NIH reduced the calorie intake of 200 monkeys by 30 percent. These monkeys were shown to have a slower metabolism rate, a longer life span, and reduced rates of cancer, heart disease, and diabetes. "We have known for seventy years that if you feed laboratory mice less food, they age slower, they live longer, and they get diseases less frequently. We find that monkeys respond in the same way as rodents and that the same biological changes may be in play here," said George Roth of the National Institute on Aging.

Scientists despair of trying to make such a spartan diet part of the American lifestyle. The faces of most Americans, in fact, would turn pale green if they saw the 940-calorie diet which might possibly increase their life span.

And Harvard biologist Steven Austad showed the harsh trade-offs one must make to achieve this long life span. After reviewing the record, he noticed that mice on this calorie-deprived diet do not have offspring. In fact, they do not mate at all! People on this kind of restricted diet may become so sluggish that they eventually lose interest in many of the things that make life worth living.

There is still room for scientific debate on the question "why?" Ron Hart, a scientist at the National Center for Toxicological Research, believes that the answer may lie in the evolutionary trade-off that mammals, and humans in particular, made by maintaining a high body temperature.

"Heat causes pieces of the molecule to split off randomly and must be repaired," Hart says. "Under calorie restriction, though, the engine runs cooler and there's less damage. Merely reducing caloric intake by 40 percent reduced this form of spontaneous DNA damage almost 24 percent!"

Furthermore, at a higher internal body temperature oxygen is being burned at a greater rate, creating more free radicals, which also speed up the aging process. Cooling the body, on the other hand, increases the amount of antioxidants in the body. Hart found a fourfold increase in catalase and a threefold increase in SOD in animals on a restricted diet.

"What's fascinating," Hart concludes, "is that reduced food intake is the only experimental paradigm ever found that enhances DNA repair." Hart is so convinced of the importance of this work that, in 1993, he began the first systematic studies of caloric restrictions on humans.

(Many scientists believe the aging process takes place in the "engine" of the cell, the mitochondria, where the chemical ATP stores much of the energy of the cell. Not surprisingly, this is where most of the oxidation takes place as well, producing the superoxide free radical in large quantities. The superoxide can be turned into hydrogen peroxide, which can turn into a hydroxl free radical, which is quite reactive. Over time, this cuts down ATP production, which reduces the efficiency of the cell. Furthermore, mitochondria, which contain their own separate DNA, lack the protein shield that helps to protect nuclear DNA from building up errors. Thus, the cell begins to degenerate on many levels, both in energy production and in cell function. The mitochondria theory of aging is attractive because it combines all three elements of aging: the oxidation theory, the gene theory, and the caloric theory.)

Perhaps once the mechanism behind caloric restriction is understood, scientists may find a way to turn it on without the caloric sacrifice. "What we want to do is to discover the mechanism that's working in these animals and then figure out how to do the same thing pharmaceutically, or by gene therapy," Hart concludes. Hart is confident that they are hot on the trail of the Master Gene of Aging. He adds: "I predict we'll have the aging gene in our hands very soon."

## From 2020 to 2050: Growing New Organs

But even if age genes do exist and we can alter them, will we suffer the curse of Tithonus, who was doomed to live forever in a decrepit body? It is not clear that altering our age genes will reinvigorate our bodies. What is the use of living forever if we lack the mind and body to enjoy it?

A recent series of experiments show that it may one day be possible to "grow" new organs in our body to replace worn-out organs. A number of animals, such as lizards and amphibians, are able to regenerate a lost leg, arm, or tail. Mammals, unfortunately, do not posses this property, but the cells of our bodies, in principle, have, locked in their DNA, the genetic information to regenerate entire organs.

In the past, organ transplants in humans have faced a long list of problems, the most severe being rejection by our immune system. But, using bioengineering, scientists can now grow strains of a rare type of cell, called the "universal donor cells," which do not trip our immune system into attacking them. This has made possible a promising new technology which can "grow" organ parts, as demonstrated by Joseph P. Vacanti of the Children's Hospital in Boston and Robert S. Langer of MIT.

To grow organs, scientists first construct a complex plastic "scaffolding" which forms the outlines of the organ to be grown. Then these especially bioengineered cells are introduced into the scaffolding. As the cells grow into tissue, the scaffolding gradually dissolves, leaving healthy new tissue grown to proper specifications. What is remarkable is that the cells have the ability to grow and assume the correct position and function without a "foreman" to guide them. The "program" which enables them to assemble complete organs is apparently contained within their genes.

This technology has already been proven in growing artificial heart valves for lambs, using a biodegradable polymer, polyglycolic acid, as the scaffolding. The cells which seeded the scaffolding were taken from the animals' blood vessels. The cells "took" to the scaffolding like children to a jungle gym.

In the past few years, this approach has been used to grow layers of human skin for use in skin grafts for burn patients. Skin cells grown on polymer substrates have been grafted onto burn patients, as well as the feet of diabetic patients, which must often be amputated for lack of circulation. This may eventually revolutionize the treatment of people with severe skin problems. As Marie Burk of Advanced Tissue Sciences says: "We can grow about six football fields from one neonatal foreskin."

Human organs such as an ear have actually been grown inside animals as well. The scientists at MIT and the University of Massachusetts recently were able to overcome the rejection problem and (painlessly) grow a human ear inside a mouse. The scaffolding of a life-sized human ear was made of a porous, biodegradable polymer and then tucked under the skin of a specially bred mouse whose immune system was suppressed. The scaffolding was then seeded with human cartilage cells, which were then nourished by the blood of the mouse. Once the scaffolding dissolved, the mouse produced a human ear. Eventually, scientists should be able to grow this ear without the aid of the mouse. This could open up an entirely new area of "tissue engineering."

Already other experiments have been done which show that noses can also be generated. Scientists have used computer-aided contour mapping to create the scaffolding and cartilage cells to seed the scaffolding.

Now that the technology has been shown to be effective on a small scale, the next step will be to grow entire organs, such as kidneys. Walter Gilbert predicts that within about ten years, growing organs like livers may become commonplace. One day, it may be possible to replace breasts removed in mastectomies with tissue grown from one's own body.

Recently, a series of breakthroughs were made to grow bone, which is important since bone injuries are common among the elderly and there are more than two million serious fractures and cartilage injuries per year in the United States. Using molecular biology, scientists have isolated twenty different proteins which control bone growth. In many cases, both the genes and the proteins for bone growth have been identified. These proteins, called bone morphogenic proteins (BMP), instruct certain undifferentiated cells to become bone. In one experiment, twelve dental patients with severe bone loss in the upper jaw were successfully treated with BMP-2. (Normally, doctors would have to harvest bone from the patient's own hip, a complicated procedure which requires surgery.)

The ultimate goal of this technology would be to grow a complex organ, such as the hand. Although this may still be decades away, it is within the realm of possibility. The step-by-step outline of such a complex process has already been mapped out.

First, the biodegradable scaffolding for the hand must be constructed, down to the microscopic details of the ligaments, muscles, and nerves. Then bioengineered cells which grow various forms of tissue would have to be introduced. As the cells grow, the scaffolding would gradually dissolve. Since blood is not yet circulating, mechanical pumps would have to provide nutrients and remove wastes during the growing process. Next, the nerve tissue would have to be grown. (Nerve cells are notoriously difficult to regenerate. However, in 1996 it was demonstrated that the severed nerve cells in mice's spinal cords can actually regenerate across the cut.) Last, surgeons would have to connect the nerves, blood vessels, and lymph system. It is estimated that the time needed to grow such a complex organ as the hand may be as little as six months.

In the future, we may therefore expect to see a wide variety of human replacement parts becoming commercially available from now to 2020, but only those which do not involve more than just a few types of tissue or cells, such as skin, bone, valves, the ear, the nose, and perhaps even organs like livers and kidneys. Either they will be grown from scaffolding, or else from embryonic cells.

From the period 2020 to 2050, we may expect more complex organs and body parts containing a wide variety of tissue cells to be duplicated in the laboratory. These include, for example, hands, hearts, and other complex internal organs. Beyond 2050, perhaps every organ in the body will be replaceable, except the brain.

Of course, extending our life span is only one of many ancient dreams. Yet another, even more ambitious one is to control life itself, to make new organisms that have never before walked the earth. In this area, scientists are rapidly approaching the ability to create new life forms.

# 11

# Playing God

## *Designer Children and Clones*

"And the LORD God formed man of the dust of the ground, and breathed into his nostrils the breath of life; and man became a living soul . . . and He took one of his ribs, and closed up the flesh instead thereof; and the rib, which the LORD God had taken from man, made He a woman and brought her unto man."

—Genesis 2:7, 21–22

"Are we going to control life? I think so. We all know how imperfect we are. Why not make ourselves a little better suited for survival? That's what we'll do. We'll make ourselves a little better."

—JAMES WATSON

EVERY CULTURE has ancient myths and tales of fantastic creatures made out of clay or mud. In Genesis, God himself breathed life into dust, thereby creating Adam, and then, from Adam's rib, Eve. In Greek legend, Venus took pity on Pygmalion, a sculptor who fell in love with Galatea, his beautiful marble creation, and brought the statue to life.

Throughout mythology, we hear of strange creatures which are half human and half animal, such as centaurs, harpies, minotaurs, and satyrs (which were half horse, bird, bull, and goat, respectively).

In modern fiction, the power of the gods has been replaced by the science of mortals, who must now bear the moral responsibility for breathing life into their creations. In Mary Shelley's *Frankenstein*, the scientist Dr. Victor Frankenstein wrestled with the moral dilemma of creating a mate for his monster. If the union produced children, then he would be creating a new hideous species of life on earth to rival humanity. He agonized that "a race of devils would be propagated on the earth who might make the very existence of the species of man precarious and full of terror. Had I the right, for my own benefit, to inflict the curse upon everlasting generations?"

(Shelley did not realize that splicing together different body parts does not affect the genetic makeup of the final product; the children of Dr. Frankenstein's creations would have normal human genes.)

With the revolution in recombinant DNA, however, we have to reanalyze many of these ancient myths from an entirely different perspective. The ancient dream of being able to control life is gradually becoming a reality via the biomolecular revolution. But this raises the question: what are the scientific limits in creating new life forms? Can science one day create new races of animals like chimeras, or even a new race of humans, "metahumans" or "homo superior," with superhuman abilities?

## Manipulating the Genes of Animals and Plants

Manipulating the gene pool of plants and animals to create new life forms is nothing new. Humans have been playing with the genes of other species for over ten thousand years, creating many of the familiar plants and animals that we see around us. But as we attain the ability to manipulate the genome of other life forms, we've learned from the breeding of plants and animals that this may have unforeseen effects.

Historically, the genetic manipulation of plants to create new food crops was considered so valuable that Thomas Jefferson once said, "The greatest service which can be rendered to any country is to add a useful plant to its culture."

Charles Darwin honed many of his ideas about natural selection by interviewing animal and plant breeders. When they crossbred animals with certain desired characteristics, Darwin saw how certain traits could be passed on to their offspring. Over several generations, this trait could then be magnified and refined, until it became a central characteristic of a new strain of plant or animal. It further supported Darwin's conclusion that this happened in the wild, with natural selection replacing the hand of the breeder.

Dogs, for example, were probably first domesticated about 12,000 years ago from the gray wolf, *Canis lupus*, eventually becoming the familiar *Canis familiaris* of today. Intensive crossbreeding has produced a bewildering variety of dogs bred for specific purposes, such as hunting, shepherding, security, retrieving, and companionship. Selective breeding has split the original *Canis lupus* into the 136 distinct breeds currently recognized by the American Kennel Club, as well as hundreds more which are not.

Cats, on the other hand, were domesticated relatively recently, probably by the Egyptians about 5,000 years ago from the wildcat, *Felis silvestris*. They were bred probably to protect grain stocks from rats. The cat is the only domesticated animal which descended from solitary animals; all other ancestors of domesticated animals are social animals (which may explain why cats are more detached and reserved than dogs).

One lesson taught by the domestication of animals is that everything comes with a price. For the dogs, one great advantage of domestication is that they have thrived: in North America, there are 50 million dogs, whereas the wolf population has dwindled to 38,000. But dogs also paid a price for living in the lap of luxury: intensive inbreeding has greatly magnified a host of genetic defects, such as blindness, hip deformities, and clotting disorders. By creating new life forms with biotechnology, we may also unwittingly inflict unforeseen damage.

Similarly, tinkering with the genome of plants occurred soon after the Ice Age ended 10,000 years ago, when human society made the transition from hunter-gatherers to an agricultural society, with villages, cities, and eventually civilization. In the process, we created the present strains of corn, beans, tomatoes, potatoes, wheat, rice, etc., that we see in the supermarket, all of which differ markedly from their ancestors.

As with the dog, there was a price paid by these crops. Centuries of selective breeding of maize by Native Americans made it so dependent on humans that it cannot survive on its own. To make harvesting easier, they bred the maize so that the kernels are firmly embedded in the cob, making it impossible for the kernels to disperse by themselves to reproduce. Thus maize depends entirely on humans to release the kernels and plant them in soil. Without humans, it may become extinct.

## To 2020: Transgenic Animals and Plants

Although the genetic manipulation of plants and animals has been going on for 10,000 years, only in the last twenty years have scientists been able to crossbreed across different species, putting genes taken from one species of plant or animal into another. Since all life on earth probably evolved from an original ancestral DNA or RNA molecule, it is not surprising that DNA from one species can propagate so easily within the genome of another.

Within a matter of minutes, it is now possible to short-circuit hundreds of millions of years of evolution and create entirely new species of "transgenic" animals which have never before walked the surface of the earth.

From now to 2020, the pace of creating transgenic animals will vastly accelerate because we will have the complete genome of thousands of life forms on the earth to guide us. For plants, in particular, the possibilities seem endless. "We can put just about any molecule that has therapeutic value into plants," comments Andrew Hiatt of the Research Institute of the Scripps Clinic in La Jolla, California. Scientists are currently only touching the surface of transgenic organisms. This process will greatly speed up in the twenty-first century, as the genomes of various crops and animals are decoded more or less simultaneously with the human genome.

So far, most successful gene transfers have involved injecting a single gene which produces a single enzyme from one animal or plant into another. (Traits of one species can be transplanted into another whenever the protein which controls a certain chemical process produced by one organism can operate in a similar manner in the other.)

This simple process has created valuable, lifesaving hormones and chemicals, by clipping certain human genes with restriction enzymes and injecting the genes into bacteria. Since 1978, insulin, once available only from the pancreas of pigs, has been produced by injecting the human gene for insulin into *E. coli* bacteria. In a process somewhat like that of fermentation, which produces alcohol, these modified *E. coli* can produce unlimited quantities of human insulin. Four million diabetics depend on this crucial process.

Similarly, human growth hormone, once available only in minute quantities from the pituitary glands of human cadavers, can now be produced cheaply in the laboratory via the same process. Since then, scores of other rare, valuable medicines have been made by genetic engineering, including interleukin-2 (for treatment of kidney cancer), factor VIII (for hemophilia), hepatitis B vaccine, erythropoietin (for anemia), whooping cough vaccine, and somatotropin (for pituitary dwarfism).

This is, in fact, one of the great successes of the biomolecular revolution. From now through 2020, when personalized DNA sequencing becomes possible, virtually all the exotic hormones and enzymes found in the body may be reproducible in quantities by inserting the human gene for that chemical into bacteria and allowing it to "ferment."

Already, scientists are unraveling the precise chemical mechanism by which certain narcotics release neurotransmitters, causing the nerves in our brain to fire all at once and creating the sensation of being "high." Being able to manufacture these rare neurotransmitters may one day ameliorate the drug problem by reducing people's urge to take drugs.

Progress in developing transgenic animals will also accelerate through 2020. The first breakthrough in creating transgenic mammals occurred in 1976 when scientists at the Fox Chase Cancer Institute in Philadelphia created a new form of mice by injecting a leukemia virus into embryonic mice cells. The technique was further refined in 1980 with the development of "microinjection." First, fifteen to twenty newly fertilized eggs were removed from a female mouse. Under a microscope, a technician manipulated a joystick that controlled an extremely thin, hairlike glass tube, which contained minute quantities of a foreign gene. The glass tube pricked the eggs and injected the foreign gene into the fertilized eggs under a microscope. The eggs were then inserted into a surrogate mouse, which delivered pups twenty days later. Analyses of the pups confirmed that their mice genome was permanently altered.

Since then, microinjection has been used successfully on rabbits, pigs, goats, sheep, and cows to produce a variety of transgenic animals. In 1982, scientists created a race of "supermice" in the laboratory. Richard Palmiter and his colleagues at the Howard Hughes Medical Institute announced that they had used microinjection to put the rat growth hormone gene into mouse eggs, creating mice who grew two to three times faster than normal and ended up twice the normal size. As expected, the mice were able to pass this new gene on to their offspring.

While transgenic animals promise a bonanza of valuable medicines, transgenic plants give the promise of increasing our food crops. At present, human society is dangerously vulnerable to even small disruptions in its food supply. Approximately 250,000 species of flowering plants exist on the earth, but only 150 plant species are cultivated for agriculture. Of these, a mere *nine* (wheat, rice, corn, barley, sorghum/millet, potato, sweet potato/yam, sugarcane, and soybeans) constitute three-quarters of our food energy. Thus a blight or infestation attacking certain key food crops could cause widespread famine.

Because the population of the earth, currently at 5.7 billion, is expected to double within the next fifty years or so, there will be even more demands placed on the earth's limited arable land (which is continuing to vanish because of rapid industrialization and urbanization). Biotechnology will not solve these enormous demographic problems, but they might alleviate some of them.

Genetic engineering of plants was demonstrated in 1983 when the DNA of the bacterium *Agrobacterium tumefaciens* was injected into plants. In 1987, scientists showed that a .22 caliber blank cartridge can serve as a "DNA pellet gun" to literally shoot DNA into plant cells. Tungsten or gold pellets are now routinely coated with DNA and then propelled into the cell. This method has had a powerful effect on agriculture.

Industry spokespeople claim that early in the twenty-first century, about half the acreage of the major crops in the United States will have at least one foreign gene in them. "Sales for such products will be about $2 billion by the year 2000, $6 billion by 2005, and perhaps $20 billion by 2010," claims Simon Best of Zeneca Plant Sciences. Pioneer Hi-Bred International, the largest seed company in the United States, predicts that fully one-third to half of its seed line will be bioengineered by 2000. "It may be as important as the first plow for agriculture," says Rick McConnell, a senior VP at Pioneer.

These techniques are being used with great success in producing new strains of plants. From now to 2020, we should see progress in the following areas:

**Pesticide-producing plants.** Genetically engineered plants can now produce their own naturally occurring pesticides. *Bacillus thuringiensis*, for example, commonly called Bt, produces a protein which kills many pests, such as the bollworm and tobacco budworm. We can embed the Bt gene into crops so they can essentially produce natural pesticides on their own. Cotton plants can now fight off bollworms and tobacco budworms, while corn can kill the European corn borer.

**Disease-resistant plants.** New plants can be created which are resistant to blights and viruses. Scientists have now isolated a gene in rice called *Xa21*, which makes a protein that protects the plant from leaf blight, a fungus that sometimes destroys up to 50 percent of the rice grown in parts of Asia and Africa. This gene can now be inserted into wheat, corn, and a host of other crops to make them resistant as well.

**Herbicide-resistant plants.** A gene from the petunia plant can be inserted into soybean plants, making them more resistant to chemical herbicides like Roundup.

**Creating valuable drug-producing plants.** Plants are actually easier to use than bacteria or yeast to manufacture certain drugs to combat disease in humans. Human genes inserted into these plants make them, in effect, factories for drugs and medicine.

## Clones

How far can we develop this technology? One active area of research is cloning. Although cloning evokes fearful images from *Brave New World*, cloning is actually found everywhere. In our gardens, the use of cuttings to create genetically identical copies of prized plants is commonplace and dates back several thousand years. In our supermarket, we find familiar fruits and vegetables, many of which are clones of specially bred plants.

Although plants are easily cloned, the cloning of mammals has always eluded scientists.

In principle, cloning of higher organisms can be performed in two ways. The first is to remove cells from an embryo (before they have differentiated into cells for skin, muscle, neurons, and so on), alter them and culture them in a laboratory or insert them into a surrogate mother. The second method is far more difficult and interesting, taking mature cells which have already differentiated and somehow coaxing them to revert back to their embryonic state. Until recently, it was believed to be impossible to clone an adult mammal.

In principle, mature cells contain all the DNA necessary to create an entire organism, but scientists had been unable to get these differentiated cells to revert back to their embryonic state. For over a decade, scientists abandoned hope that they could coax a skin cell, say, to regenerate an entire animal.

All this changed with the work of Ian Wilmut of the Roslin Institute outside Edinburgh, Scotland. The world was unprepared for his announcement in 1997 that he successfully completed the second method, the cloning of sheep from an adult cell, by extracting a cell from the mammary gland of an adult sheep. After 277 unsuccessful tries, Wilmut's team produced the world's first cloned mammal of an adult sheep, which they called Dolly. "Not since God took Adam's rib and fashioned a helpmate for him has anything so fantastic occurred," hailed *Newsweek.*

Wilmut's work had begun quite conventionally. First, his team extracted the nucleus of a cell from an adult sheep. As before, they used an electric pulse to fuse this with an embryonic cell whose nucleus had been removed. Normally, this hybrid will refuse to become an embryo. The key to Wilmut's breakthrough was in coaxing this cell to "wake up" long-dormant genes which are stored in their nuclei.

Much of the DNA in mature cells is often folded up and hence becomes inaccessible, which ensures that skin cells, for example, do not suddenly become liver cells. Previously, scientists knew that the protein scaffolding surrounding a mature cell's DNA is somehow responsible for turning off these genes. The twist that Wilmut's team devised was to "starve" the cells for a week, depriving them of nutrients, which somehow altered their protein scaffolding. This tricked the cells into reactivating their dormant genes and reverting back to an embryonic state.

In doing so, Wilmut's team disproved a "law" of nature often quoted in textbooks, that mature cells, once differentiated, cannot revert back to an undifferentiated, embryonic state. This may have enormous medical benefits for the future. Cells in the spinal cord, brain, and heart are notoriously difficult to regenerate because they have "forgotten" how to multiply as they did in their embryonic state. If one can coax these cells to multiply, it could allow doctors to repair spinal cords broken in severe back injuries, regenerate brain tissue after strokes, and repair damaged hearts after heart attacks. Millions of people imprisoned in wheelchairs or wasting away paralyzed in hospitals and nursing homes could be cured if these cells can be rejuvenated. Other benefits include the development of replacement organs for transplant patients, such as livers, and cloning endangered species that have difficulty reproducing in captivity.

Numerous hurdles still exist. The experiment has yet to be duplicated in other laboratories. In addition, Dolly's cells, cloned from a six-year-old adult, may show signs of premature aging. Genetic damage is also a distinct possibility in cloning. More important, the precise mechanism that makes cells "remember" long-forgotten genes still has to be elucidated.

Ron James of PPL Therapeutics, which provided a third of the funding for Wilmut, says that practical applications include creating herds of sheep which produce milk laced with beneficial enzymes and drugs. But when asked how long it will be before this technique is used to clone humans, he replied, "Hopefully, an eternity."

It's always difficult to make the leap from animals to humans, but bioethicist Arthur Caplan foresees the first human clone within seven years. Even if human cloning is banned, it is possible that an underground cloning industry may develop over time.

Cloning, of course, raises thorny ethical questions which I will turn to in the next chapter. But the moral dilemmas raised by cloning pale in comparison to those raised by genetic engineering of humans. Cloning only produces a carbon copy of an individual; genetic engineering promises the ability to change the human genome and hence the human race. To give an analogy: it is a relatively simple matter to Xerox a work of Shakespeare, but infinitely more difficult to improve on it.

## Beyond 2020: Polygenic Traits

As we have seen, the development of biotechnology will continue to accelerate to the year 2020 because of computerized and automated DNA sequencing. And as long as we focus on single gene transfers to plants and animals, we should be able to create any number of desired effects.

However, beyond 2020 progress will likely slow down considerably. First, the huge advances ushered in by computerized DNA sequencing will have been completed. Second, more complex traits are usually polygenic, involving interactions with several genes and with the environment. Third, to determine the structure and hence the properties of many complex proteins may require solving the notoriously difficult "protein folding" problem. (X-ray crystallography, we recall, is useless in determining the three-dimensional structure of a protein if we cannot crystallize it. This forces scientists to use quantum physics and computers to mathematically calculate how the protein molecule folds up.)

Just as possessing the telephone numbers of everyone in the United States does not mean we know what each person does for a living, what he or she thinks, or how U.S. society functions as a whole, possessing the location of the 100,000 genes in our body does not imply that we know their function or how they interact with each other and with the environment.

As we approach 2020, the emphasis in biotechnology will therefore gradually shift from DNA sequencing to determining how genes function and interact with each other, called "functional genomics." Unlike DNA sequencing, which has advanced exponentially because it can be computerized and roboticized, the task of functional genomics will be painstakingly slow. This process, in fact, may never be completely computerized.

One way to determine gene function is to analyze the genes of other animals. For years scientists have been tediously mutating or removing specific genes within fruit flies and mice to see how the resulting progeny were affected. Because of the rapid reproductive cycle of these animals, scientists could get useful laboratory data relatively quickly. Once the function of an animal gene was determined, scientists could hope to find homologues in the human genome by computer searches. But the human homologues often differ from their animal counterparts by many key DNA sequences; as a result, one can never be sure of the precise function of a human gene. This labor-intensive effort is not easily adapted to computerization.

Because of the difficulty in unraveling polygenic traits, from now to the year 2020 scientists probably won't be able to do more than isolate many of the key individual genes which contribute to polygenic traits. By analogy, scientists won't be able to see a tapestry in its entirety, but they will be able to isolate and examine individual threads from which it is woven. For example, scientists should be able to isolate the individual genes which control certain polygenic traits, such as the shape of our bodies, including our face, and rudimentary forms of behavior.

From 2020 to 2050, scientists should be able to weave together these individual threads and determine how they create the complete tapestry. Progress will be slower, but certain polygenic traits, especially those controlled by a small collection of genes, will probably be decoded during this period.

However, manipulating more than a handful of genes will remain beyond our reach for many decades. Even beyond the year 2050, science will be incapable of performing the genetic feats sometimes portrayed in science fiction. It will likely be powerless to manipulate the genes which control the development of entire organs. Several thousand genes are required to control the key organs of the body, which will probably be outside genetic manipulation (except in a trivial way) even late in the twenty-first century. The idea of transplanting entire organs (such as wings) across different species will almost certainly be beyond anything attainable in the twenty-first century.

In sum, progress in transgenic organisms will be explosive early in the next century as long as it involves a single protein transferred from one life form to another, giving us the ability to create new organisms and clone others. But the technological wizardry featured in Hollywood movies (i.e., the creation of "metahumans" or "homo superior" involving the manipulation of thousands of genes) is perhaps centuries away, if possible at all. At present, we can barely move snippets of DNA from one organism to another, let alone modify the hundreds if not thousands of genes which control our basic bodily functions.

## 2020 to 2050: Designer Children?

Among polygenic traits controlled by a small handful of genes are those which determine the general shape of the human body and simple forms of behavior. Can these techniques be used to create "designer children," such that parents decide their children's genes?

In the near future, unless restricted by law, science will have the capability of changing the genes of our progeny. Already it is possible to control the height of our children via genetically engineered growth hormone. Scores of other traits that are controlled by a single protein will soon follow.

### GENES FOR BODY SHAPE

Scientists are now closing in on the handful of genes which control body weight, the "fat genes." Already, five such genes have been found in mice. The homologues in humans have also been found, and scientists think they control our body weight as well. It is expected that many more individual genes will be found in the coming decade, although it may not be until 2020 that scientists have a complete description of how these myriad genes interact to control our metabolism and body shape.

The control of the handful of genes which govern our body weight could have significant consequences. According to a 1995 report of the Institute of Medicine, 59 percent of the adult population in the United States is obese. Not only has this generated a huge industry producing diet books, tapes, and programs, but the economic health burden caused by obesity is staggering. According to epidemiologists at the Harvard Medical School, obesity cost the nation $45.8 billion in health bills in 1990, in addition to $23 billion from lost work, helping to drive health costs through the roof and killing as many as 300,000 Americans. Diabetes, heart disease, strokes, and colon cancer are some of the severe complications of obesity.

The fact that every industrialized society is undergoing this explosive growth in girth indicates not only the widespread availability of high-fat foods but also perhaps a genetic predisposition to obesity. By analyzing twins reared in different environments, one can quantify the extent to which height and weight are governed by a genetic component. Most studies show that 50 percent of our weight and height are genetically controlled (although a few have found a correlation as high as 80 percent). This study on twins shows that body weight is influenced by genetics, although it is not controlled by a single gene. Several genes are involved, some of them triggered by cues from the environment.

Scientists are beginning to appreciate how this handful of genes interact. First, the hormone leptin, which is controlled by these genes, seems to govern appetite. The more you gain weight, the more you make leptin, which increases metabolism and shuts off appetite. If you get too thin, then leptin levels begin to fall, appetite increases, and you burn less fat.

Second, since this feedback loop involves the brain, one can create drugs which influence neurotransmitters involved with feelings of satiety and well-being. The appetite suppressant dexfenfluramine (which was approved in 1996 by the FDA under the name Redux, but which is already available in sixty-five other countries) is one of several drugs which control the neurotransmitter serotonin, which quells appetite.

By 2020, this handful of individual genes will grow in number until scientists have a complete understanding of the individual genes which control not only fat but the overall body shape as well, including its muscles and skeletal system. But it may not be until after 2020 that we understand how these individual genes interact to orchestrate these polygenic traits and shape a human body.

### GENES FOR THE FACE AND SCALP

Similarly, early in the next century, the individual genes which control the features of our face and scalp should be isolated. In 1996, it was announced that a gene for hair growth was isolated by a team operating primarily out of the University of Washington in St. Louis. A defect in the X chromosome of this gene can cause anhidrotic ectodermal dysplasia, which affects about 125,000 Americans. One of the symptoms of the disease is baldness or severe loss of hair (as well as missing teeth and poorly developed or missing sweat glands; women are not affected). The hope is that by studying this gene, scientists will be able to control baldness.

At the opposite end, scientists may have found the gene on the X chromosome which causes uncontrolled hair growth, nicknamed the "werewolf syndrome." People who suffer from congenital generalized hypertrichosis have hair covering most of their upper body and face and unfortunately in the past ended up in circus sideshows. (Scientists believe that humans once had a protective hairy coat, but sometime in our evolutionary past a mutation turned this gene off. This, in turn, probably means that we still possess hundreds of genes that have been turned off over millions of years which represent the bodies of our primitive ancestors.)

Scientists have also been able to isolate some of the genes which affect the shape of the face. Increasingly, scientists analyzing rare genetic defects are finding that a gene codes not just for a single organ but also for features and organs such as the face, heart, and the hands as well.

"People are now finding these single genes that give the face its appearance. Quite often it turns out to be a surprise. Either they hadn't heard of the gene before or they hadn't a clue that it was involved with the face," notes Robin M. Winter of the Institute of Child Health in London.

Other genes that have been found to code for the human face include the genes for Williams syndrome, Crouzon and Pfeiffer syndromes, Waardenburg syndrome, and Treacher Collins syndrome.

By isolating the handful of individual genes that control our body weight, scalp, and face, scientists may ultimately be able to cure the gross abnormalities which afflict people suffering from these genetic diseases. By 2020, this small collection of genes should increase into an almost complete set of genes that code for the face and scalp. But it may not be until after 2020 that we understand the intricate ways in which these polygenic traits interact to control body shape and the appearance of the face.

For the foreseeable future, the best way to be beautiful, as Candice Bergen once said, is to "choose your parents well."

### GENES FOR BEHAVIOR

Scientists have long suspected that certain types of behavior are influenced by genetics, involved in an intricate web of genes and complex cues from the environment. But for the first time in history, a few of the individual genes that code for certain types of behavior are being isolated.

(I will discuss the important social implications of this research in the next chapter.)

One interesting development was the discovery in 1996 that a single gene, called *fru*, controlled almost the entire courtship ritual of the male fruit fly. For the first time, scientists discovered a single gene that controlled complex brain functions. Scientists from four universities (Stanford University, the University of Texas Southwestern Medical Center, Oregon State University, and Brandeis University) showed that this single gene was responsible for a male fruit fly recognizing a female, nuzzling her, singing by vibrating its wings, and mating if the song was approved. Since the brain of a fruit fly is relatively simple, containing only 10,000 cells (10 million times fewer than the human brain), many behavior patterns are probably hard-wired into its brain.

The entire genome of the fruit fly is expected to be unraveled around the year 2000. Since the entire behavioral repertoire of the fruit fly is not very large, we might expect that scientists will identify which genes control which behavior patterns as early as 2010.

The genes that contribute to the behavior of mice are much more complex, yet a few of these have also been discovered; this could have direct effects on human health. In 1997, for example, scientists isolated a gene that influences memory, a milestone in genetic research. Biologists have long suspected that the brain first processes many of its experiences in the hippocampus, a tiny cashew-shaped structure located deep within the brain. The hippocampus, which helps in constructing a three-dimensional "mental map" of our surroundings, is crucial for our ability to move about in the real world.

Memories are probably first processed and kept in the hippocampus for several weeks, before they are transferred to the cerebral cortex for permanent storage. This may explain why people with brain damage to their hippocampus retain previous memories of faces and places, which are stored in their cortex, but have difficulty forming new short-term memories.

The hippocampus of the mouse brain contains about a million large nerve cells, called "place cells," which enable mice to record where they are located in space. These place cells consolidate their memory via the protein kinase.

Scientists at MIT and Columbia University independently announced in 1996 that they could change the memory of mice by altering a gene that codes for kinase. (The Columbia group was able to create a strain of mice which produced a defective kinase molecule. The MIT scientists knocked out the receptor kinase gene from hippocampal cells, creating mice which lacked the protein entirely. In both cases, the mice appeared perfectly normal, but were severely impaired in their ability to find their way around. The connections between their nerve cells, called synapses, did not form correctly; hence the mice could not learn to adjust to new environments.)

The fact that the homologous genes for this particular memory gene can be found in humans may have a tremendous impact on medicine as well. Walter Gilbert believes that by 2020 this could enable scientists to create drugs that will help people with impaired memory. Alzheimer's disease, for example, starts with the hippocampus and causes short-term memory loss. But Gilbert also believes that this development will lead to drugs that will help sharpen the memory and learning experiences of normal people as well. In the near future, it may be possible to increase one's ability to assimilate new experiences by ingesting a protein that helps us form new synapses. Although memory in humans is probably influenced by a very complex interaction of genes, Gilbert foresees a time about a decade from now when we will be able to create a class of memory-enhancing drugs.

Another form of behavior that may have genetic roots is alcoholism, which contributes to half of all traffic accidents and violent acts in the country and costs the country roughly $1 billion a year. Since an identical twin of an alcoholic is twice as likely to become an alcoholic as a fraternal twin, there is a definite genetic link to alcoholism. Unfortunately, since the correlation is not perfect, it must be that many genes, including triggers from the environment, are involved as well. So far, scientists have found evidence for alcoholism genes on chromosomes 1, 4, 8, and 16.

In 1996, a gene that contributes to "anxiety" was discovered. The gene codes for serotonin, the same neurotransmitter that is targeted by the antidepressant Prozac. The gene occurs in both long and short varieties, which are inherited from the parents. People with long/long genes (about a third of the population) tested optimistic in their view of the future on a personality test. The rest of the people, who inherit the short gene, scored higher in terms of anxiety, worry, and neuroticism.

The same year, psychologists at the University of Minnesota announced that the feeling of "happiness" may be genetically based. Although they did not locate a gene controlling "happiness," they claimed that, after analyzing 2,000 twins born between 1936 and 1955 in Minnesota, there is a preset "happiness set point" that seems to be hard-wired into our genes. Whether we face good fortune or bad times, we eventually return to that set point.

Time will tell whether these genes will be confirmed by independent analysis. The point is that we are, for the first time, beginning to isolate individual genes that contribute to the complicated mix which creates polygenic traits. Finding these individual genes will accelerate in the future. Individual genes contributing to a wide variety of behavior patterns will be discovered by 2020, but it may take many years after that before we understand how they are orchestrated and how they take their cues from the environment.

## Beyond 2050

In the remake of the movie *The Fly*, Jeff Goldblum plays a brilliant physicist who designs the world's first teleporter, which will revolutionize transportation by disassembling people's molecules, transporting them across space, and reassembling them. But when he enters the transporter chamber, he fails to notice that a tiny fly has entered with him. When he teleports across space, the fly's DNA is irreversibly scrambled with his own. He is stunned to find that the fly's DNA in his body is slowly beginning to change his metabolism, his appetite, his sex drive, his physical powers, and his body shape, as he slowly and hideously mutates into a giant fly.

We now know enough about genetic engineering to make reasonable conclusions about some of the scenarios found in such science fiction. For example, it is unlikely that a random merging of thousands of human and fly genes will convert one species into the other. The reality is much more complicated. Many of the genes controlling the general shape of our body are activated only at the embryonic level. In adulthood our cells have already differentiated into our present organs and hence do not respond to new instructions that would change their function. Thus, merging DNA from other animals does not cause one's body to change into that animal. For the most part, nothing happens at all.

But randomly mixing up genes will shut down many of the biochemical processes of the human cell. The purpose of our genes is to produce proteins, which then control chemical reactions and create tissue. If the genes are mixed up, these proteins are no longer being produced; as a result, the cell may eventually cease to function and may die. So instead of gradually turning one organism into another, randomly mixing up DNA will probably cause the organs of the body to slowly malfunction and perish. Thus, many of the plots featured in science fiction are probably not possible.

At first glance, it would seem to be a hopeless undertaking to try to decipher the thousands of genes necessary to create a single organ of the body. For example, it is believed that our nervous system may involve almost half of our genome, or almost 50,000 genes. Untangling this many genes seems unfeasible. But the period after 2050 does not necessarily have to be mired in total confusion. One key to understanding polygenic characteristics is to concentrate on "master genes" and "master architect genes" which developed over hundreds of millions of years of evolution. Already, scientists have isolated the master gene that controls the eye of the fruit fly, which in turn can be manipulated almost at will.

### Master Genes for the Eye

Evolution, scientists have found, usually builds on top of previous structures. Although there are exceptions to this rule, we usually see nature adapting and co-opting previous structures to build new ones. Thus even in our bodies there are still remnants of genes that controlled the bodily functions of our earliest ancestors. (It's staggering to realize that our own genome contains fragments of the early evolutionary history of our species, going back to fish, worms, and even the earliest bacteria.)

To some extent, this can be seen graphically in the development of the human embryo, which in the early stages echoes the development of worms, fish, and mammals. As Darwin said in *The Origin of Species:* "The embryo is the animal in its less modified state; and in so far it reveals the structure of its progenitor." One theory holds that the gills found in the human embryo are remnants of the ancient gills that our fishlike ancestors once possessed.

But the DNA revolution is also springing a few surprises, forcing us to revise the biology textbooks. For example, for decades biologists have believed that evolution independently "discovered" the eye time and time again on widely differing branches of the ancestral tree. The eye of a mammal, with a single retina, differs so much from the eye of a fly, with 750 hexagonal facets, that it was always assumed that evolution, by trial and error, independently came up with the idea of an eye forty distinct times in the animal kingdom.

Some biologists have wondered, however, why the pigment rhodopsin, which is vital to capturing light radiation, is found in all of those eyes if it was "discovered" independently. Does this common fact point to the existence of an "Ur-eye," the mother of all eyes, they wondered, if we go back far enough in time?

Molecular biologists now believe so. In 1995, Walter Gehring and his colleagues at the University of Basel in Switzerland discovered that there is a "master gene" that controls whether or not an eye develops in a fruit fly, although as many as 5,000 genes may separately control the precise development of the millions of cells in the eye. The gene is called *eyeless* (because its absence deletes the presence of an eye). By placing this *eyeless* gene on different parts of the fly, Gehring's team was able to create a complete set of eyes on the wings, the legs, even the antenna of the fruit fly. One fly (as a result of their experiment) had fourteen separate eyes on its body.

This result was so remarkable that the normally staid journal *Science* emblazoned its cover with flies sprouting eyes all over its body. "It's the paper of the year. This is Frankensteinian science at its best," joked Charles Zuker of the Howard Hughes Institute in San Diego.

Gehring found that this gene repeats itself throughout the animal kingdom, in flatworms, squid, sea squirts, mice, and even humans. "Everywhere we look, we're finding it," says Gehring. "All this points to a common, ancestral eye," says Russel Fernald of Stanford University.

On one level, this means that the *eyeless* gene is ancient and fundamental to many branches of the animal kingdom. It indicates that we are probably all descendants of some sea-dwelling worm that lived 500 million years ago which first developed the original "Ur-eye" gene. That organism, which developed the first rhodopsin necessary for a functioning proto-eye, probably transferred its genes to all creatures on earth with eyes. "It means that we are all basically just big flies," jokes Zuker.

On the other hand, it also means that there are likely to be other "master genes" lurking in our chromosomes which control the development of entire organs of our body. Scientists have often wondered how a mere 100,000 genes can contain all the instructions needed to put together a human being. It makes sense that some genes are more important than others.

In building a modular house, for example, there is a "master set" of instructions that tell one how to order the various rooms and furniture. Similarly, "master genes" trigger thousands of other genetic instructions, such as how to build the organs of the body. Like instructions for a construction foreman assembling a modular house, some instructions tell the genes how to arrange the general outline of a structure, and other instructions tell them how to arrange the details.

Once we find the master set of genetic instructions, it should be much easier to figure out the blueprint of the entire body with its various organs.

What does all this mean? There are thousands of genes involved in creating various organs of the body—simply too many to track. One key to studying complex clusters of genes beyond 2020 may be these "master genes." By isolating master genes, scientists can study how thousands of other genes are successively triggered to activate the development of an entire organ.

This does not mean that one day we will be able to transfer the eye of a fruit fly into other organisms, as in *The Fly*. These master genes instruct undifferentiated, embryonic cells, not adult cells. But homologous master genes have already been shown to exist in humans, and it is logical to assume, therefore, that human embryos develop genetically in ways similar to fruit flies.

If all the master genes in mammals can be found, it would significantly speed up our learning about how thousands of genes interact to produce a single organ. Thus, in the period beyond 2050, when the focus turns to deciphering polygenic traits involving thousands of genes, master genes may be one of the important tools for decoding the 100,000 genes in our genome.

### "Master Architects" for the Body

Another important discovery which may eventually simplify our understanding of how thousands of genes act in concert involves the "master architects" helping to supervise the design of our overall body shape.

The existence of these universal genes can be deduced by analyzing animal body shapes, most of which share the same head-to-tail axis and bilateral symmetry. Our basic body shape—the head at one end, the body in the middle, and the tail at the other end, with appendages sticking out from the side, all with a line of symmetry traveling down from head to tail—goes back hundreds of millions of years. Think of dinosaurs, insects, sharks, crocodiles, rabbits, etc. All have the same basic body design.

Recently, scientists have isolated the genes partly responsible for the body plan. This remarkable set of genes control the way in which uniform embryonic cells gradually differentiate into our head, arms, torso, and feet. These "homeobox" genes, as they are called, help to control the grand blueprint of the body's architecture for a wide variety of animals—everything from flies and mice to humans. (These genes are called the HOM genes in invertebrates; their counterparts in vertebrates are called Hox genes.) Many, though not all, of the homeobox genes of fruit flies, mice, and humans have a one-to-one relationship. Furthermore, the order in which these homeobox genes occur along the chromosome matches exactly the order in which they occur in the organism itself, starting with the head, which has greatly simplified the identification of the homeobox genes.

The fact that we can freely interchange some of these genes between widely separated species indicates how ancient these genes are. For example, the *Pax-6*, *Dlx-1*, *Hox-7*, and *wnt-7a* genes in vertebrates are homologous to the *eyeless*, *Distal-less*, *msh*, and *wingless* genes found in fruit flies, which, respectively, control the development of the eye, the appendages, the muscles, and the wings.

That these HOM genes control the body's architecture was shown by mutating fruit flies. By analyzing what happens when certain HOM genes are mutated, scientists have been able to determine the function of the genes. Mutating the *antennapedia* gene, for example, can cause the antennae of the fruit fly to be substituted for an extra pair of thoracic legs.

A team of scientists at the Harvard Medical School showed that a mutation in the homeobox genes can cause a deformity in humans called synpolydactyly, which causes webbing between our fingers and extra fingers. Another group found bone morphogenetic protein (BMP), the chemical signal that determines whether webbing will occur or not. Blocking the BMP, they could induce a chicken which had webbing like a duck, or feathers on its legs rather than scales.

Scientists now think that these homeobox genes produce chemicals, perhaps like BMPs, which control other genes, telling when to shut off or turn on certain processes. Some of these BMPs may even tell certain cells (like those for webbing) to commit cell suicide, thereby suppressing the webbing of feet.

Such homeobox genes may be the key to unraveling the mystery of how the body plan is laid out at the embryonic level. Obtaining a complete map of the homeobox genes, which should be possible in the next decade, may help to guide us through the dense thicket of polygenic traits, some involving thousands of genes, in the period after 2050.

## Beyond 2050: Angels in America

We have seen how primitive our understanding of gene transfers is: scientists usually transfer snippets of DNA, for the most part involving a single gene, from one organism to another. We have seen how notoriously difficult polygenic traits (e.g., metabolism and body shape, hair growth, face, body organs) are to understand at the genetic level, because they involve scores to thousands of genes. But if one day we are able to manipulate the master genes that control entire organs of the body, the question naturally arises: will this knowledge give us the ability to create "metahumans" or "homo superior" in the future?

From 2020 to 2050, we will probably be able to decipher many of the thousands of genes involved in shaping crucial body organs. However, it may be decades after that before we are able to manipulate them.

To understand the difficulty of manipulating polygenic traits, consider the example of human flight, which has excited the imagination of mystics and theologians since the beginning of history. Angels have appeared in religious mythology for thousands of years.

Creating people who can fly requires manipulating perhaps thousands of genes that control the development of the wings, all acting in synchronization to create the necessary tissue and bones, which is far beyond modern biotechnology. Ultimately, it may be possible to perform such miraculous feats (evolution, after all, has created even greater feats of biological magic), but it may take centuries to master such a technology.

To see this, we first realize that master genes for bird wings may be of no use. Wings, like the fruit fly's eyes, may in fact be controlled by master genes, but only the birds and select flying animals have the proper gene clusters which can be activated by these master genes. Inserting the master gene for wings into a human may do nothing (or may activate homologous organs, like arms).

Next, consider the series of difficult steps involved in creating an animal with wings. Although the genome for birds may be sequenced early in the twenty-first century, identifying the vast collection of genes necessary for wings may take many decades after that. Determining how they act in concert to produce the proper bones, muscles, tendons, feathers, blood supply, immune system, etc., may take more decades still.

Then one would have to worry about the aerodynamics. One reason why birds can fly is that they have hollow bones. Because humans have solid bones and are quite heavy in relation to birds, our wingspan would have to be huge. According to Bernoulli's principle, which governs the aerodynamics of flight, a human wingspan would have to be on the order of twenty feet, or comparable to the wings of a hang glider. But then the muscle power necessary to flap wings of this size would exceed anything possible with the human frame. Such wings would require a major reengineering of the human body, developing massive back muscles and lighter but stronger bones.

Then there is the problem of splicing the genes for wings onto a human's back. First, one would have to modify the genes for bird wings to create a wingspan of twenty feet. This alone poses problems, since the process of expanding a bird's wing by over ten times involves increasing the blood supply, strengthening the muscles, and hardening the bones.

Then one has to modify the human genome to accept wings. It might be easy to find the location among our homeobox genes where the wings should be inserted. But the problem is that our body organs must be radically altered: our muscles must become stronger and our bones lighter. Moreover, to splice the muscle, bone, and tissue of a bird, which requires thousands of genes, to ours, also requiring thousands of genes, would necessitate decades of experimentation and involve a microinjection technique that is beyond anything attainable for the foreseeable future.

The point here is that the relative success of creating transgenic animals and plants, based on transferring single genes, does not translate at all into creating chimeras which have the polygenic traits of other animals. To transfer the organs of one animal (e.g., wings of a bird, fins of a fish, trunk of an elephant) into another will require techniques not available until long after 2050, perhaps after the twenty-first century.

In other words, most of the fantastic creations from ancient myths may remain just that: myths. True chimeras are probably beyond the reach of biotechnology for perhaps a century or more, if they are attainable at all.

But the overriding question this very discussion raises is: is it ethical to manipulate the human genome? If so, under what guidelines?

In the next chapter, we will examine some of the touchy moral and ethical issues raised by the biomolecular revolution, which promises not only to give us health and prosperity but also to challenge our moral principles and perhaps force us to redefine who we are.

# 12

# Second Thoughts

## *The Genetics of a Brave New World?*

"There are some things about which we must simply say you can't do."
—JAMES WATSON

"Any attempt to shape the world and modify human personality in order to create a self-chosen pattern of life involves many unknown consequences. Human destiny is bound to remain a gamble, because at some unpredictable time and in some unforeseeable manner nature will strike back."
—RENÉ DUBOS, *Mirage of Health*, 1959

"The very reductionism to the molecular level that is fueling the medical revolution also poses the greatest moral challenge we face. We need to decide to what extent we want to design our descendants."
—ARTHUR CAPLAN, University of Pennsylvania Center for Bioethics

THE DNA REVOLUTION gives us at least two startling divergent visions of the future. One vision, promoted by the biotech industry, is that of health and prosperity: gene therapy will eliminate hereditary diseases and possibly cure cancer, bioengineering will create new drugs to vanquish infectious diseases, and gene splicing will create new animals and plants which will feed the world's exploding population.

However, a much darker vision of the future was painted by Aldous Huxley in his unsettling yet prophetic book *Brave New World*, written in 1932, with the world still reeling from the unremitting savagery unleashed by World War I and from the grinding poverty of the Great Depression.

The novel is set six hundred years in the future, when a similar series of disastrous wars have convinced the world's leaders to impose a radical new order. Recoiling from the chaos of the past, they decide to impose a Utopia based on happiness and stability, rather than concepts that have proven to be inherently unstable and messy, such as democracy, freedom, and justice. To be unhappy is against the law of the land. And the key to this state-mandated paradise is biotechnology.

Babies are mass-produced in huge embryo factories and are cloned to produce a caste system of Alpha, Beta, Gamma, Delta, and Epsilon human beings. By restricting the oxygen given to the embryos, scientists can cause selective brain damage and clone an army of obedient workers. The most brain-damaged are the Epsilon morons, subhumans who are carefully brainwashed into happily doing all the menial labor in society. The highest caste are the Alphas, who are, by contrast, carefully groomed and educated to become the ruling elite. Happiness is guaranteed by incessant, numbing brainwashing and unlimited access to mind-dulling drugs and sex.

The world was scandalized by Huxley's outrageous novel, and many attempts were made to censor it. Ironically, events have outpaced even Huxley's fertile imagination. In the 1950s he wrote: "I projected it six hundred years into the future. Today, it seems quite possible that the horror may be upon us within a single century." But even a century may seem too long; already, many of his technological predictions are within grasp.

Huxley's predictions were certainly prophetic. He wrote in a time when the laws of embryonic development were largely unknown. Less than forty years later, however, Louise Brown, the first "test tube baby," was born. By the 1980s, parents had a wide selection of commercially available birthing options: embryos can be frozen and then thawed out years later, infertile couples could employ surrogate mothers to bear their children, and even grandmothers could give birth to their own grandchildren (by having the fertilized egg taken from their grown daughter implanted into their uterus). And with the coming of the biomolecular revolution, many of his other predictions may also be within reach—human cloning, selective breeding, and so on.

Therefore the question must be asked: which future will we choose?

In this chapter, I will look at how the biomolecular revolution will impact on society, for better or for worse. Few will dispute the tremendous accomplishments and potential of the biomolecular revolution. However, even the creators of the revolution have expressed reservations about the moral and ethical direction of this revolution if its excesses are not checked. In a democracy, only informed debate by an educated citizenry can make the mature decisions about a technology so powerful that we can dream of controlling life itself.

## Nuclear Energy vs. Genetic Revolution

The awesome scientific knowledge that will be unveiled early in the next century must be tempered by the enormous ethical, social, and political questions that it raises. One framework in which to discuss the implications of the biomolecular revolution is to compare it with the nuclear revolution.

Biomolecular scientists are determined to avoid the kind of blunders committed in atomic energy research, which was originally conducted in total secrecy under the cloak of "national security." Because there was little democratic discussion of the implications of atomic energy, the United States is now faced with seventeen leaking nuclear weapons dumps, which may cost upward of $500 billion to clean up. The human price is incalculable; unethical radiation experiments were conducted on 20,000 unsuspecting human subjects since the 1940s, including injecting plutonium into the veins of innocent patients, releasing radioactive materials over populated areas, and exposing pregnant women.

Mindful of this parallel, the originators of the Human Genome Project set aside 3 percent of the budget for what they called the Ethical, Legal, and Social Implications Branch (ELSI) of the Human Genome Project. It is the first time in history that a crash government project has ever devoted even a fraction of its resources to larger societal questions.

One danger that both supporters and critics of the technology fear is the equivalent of a Three Mile Island—i.e., a catastrophic accident due largely to human error, design flaws, or inadequate testing that could endanger the lives of millions and give the entire industry a black eye.

But there is an important difference between the atomic and biomolecular revolutions. It is possible, to some degree, to control the proliferation of nuclear weapons because of the tens of billions of dollars in resources necessary to develop a large nuclear infrastructure, complete with enrichment facilities, reactors, and top nuclear scientists. One cannot simply start up a nuclear program in one's basement. For example, the flow of enriched uranium and plutonium is restricted via stringent security measures, which has been one of the principal reasons only a handful of nations possess nuclear weapons today. The genie cannot be put back into the bottle, but we can limit the number of genies set loose on the world.

The nature of bioengineering is radically different. With only a modest $10,000 investment, one can conduct biotech experiments in one's living room and begin to manipulate the genome of plants and animals. With a few million, one can create a fledgling biotech industry. The low initial investment, high return, and potential for feeding its people are some reasons why a poor nation such as Cuba has decided to jump into biotechnology.

But this also means that biotechnology is impossible to contain. One cannot restrict the flow of DNA; it's everywhere. Because the technology can never be entirely banned, it is important to discuss and decide which of the various technologies should be allowed to flourish and which ones should be restricted, either via governmental fiat or by social and political pressure.

### You Can't Recall a Crop

Jane Rissler of the Union of Concerned Scientists worries that the lack of proper oversight may release a seemingly harmless gene into our food supply which may cause life-threatening allergies to the unwary customer. (Banana genes, for example, have been inserted into tomatoes; such tomatoes could be unwittingly eaten by children with severe allergies to bananas.) Rebecca Goldburg, senior scientist at the Environmental Defense Fund, points out that there are 5 million people with food allergies ranging from mild to life-threatening. Goldburg recounts the recent case of a soybean that was engineered to contain a gene from the Brazil nut. Subsequent testing by the company showed that it was allergenic and could have caused life-threatening shock if the product had been prematurely released to the public.

Critics also worry that the FDA is approving new foods for the supermarket without adequate testing, while the Department of Agriculture allows companies to do field testing without permits. "I think they've taken a good idea and taken it too far," concludes Rebecca Goldburg.

Under pressure from powerful agribusiness interests to cut red tape, the Department of Agriculture streamlined the process of field testing these plants. From 1987 to 1995, 500 field test permits covering forty new species were granted by the Agriculture Department (including barley, carrots, chicory, soybeans, peanuts, broccoli, cranberry, various berries, and watermelons) with minimal oversight.

What worries Goldburg is that the same corporations that are pushing pesticides are now pushing genetically altered plants that are more resistant to pesticides. This, she thinks, smacks of self-interest. The net result will be that farmers buy more pesticides, thinking that their crops can handle the increased load, which means more pesticides in our foods, and potentially the creation of a new generation of pesticide-resistant bugs. This could spark a new "arms race" between insects and bioengineered products, creating a host of "super-bugs" that are resistant to potent levels of pesticides and also leaving more pesticides in our food.

But the primary fear is that entirely new plants never seen before in nature may escape into the wild once the floodgates are opened, where they may displace native plants and take over whole ecosystems, with unforeseen results.

"If the plants are raised outdoors and the new genes get into the wild gene pool, it could have a potentially destabilizing effect on the ecological system," says Jeremy Rifkin of the Foundation for Economic Trends, one of the leading critics of the biotech revolution.

The worry over transgenic crops is summarized by the phrase: "bioengineered crops can't be recalled." Many critics point out the unforeseen consequences of alien species being introduced, deliberately or accidentally, into new environments, as has happened with zebra mussels, Dutch elm disease, kudzu, and chestnut blight. The delicate ecological balance can be severely affected by a new species.

A case in point is the African bee *(Apis mellifera adansonii scrutellata,* sometimes referred to as the "killer bee" by the press), which was deliberately imported into Brazil in 1957 to replace the European honeybee *(Apis mellifera),* which did not adjust well to the daylight cycle of the country's equatorial climate. When scores of queen bees escaped, this highly aggressive species spread out of control and wreaked havoc with the bee industry. Unlike the gentler European bee, the African bee is easily aroused and attacks and swarms by the thousands. It has already killed 1,000 people and caused millions of dollars in losses.

Today, the African bee is the dominant bee species over 20 million square kilometers of the Western Hemisphere, including all of South and Central America. The bees reached Texas in 1990, Arizona in 1993, and are expected to colonize most of the American Southwest until they are stopped by the colder climate of the North around the year 2000.

This is a telling example of how humans have upset a natural ecosystem by unwittingly introducing a new life form which is aggressive enough to displace the milder domestic one.

Instead of pesticides, Rissler advocates an alternative vision of the future, something called "sustainable agriculture," which involves using natural enemies of certain insects to control their population, which can be done without pesticides. By balancing the ecology of insects in a field, one may be able to keep certain insect populations within limits by importing its natural enemies.

### "Who Owns the Genome?"

Critics point out that the secret of life is being unraveled by companies with the freewheeling morality which prevailed in the Wild West. *Science* even devoted its cover to the issue with the title "Who Owns the Genome?"

Daniel Cohen, of the Center for the Study of Human Polymorphisms in Paris, has compared the patenting process to "trying to patent the stars. . . . By patenting something without knowing the use of it, you inhibit industry. That could be a catastrophe."

In 1996, Jeremy Rifkin led a coalition to protest the patenting of the tumor suppressor breast cancer gene *BRCA1*. Myriad Genetics of Salt Lake City, whose scientists isolated the gene in 1994, patented it and began marketing commercially available genetic tests. The coalition argued that patenting the gene would jeopardize women's privacy, especially if the information wound up in the hands of insurance companies. It also argued that patenting genes constricts scientific competition, drives up prices, and allows private industry to reap profits from publicly funded research.

But Collins believes that this thorny problem will gradually disappear with time. "When the Human Genome Project is done, all the sequences are going to be publicly available; no one is going to be able to patent the sequence anymore. At that point, the patenting (if there is patenting, and I think there will be) will be on uses of the sequence, showing that this particular region can be used to make a product that benefits people. And that will probably be for the best."

### Genes and Privacy

"Should you have a legal right to demand someone accused of rape to give a DNA sample?" asks James Watson. "Should you, if you're running for President, have to say what your DNA constitution is?" What might have happened if J. Edgar Hoover, the pugnacious and ruthless director of the FBI, had the genetic profiles of politicians in his drawer? Watson wonders. For many decades, J. Edgar Hoover bullied politicians because he had their sexual peccadilloes and drinking habits on file. How much more pressure might he have applied if he knew the complete genetic history of Washington's sometimes wayward politicians?

What is to prevent someone from swiping strands of hair from a presidential candidate and having them genetically analyzed? John Kennedy, for example, may never have been elected President had it been known that he had a serious medical problem with his adrenal glands, which became known only after his death. Recent studies of preserved DNA samples taken in 1967 from former Vice President Hubert Humphrey showed they possessed a cancerous *p-53* mutation associated with bladder cancer (Humphrey died of cancer in 1976). With modern techniques, he could have been diagnosed as being predisposed to cancer before the 1968 presidential elections, possibly eliminating him from the race. David Sidransky of Johns Hopkins, who led this study, says, it "could have changed the course of political history."

A related question concerns mandatory testing. Already, DNA data banks in this country are being formed by testing prisoners. But should the government be allowed to force people to be tested against their will? Arthur Caplan of the Center for Bioethics believes that thirty years from now, health costs in the United States will be so exorbitant that some in the government may be tempted to call for mandatory testing for genetic diseases and simply refuse to pay the health costs for a baby whose genetic disease was preventable if it had been tested.

Caplan believes that within fifteen years the debate over whether or not to have your future baby genetically tested will become even more raucous than the abortion debate today. Are you irresponsible if you have children without genetic testing? And if so, should the government pay for such genetic irresponsibility? Eventually, he believes, people who have children without genetic testing may be treated as pariahs.

Next, what happens if our genome is leaked out publicly, to our employers, our insurance company, our fiancé, especially for those who harbor potentially deleterious genes? Since time immemorial, societies have committed some form of genetic discrimination. People with obvious deformities or diseases were taunted, labeled witches (as in Huntington's disease), systematically isolated from society, or even killed. What is new, however, is that today it will be possible to screen individuals for a genetic disease even if the disease never appears. Someone who may never suffer from a particular genetic disease may be denied insurance or a job if the person has a high probability of developing a genetic disease.

Nancy Wexler, head of the Ethical, Legal, and Social Implications Branch (ELSI) of the Human Genome Project, says, "Genetic information itself is not going to hurt the public. What could hurt the public is existing social structures, policies, and prejudices against which information can ricochet. We need genetic information right now in order to make better choices so we can live better lives. We need the improved treatments that will eventually be developed using genetic information. So I think the answer is certainly not to slow down the advancing science, but to try, somehow, to make the social system more accommodating to the new knowledge."

According to the now defunct Office of Technology Assessment, the former investigative arm of the U.S. Congress, 164,000 applications for medical insurance are being turned down for medical reasons. The OTA report stated: "Applicants for insurance plans are already being asked to provide information to prospective insurers related to genetic conditions like sickle cell anemia. Some experts fear that individual policies will become increasingly difficult to acquire as more genetic screening tests become available."

A recent study done by Harvard and Stanford universities identified 200 cases in which people were denied insurance, fired from their jobs, or prevented from adopting children because of their genes.

Four bills have been introduced in Congress and twenty bills in various state legislatures to prohibit genetic discrimination. Fourteen states have so far passed laws against it. President Clinton in 1996 signed a bill that prohibits insurance companies from discriminating on the basis of a "pre-existing condition."

Genetic discrimination could also affect your marriage prospects. Since everyone has some genetic disease in their genome, this could play havoc with dating rituals. Already, there are dating services exclusively for singles who have tested negative for AIDs. In the future, there may be dating services for people who have tested negative for potentially fatal diseases, like cancer.

But Collins thinks some of these fears of "genetic wallflowers" are exaggerated. "Those concerns are somewhat lessened when you realize that we are all walking around with four or five genes that are pretty badly screwed up, and maybe another twenty or thirty that are moderately defective. So if you're going to wait around for the perfect genetic specimen to walk in, to be your mate, you're going to be single for the rest of your life. It's not going to happen. And you're not going to be able to offer them a perfect genetic specimen either. So we are all flawed. That's the way it is."

## Are Genes Us?

One area that could cause considerable misunderstanding is the link between genes and human behavior. Although human behavior is influenced by genes in complex ways, to say that we have a gene for this or that behavior goes too far. Caplan thinks that for the next thirty to forty years this link between genes and behavior will be a "ticking time bomb." While he believes that math ability, personality types, mental illness (depression and schizophrenia), homosexuality, alcoholism, and obesity all have genetic roots, he cautions, "It would be silly to just equate our behavior with genes. It's obvious that even in the same family twins don't turn out to have precisely the same behavior." Genes are only one ingredient in the mix.

Christopher Wills of the University of California at San Diego says, "Simply determining the sequence of all this DNA will not mean we have learned everything there is to know about human beings, any more than looking up the sequence of notes in a Beethoven sonata gives us the capacity to play it. In the future, the true virtuosos of the genome will be those who can put this information to work, and who can appreciate the subtle interactions of genes with each other and with the environment."

Mistakes have already been made. In 1996, the announcement that the gene *D4DR* controls "novelty seeking" in humans made the front page of the *New York Times*. But an exhaustive study of 331 people failed to show any such link.

A more controversial claim which has not held up under scrutiny is the genetic link to violence. The initial controversy was sparked in 1965 by a study which reportedly found that out of 197 patients in a high-security mental hospital in Scotland, 3.5 percent had an unusual XYY chromosome. XYY males were widely stereotyped as being violent and subnormal. The press dubbed the Y chromosome the "criminal chromosome." (Actually, later studies showed that males with XYY chromosomes are known to be more widespread than previously thought, and 96 percent of them lead perfectly normal lives. The most common traits among XYY males seem to be tallness, higher IQ than normal, and slight slurring of speech.)

There is a lesson for the future. By 2020, when personalized DNA sequencing is widely available, there will be many claims to have isolated the "violence gene." By then, it will be a simple matter to correlate prison populations with any number of genes. Genes will certainly be discovered which will, superficially at least, appear to be associated with violent individuals. For example, genes will be found which influence the production of male hormones like testosterone, which some believe may increase aggression under certain circumstances. However, to claim that the "violence gene" has been discovered might be a gross error. Although these genes may in fact be found in a tiny fraction of violent individuals, the vast majority of violent individuals may be linked to totally unrelated factors (e.g., poverty or racism).

This controversy erupted again in 1992, when preparations were made for a major government-funded conference on violence and genetics. African-American critics charged that the conference was unbalanced, giving the impression that genetics was a driving feature of violence, rather than one among a host of contributing factors. Said psychiatrist Peter Breggin, "If you think back, the genetic policy argument used to be that blacks were docile—in one generation they are now genetically violent. This is not science. This is the use of psychiatry and science in the interest of racist social policy."

One controversy about the genetic roots of behavior which will continue for many decades into the future concerns one of the touchiest issues in modern society, race and IQ. In general, most molecular biologists avoid simplistic comments about genes being the sole source of human behavior. However, there is a tendency of others, especially those with a hidden political agenda, to use the results of genetic research to support their often exaggerated claims.

The whole question of DNA, genetics, and race burst upon the national scene in 1995 with the publication of *The Bell Curve* by Richard Hernstein and Charles Murray. It soon ignited a national controversy and opened deep wounds.

Some facts are indisputable. African-Americans consistently score about 10 percent lower than white Americans on the IQ exam. And Asian-Americans consistently score a bit higher than Caucasians. But does that mean that Asian-Americans are a bit smarter than Caucasians, who are in turn 10 percent smarter than African-Americans?

From an evolutionary point of view, it seems unlikely that race and intelligence are strongly linked. The various races of the earth began to diverge about 100,000 years ago in waves of migration from Africa, long after humans had evolved their large brains, which took millions of years of evolution. So the races of the world are a relatively new phenomenon, whereas human intelligence is much more ancient. (DNA analysis, furthermore, clearly shows that the greatest genetic variations exist not between races, but within the races. So the genetic distance between Richard Hernstein and Nelson Mandela, for example, may be much smaller in principle, than the genetic difference between Hernstein and Murray.)

Caplan, echoing the comments of many other scientists, believes that intelligence is actually multidimensional, involving many facets that are totally neglected by IQ exams. He concludes: "*The Bell Curve* was pretty dopey. Psychiatrists and geneticists know that intelligence is a very complex trait made up of many different things. We all know people who can compute well but can't interact socially with anyone, or people who are very good at finding their way around a neighborhood and others who cannot seem to locate the door in their house. Many different things contribute to intelligence. *The Bell Curve* didn't reflect any of this."

One lesson for the future is this: Commentators have noted that the issue of race and IQ usually surfaces during times of economic hardship. Inevitably, because of the business cycle, there will be many periods in the twenty-first century when the economy descends into recession. Demagogues looking for scapegoats will find a receptive ear among the millions who are thrown out of work. In American history, racial theories of intelligence usually receive widespread publicity during times of economic crisis, when people feel threatened by new waves of immigrants. In 1923, for example, Carl Brigham published *A Study of American Intelligence*, using IQ tests to prove that Alpine and Mediterranean "races" were inferior to the Nordic "race" and that Africans were inferior to both. This fueled the movement to exclude people from Southern and Eastern Europe, especially Italians and Jews. One congressman said, "The primary reason for the restriction of the alien stream . . . is the necessity for purifying and keeping pure the blood of America." President Calvin Coolidge, who signed the Immigration Act enforcing quotas on certain nationalities, was on record as stating, "Biological laws show . . . that Nordics deteriorate when mixed with other races."

## After 2020: Manipulating Our Germ Line?

Beyond 2020 new ethical questions are likely to be raised. Gene therapy, which I discussed in Chapter 8, relies on manipulating the genes of somatic cells. Thus the new genes cannot be passed on to succeeding generations. The new genes die when the patient eventually dies. But germ-line therapy would change the genome of our sex cells, so the new gene can be passed on permanently to our offspring. As with transgenic mice, using microinjection of human embryos to permanently change the genetic heritage of the patient could, for example, eliminate cystic fibrosis from a family forever.

Although the idea of eliminating genetic diseases from one's germ line is appealing, there is also enormous potential to misuse germ-line therapy. By and large, the scientific community is against the idea of germ-line therapy. In 1988, the European Research Council stated flatly, "Germ-line gene therapy should not be contemplated." There are, however, some disagreements among scientists.

Would parents opt for germ-line gene therapy, if it were available, to choose the height, sex, strength, eye, and hair color of their children?

"Are you kidding? Yes!" claims Arthur Caplan. There is ample evidence that some families, given the chance, would readily pay to have this done. Parents already try to shape their children in hundreds of ways, such as giving them lessons in piano, languages, sports, etc. "I think there is no doubt that many parents will want to use genetic information to design their kids," Caplan says.

But is this a good thing? The question is: what should be the role of doctors? Are they servants who are expected to simply carry out the wishes of the consumer? Or do we want them to be ministers and guardians of morality, deciding what forms of treatment are unethical? Caplan predicts that it's going to be "one whopping moral debate."

Yet banning such therapy could create a thriving black market in germline therapy, especially in Third World countries. Even a simple test such as determining the sex of the unborn infant is causing a major demographic earthquake.

"The use of this technology for sex selection insults the reasons I went into genetics in the first place. Sex is not a disease but a trait!" declares Francis Collins.

Will parents, for example, primarily ask for children who are male, tall, strong, and handsome? The answer, unfortunately, in many countries and in many families, is yes. The laws of evolution dictate that animals will try to give every possible genetic advantage to their unborn children. And humans are no different. Consciously or unconsciously, we want our children to have a head start on life.

To poor families in the Third World, the idea of tinkering with their unborn children seems one way out of poverty. Even before gene modification becomes a reality in the next century, the introduction of a simple device like the sonogram is creating a major demographic shift in China and India, with grave implications for the next generation.

In large portions of the developing world, peasant families place an inordinate emphasis on male children. Not only do male children carry on the name and enjoy numerous feudal privileges, but families with females are required to prepare an expensive dowry at the time of marriage, which is a drain on a poverty-stricken family's finances. According to Monica Das Gupta of Harvard University, from 1981 to 1991 a million girls in India were lost to selective abortions when sonograms were introduced. Four million other girls simply "disappeared" during their first four to six years of life. In other words, 3.6 percent of the female population for that age bracket disappeared.

China's one-child-per-family policy, which has finally brought its population explosion under control, had the unintended side effect of fostering female infanticide. Informal estimates of the young female population in rural areas have shown that up to 10 million female children are "missing." On the southern coast of China, the normal sex ratio, which is 100 females to every 103 male babies born, became skewed in 1995 to 100 females for every 115.4 males.

If the introduction of the simple sonogram could unleash this demographic nightmare, think of the social upheaval that could result from the ability to genetically control our progeny. To give a mild example, in the case of the genetically engineered human growth hormone (HGH)—which only children who suffer from HGH deficiency or chronic kidney failure can quality for—a recent study found that 60 percent of the children receiving HGH did not qualify. Apparently, anxious parents, concerned about the height of their children, have been pressuring doctors to administer HGH, even at costs of up to $16,000 a year for treatments.

Inevitably, science in the twenty-first century will require the passing of certain laws to prohibit rampant meddling with the human genome, and certainly the human germ line. Some argue that pernicious genetic defects which have caused excruciating pain and suffering for generations should be eliminated from our germ line forever. Others argue the Law of Unintended Side Effects—that by playing God we will inadvertently cause even more suffering later. The question that will dominate ethical battles in the next century is precisely where this fine line should be drawn. Many scientists believe genetic manipulation of our germ line for strictly cosmetic reasons should be banned, as this is a frivolous (and potentially dangerous) application of a powerful technology. However, if it can be proven that grotesque diseases like Huntington's serve no practical purpose, then an equally powerful case can be made to eliminate them forever from one's germ line.

There may be no definitive answer to where to draw the line, as public perceptions and scientific advances change over the decades. However, since random chemical errors, cosmic rays, chemical pollution, poor diet, and other environmental insults continually create new mutations in our genome, it is a question that will be with us for centuries to come.

## To Clone a Human

Some predictions from Huxley's *Brave New World* remain firmly in the distant future. At present, it is impossible to bring a fertilized egg to full term in a test tube. Thus, Huxley's prediction that birthing might be replaced by huge embryo factories is far beyond today's technology. Reproducing the delicate, complex chemical environment found in a womb necessary to nurture a human fetus for nine months will probably remain technically out of reach for several decades.

The cloning of humans, however, is now distinctly within the realm of possibility. The astonishing announcement by Ian Wilmut of the successful cloning of an adult sheep has opened up enormous ethical and social questions. Many biologists now believe that only technical and legal barriers prevent the cloning of humans.

The ramifications of human cloning are considerable, ranging from the silly and the humane to the fantastic:

- Prominent athletes from different sports and even different decades may be cloned to create a lucrative "dream team."
- Wealthy individuals and aging monarchs without children might bequeath their fortunes and thrones to their clones.
- Parents might want to clone a child who died from a fatal disease or accident.
- Cells may be stolen from famous or glamorous figures and then sold to people who want these individuals for children.
- Graves of famous people may be raided to obtain DNA samples capable of being cloned.
- Dictators may create armies of cloned soldiers or slaves with great physical strength but limited mental capacity, or human hybrids resembling the nightmares from *The Island of Dr. Moreau.*

Other possibilities, such as cloning individuals to perform the undesirable menial tasks necessary for society, as in *Brave New World*, are not such farfetched concepts, given the fact that industrialized societies already import cheap immigrant labor to perform these duties.

Some have even speculated about a mythical society based entirely on clones, in which males would be superfluous. Parthenogenesis, whereby a female produces offspring without a male, could become the dominant mode of human reproduction. (In the long term, such a society would probably be unstable, since one of the evolutionary purposes of sex is to ensure genetic diversity, which is essential to survival in a constantly changing environment.)

There will certainly be a demand for this technology, legal or not. If some parents yearn for a "chip off the old block," then why settle for anything less than an exact copy? Some see clones as fulfilling a deep-seated wish for immortality. After all, the search for immortality probably led the Pharaohs of Egypt to build the pyramids and dying kings to build opulent tombs. Cloning offers a kind of immortality that would be infinitely cheaper.

Cloning also raises a host of other unresolved questions. Theologians have debated whether a human clone has a "soul." If humans can be cloned without limit, then what determines their individuality and essence? Ethicists have asked whether it is morally right to force our own genetic desires onto our offspring, who have no say in the matter. Moralists have been disturbed at the thought of the hundreds of embryos that may be sacrified in order to produce a single successful clone. Lawyers have asked what are the legal rights of clones—can they assume the legal rights, privileges, and debts of their predecessor? If clones are produced in order to "harvest" their organs, what happens if they refuse to be sacrificed?

Certain things are clear. There is no guarantee that cloning well-known figures will produce equally great offspring. In the movie *The Boys from Brazil*, for example, neo-Nazis cloned young versions of Hitler to resurrect the Third Reich. However, many historians have argued that it was the economic collapse of the German middle classes in the 1930s that set the stage for fascism and gave rise to Hitler. A social or political movement is rarely created by one man alone. Cloning Hitler may do no more than produce a second-rate artist. Similarly, cloning an Einstein does not guarantee that a great physicist will be born, since Einstein lived in a time when physics was in deep crisis; many of the great problems in physics today have already been solved. Great individuals are probably as much the product of great turmoil and opportunity as the product of favorable genes.

It may be that human cloning will be banned in most countries. Even before Wilmut's announcement, the United Kingdom passed the Human Embryo Act prohibiting experimentation on human embryos. President Bill Clinton previously restricted federal funding for human embryo research. In 1997, a federal panel appointed by President Clinton recommended legislation to restrict both public and private research for at least three years.

Ironically, barring some unforeseen technical problem, it is likely that human cloning will soon become a fact of life. Laws banning cloning will simply push cloning research into private, foreign, and underground laboratories, which will be able to continue this line of research because start-up costs are so low and the economic incentives so attractive.

Because of the laws of the marketplace, some predict a small but bustling underground economy based on cloning. "I don't see how you can stop these things. We are at the mercy of these technological developments. Once they're here, it's hard to turn back," says bioethicist Daniel Callahan of the Hastings Center in Briarcliff Manor.

In the future, it is likely that a small fraction of society will, in fact, be clones, given the demand for cloning. For the most part, society may eventually learn to accept the presence of small numbers of clones, in the same way that society has already accepted the presence of test tube babies from surrogate mothers and other unorthodox birthing options.

For all the controversy generated by cloning, the ultimate social impact from these clones could ultimately be negligible. People will learn that the few clones that exist will probably not pose a threat to society. After all, we already live in a world with twins; the more insidious possibility is that cloning may revive the eugenics movement.

## The Eugenics Movement

We sometimes forget that the eugenics movement in the United States has a long and unsavory history with deep roots in our culture.

The movement's founder and chief propagandist was Francis Galton, a cousin of Charles Darwin. Inspired by Darwin's work, Galton spent several decades studying the ancestral trees of eminent writers, scientists, philosophers, artists, and statesmen and became convinced that their great abilities were passed down from generation to generation. (Coming from a wealthy family, Galton was apparently blind to environmental influences. He could not admit that perhaps poor people rarely produced great statesmen because they spent most of their time trying to survive.)

Galton concluded that it would be desirable "to produce a highly gifted race of men by judicious marriages during several consecutive generations." In 1883, he coined the word "eugenics" from the Greek to mean "endowed by heredity with noble qualities." Attempts were even made to breed the perfect race. In 1886, Elisabeth Nietzsche, the sister of the philosopher, selected a group of pure-blooded individuals and set sail for Paraguay to create Nueva Germania (New Germany). According to geneticist Steve Jones: "Today the people of Nueva Germania are poor, inbred, and diseased. Their Utopia has failed."

One of Galton's disciples was Charles Davenport, a professor at the University of Chicago. He used his influence to launch a major institution at Cold Spring Harbor on Long Island to collect a massive database on family hereditary histories. His popular book *Heredity in Relation to Eugenics* helped to inspire the eugenics movement in the United States. In the book, not only did he call for selective breeding to enhance the intellectual qualities found among artists, musicians, scientists, etc., but he also said it may be necessary to use forcible methods to eliminate undesirables with unwanted characteristics. "Society must protect itself," he wrote. "As it claims the right to deprive the murderer of his life so also may it annihilate the hideous serpent of hopelessly vicious protoplasm."

In 1927, this was given legal stature when the U.S. Supreme Court upheld the constitutionality of sterilization in the case of *Buck* v. *Bell*, involving a Virginia sterilization statute. Justice Oliver Wendell Holmes wrote: "It is better for all the world, if instead of waiting to execute degenerate offspring for crime, or to let them starve for their imbecility, society can prevent those who are manifestly unfit for continuing their kind. The principle that sustains compulsory vaccination is broad enough to cover cutting the Fallopian tubes."

By 1930, twenty-four states had passed laws allowing for the sterilization of a wide variety of "undesirables," which included criminals, epileptics, the insane, and the retarded. By 1941, 36,000 people were sterilized in the United States.

The Nazis openly expressed a deep gratitude to the eugenics movement in the United States, which provided an inspiration for their own ideas. Eugenics was incorporated as an integral part of the Nazi ideology, based on breeding the Aryan "master race." Eventually, millions would be rounded up, thrown into camps, or gassed, victims of the abstract, theoretical ideas proposed by the eugenicists.

Many of these ideas still percolate in the United States. In the 1980s, Nobel Prize-winning physicist William Shockley, co-inventor of the transistor, called for Nobel Laureates to contribute to a sperm bank. Any eligible female could then perform her duty to humanity and improve the human race by being inseminated from this sperm bank of "geniuses."

One long-term danger for the far future is that those who are the wealthiest will be able to afford to improve their germ line, while others will not, leaving the rest of society behind, eventually creating a new biological caste system. Gregory Kavka, a philosopher at the University of California at Irvine, says, "Any such move toward genetic enhancement has the potential of reestablishing social inequality, though along new lines. Old aristocracies of birth, color, or gender may dissipate, only to be replaced by a new genetic aristocracy, or 'genetocracy.' "

The deep fracture lines of society could become chasms if only the wealthy have access to choosing their germ line (eventually creating a nightmarish, two-tiered society like the one portrayed by H. G. Wells in *The Time Machine*, when the Morlocks toiled with their machines in underground caverns while the childlike Eloi pranced and frolicked aboveground).

In the future, society must be wary of those who would use the benefits of the genetics revolution to further their own social agenda.

## Biological Warfare

But perhaps the greatest fear concerning biotechnology is the deliberate misuse of this technology, especially for warfare.

Unfortunately, biological warfare has a long and ugly history. When conquest or national survival is at stake, nations often resort to the most destructive weapons at their disposal, including biological ones.

One of the earliest recorded uses of biological warfare was in 600 B.C., when Solon from Athens contaminated the water supply of the city of Kirrah with the poisonous hellebore plant. During the fourteenth century the Tartars catapulted the bodies of dead plague victims over the walls of the Crimean town of Kaffa in order to ignite an epidemic. And in the eighteenth century both British soldiers and U.S. government agents traded smallpox-infected blankets to Native Americans, which accelerated their extermination.

During World War I, 100,000 tons of poison gases (chlorine, phosgene, and mustard gas) were used to kill 100,000 soldiers and incapacitate 1.3 million more.

During World War II, the Nazis gassed millions of Jews, Russians, Gypsies, and other "undesirables," and Japanese conducted hideous germ warfare experiments on prisoners of war (even Britain and the United States had plans, never put into effect, to use anthrax as a weapon, either stored in 200-kilogram bombs or employed as a poison to infect enemy livestock).

In March 1995, a fanatical Buddhist cult in Japan unleashed the nerve gas sarin in the Tokyo subway system, killing twelve and injuring 5,500. The only thing that prevented tens of thousands from dying was the fact that the mixture was impure. (Like other nerve agents, sarin, developed by the Germans in the 1930s, blocks the chemical acetylcholinesterase, which is necessary for the transmission of nerve impulses.) There is evidence that this same cult actively tried to obtain samples of the Ebola virus as well.

But perhaps the greatest fear is that an accidental release of an incurable virus from one of the biological warfare centers (such as Fort Detrick outside Washington, D.C.) may threaten the very existence of the human race. A mutated, airborne Ebola or HIV virus could infect most of the planet within a matter of weeks or months.

Some scientists' greatest fears were voiced by Karl Johnson of the CDC when he said, "I worry about all this research on virulence. It's only a matter of months—years, at most—before people nail down the genes for virulence and airborne transmission in influenza, Ebola, Lassa, you name it. And then any crackpot with a few thousand dollars' worth of equipment and a college biology education under his belt could manufacture bugs that would make Ebola look like a walk around the park."

Such a doomsday scenario cannot be ignored. D. A. Henderson, who helped to lead the campaign against smallpox, has observed: "Where would we be today if HIV were to become an airborne pathogen? And what is there to say that a comparable infection might not do so in the future?"

A warring nation could also use biotechnology to create a disease to destroy an enemy's crops, thereby unleashing famine. "It can easily be done. This is not science fiction," says A. N. Mukhopadyay, dean of agriculture for the G. S. Pant University in India.

Barbara Rosenberg of the Federation of American Scientists, comments: "None of the equipment is so high-tech that it could not be homemade by any nation intent on developing BW [biological warfare] capacity. No nation is immune to the dangers."

At the annual meeting of the American Society of Tropical Medicine and Hygiene in Honolulu in 1989, scientists staged an extraordinary but purely hypothetical war games exercise involving germ warfare. In that exercise, a civil war and mass chaos erupts in central Africa. An airborne Ebola virus suddenly emerges out of the squalor of a refugee camp. Within days, it begins to spread outside the camp, eventually reaching the airports and spreading to Europe and America. Within ten days, it reaches Washington, D.C., New York, Honolulu, Geneva, Frankfurt, Manila, Bangkok. Within a month, a global pandemic is unleashed, triggering worldwide panic.

Recalling that chilling exercise, Karl Johnson said, "You may say 'ridiculous,' but I don't think we can disregard that possibility. It was, and still is, a potential," he says.

Perhaps one of the most frightening forms of germ warfare to contemplate is what are called "ethnic weapons"—i.e., genetically altered germs which attack specific ethnic groups or races. Ethnic weapons were first proposed publicly in 1970 in *Military Review* magazine, which noted that certain Asian people cannot digest milk. The article used this example to demonstrate that certain races are vulnerable to certain chemicals.

Recently declassified documents reveal that back in 1951 the U.S. Navy conducted top secret tests to determine how vulnerable it was to an enemy attack which selectively affected primarily African-American defense workers by using *Coccidioides immitis*, which causes San Joaquin Valley fever, which kills ten times more African-Americans than Caucasians.

Charles Piller, author of *Gene Wars*, notes that San Joaquin Valley fever, a systemic fungal disease, was developed by the United States as a potential biological weapon back in the 1940s. Military planners once considered mutating the organism so that it would attack one specific ethnic group.

## Legislating the Genome

Francis Collins says, "I am not such a Pollyanna as to imagine that information this powerful cannot in some future instance be used in the wrong way. . . . If you believe that one of the strongest mandates of humankind is to pursue ways to alleviate human suffering, you really can't be against this research. But it's knowledge. It's not good or evil. It's just knowledge."

But knowledge is power, and power is inherently a political and social question. To help clarify the essential issues at stake with the genetics of the future, the ELSI has come out with simple guidelines for dealing with some of these thorny ethical issues. What they advocate is this:

> Fairness for all—no genetic discrimination
> The right to privacy—prevent disclosure
> The delivery of health care—services made available to all
> The need for education—raising public consciousness

These guidelines identify some of the essential issues and give cogent responses to them. However, how to implement these guidelines is still in question. Ultimately, many of these ethical questions may be solved by a combination of social pressure, legislation, and treaties between nations.

There is no viable way to completely stop the progress of science—but we must find a way to carefully control the excesses of technology. Certain aspects of genetics research may need to be banned entirely. But the best overall policy is to air the risks and potentials of genetics research in public, and democratically pass laws which will shape the direction of the technology toward alleviating sickness and pain.

Caplan thinks that some of the simpler questions can be resolved by peer pressure. For example, today many women in their late thirties or older voluntarily ask that an amniocentesis be performed, which can determine if the fetus is suffering from Down's syndrome. In the future, as testing for more genetic diseases is perfected, women may voluntarily agree to be tested during their pregnancy.

Other issues, however, will require outright legislation. Already, for example, several bills in Congress are being considered to ban insurance companies from discriminating on the basis of one's genetic makeup.

Similarly, society may have to pass legislation to decide which germ-line therapies will be banned. For example, is being short a disease? Many scientists working on gene therapy are horrified that the fruits of their work may be used for purely cosmetic reasons. They argue persuasively that germ-line therapy for cosmetic reasons should be banned, but germ-line therapy might be allowed for certain classes of debilitating genetic diseases.

Ultimately, the question of germ warfare will have to be decided by treaty. The Biological Weapons Convention of 1972, signed by the United States, the former Soviet Union, and scores of other countries, was a milestone in trying to ban or restrict germ warfare. Unfortunately, it was signed before the coming of recombinant DNA technology; hence there are many potential flaws. First, it banned the use of biological weapons for "hostile purposes or armed conflict." However, in the age of recombinant DNA, there is precious little difference between the offensive and defensive use of deadly germs. Second, it outlawed the "development" of germ weapons, but allowed for "research" on them. Unfortunately, this means that it is legal to do "research" on large quantities of deadly germs with the intent to use them in a future war. In the biotech business, there is no great distinction between researching a biological weapon and developing it.

Given that there is no easy dividing line between defensive and offensive uses of germ warfare, ultimately the entire field of biological weapons might have to be banned. In 1995, in a report by the Office of Technology Assessment, seventeen countries were said to be working on biological weapons.

Ultimately there must be tight international restrictions on this kind of weapons technology, including on-site inspections, dismantling known biological weapons facilities, monitoring the flow of certain chemicals and life forms, etc. It will not be easy, but such guarantees are necessary to prevent dangerous life forms from emerging from renegade laboratories.

Banning these weapons of war may be generally accepted once nations realize that biological weapons are unstable, unpredictable, and unreliable in actual warfare.

Ultimately, society must make democratic decisions on whether or not to restrict certain kinds of technology. Unlike nuclear technology, the debate about the risks and benefits of biotechnology is in its early stages, giving society time in which to decide which forms of the technology should be allowed to flourish and which ones should be restricted. In a democracy, what is decisive is informed debate by an enlightened electorate.

# Part Four

# The Quantum Revolution

# 13

# The Quantum Future

"Anyone who is not shocked by the quantum theory does not understand it."

—NIELS BOHR

"In the future, there are no roads!"

—DOC BROWN, *Back to the Future*

IN THE FINAL SCENE of the movie *Back to the Future*, the crackle of electrical energy and thundering bolts of lightning announce the dramatic arrival of a sleek "hover car" floating in from the future. As the hover car lands gently on Michael J. Fox's lawn, the manic scientist Doc Brown dashes out, frantically looking for some fuel for his empty gas tank. He races to the nearest garbage can and lifts the cover, revealing smelly banana peels, broken eggshells, and other waste.

Ah, yes, banana peels. He throws the banana peels into his gas tank, and is ready for takeoff.

Banana peels?

Then the camera zooms in on his gas tank, which reads: "Mr. Fusion." The fusion chamber in the hover car quickly ionizes the garbage, converting it into raw power. Energized by a healthy dose of garbage, the hover car rises into the air and then rockets off into the sky. As Doc Brown enters the time warp and sails back to the future, he loudly proclaims, "In the future, there are no roads!"

This scene so captured the imagination of millions of Americans that even President Ronald Reagan mentioned it in a State of the Union address.

*Back to the Future* was, of course, just a Hollywood creation, but cars that float in the air using powerful magnets or fusion machines that can be miniaturized to energize our machines may not be. Ultimately, it is quantum physics that will provide the answers to these questions.

It was quantum physics that first ignited the DNA and biomolecular revolutions in the 1950s. With the introduction of PET, MRI, and CAT scans, quantum physics helped to change the way medical research was conducted in the 1990s. It has also given us the laser and the transistor, which are fundamentally altering the nature of commerce, business, entertainment, and science. The smooth transition to the world of 2020, where computer power and DNA sequencing double roughly every two years, is made possible by the technology of the quantum theory. But quantum physics has not stood still in the last four decades. There has been major progress in several areas which will determine the course of the next century.

## Nanotechnology: Molecular Machines

In *Back to the Future*, Doc Brown's hover car was energized by the cosmic power of the sun in a device no bigger than a tea kettle, raising the question: how small can machines be made?

Nanotechnology is a field that promises perhaps the smallest of all possible machines: molecular machines. Although many major conceptual breakthroughs are needed to bring nanotechnology to the marketplace, it seems perfectly consistent with the laws of physics. Moreover, its promise is so amazing that it cannot be easily dismissed.

Nanotechnology may well open up a new era in our relationship to biology and technology. Because scientists have recently been able to manipulate individual atoms, it is not farfetched to believe that one day scientists will be able to manufacture gears and wheels that are no larger than several atoms in diameter. Given the enormous strides in manipulating single atoms, there is a consensus within the scientific community that atomic-size micromachines may be within reach within the foreseeable future.

What is particularly exciting (and extremely controversial) about nanotechnology is the belief that these machines may be able to scavenge molecules from their environment to *reproduce themselves*, creating an unlimited number of molecular robots that can perform feats of engineering that defy our imagination. About a tenth of a micron in size, these micromachines would be able to manipulate individual atoms, creating an atomic Lego factory. With trillions upon trillions of these molecular robots converging on a site, biological and engineering problems that are currently impossible to deal with may (at least in principle) be solved.

Like viruses or bacteria, these atomic micromachines will have the power to make duplicates of themselves, so they will multiply like living entities and reshape the environment around them.

Some of the ways in which such machines could be used include:

- destroying infectious microbes
- killing tumor cells one by one
- patrolling our bloodstream and removing plaque from our arteries
- cleaning up the environment by devouring hazardous wastes
- eliminating world hunger by growing cheap and plentiful foods
- building other types of machines, from booster rockets to microchips
- repairing damaged cells and reversing the process of aging
- building supercomputers the size of atoms

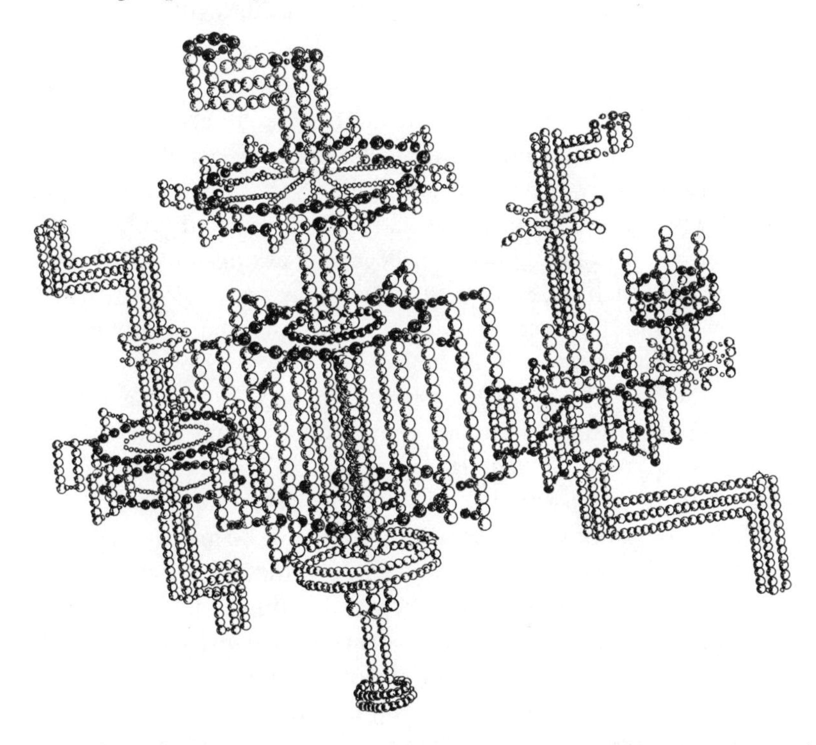

In principle, molecular machines with gears and moving parts can be made by manipulating individual atoms. If, in the future, they can be programmed to reproduce themselves, then they might be able to perform nearly miraculous feats of biological and technical engineering. (Courtesy Robert O'Keefe)

Nanotechnology, its advocates claim, may also give us a form of immortality. They believe in freezing the human body after death, but then using molecular robots to reverse the inevitable cell damage that takes place when ice crystals break the cell wall. (In fact, many of its proponents have already signed up to have their bodies frozen after death.)

Because they are self-replicating, the cost of these molecular machines would be almost nothing. One can literally dream up thousands of mind-boggling applications for such armies of molecular robots that can reproduce and manipulate the atoms around them.

The siren call of nanotechnology has echoed throughout Silicon Valley, which is used to thinking decades ahead in microelectronics. Nor have the military applications of such a powerful technology escaped the attention of the Pentagon. Admiral David E. Jeremiah, the former Vice-Chairman of the Joint Chiefs of Staff, has said, "Military applications of molecular manufacturing have even greater potential than nuclear weapons to radically change the balance of power." An enemy force could be devoured in a few hours by near-invisible hordes of trillions of such self-replicating robots.

But for all the potential of nanotechnology, scientists' enthusiasm must be tempered with the practical constraints of engineering and physics. The question remains whether nanotechnology is possible.

Nanotechnology was first proposed by Nobel Laureate and gadfly Richard Feynman in a seminal but witty article entitled "There's Plenty of Room at the Bottom." Feynman began by asking himself: how small can you make a machine, consistent with the laws of physics? Much to his surprise, he found that there is nothing in the laws of quantum mechanics that forbid machines the size of molecules.

"The principles of physics, as far as I can see," Feynman wrote, "do not speak against the possibility of maneuvering things atom by atom. It is not an attempt to violate any laws; it is something, in principle, that can be done; but in practice, it has not been done because we are too big. . . . The problems of chemistry and biology can be greatly helped if our ability to see what we are doing, and to do things on an atomic level, is ultimately developed—a development which I think cannot be avoided." He challenged the physics community by offering a prize of $1,000 to the first person who could demonstrate nanotechnology.

Some critics of nanotechnology point out that the claims of those who support it are as breathtaking as their results are meager. This criticism is true enough. A few molecular machines actually occur in nature, but not a single molecular machine has been made in the lab, and none is expected for at least a decade. And self-replicating molecular robots are yet another leap of faith far beyond that.

One acid critic of this technology is David E. Jones, a chemist and columnist with the prestigious *Nature* magazine, who asks: How do these machines know where every atom is located? How do you program such machines to perform their miraculous feats? How do they navigate? Where does their power come from?

Orchestrating the efforts of several quadrillion micromachines with perfect precision on the atomic scale may prove to be impossible. "Until these questions are properly formulated and answered, nanotechnology will remain just another exhibit in the freak show that is the boundless-optimism school of technical forecasting," Jones scathingly concludes.

Another critic is Philip W. Barth, an engineer with Hewlett-Packard, who says, "There's a plausible argument for everything, but there are detailed answers to nothing." Barth even posted a message on the Web claiming that nanotechnology was becoming a pseudoscientific political/social sect like any other religious cult.

Indeed, nanotechnology has become a trendy subject in science fiction novels. Istvan Csicsery-Ronay, Jr., an editor of *Science Fiction Studies*, says, "It seems like nanotechnology has become the magic potion, the magic dust that allows anything to happen with a pseudoscientific explanation."

## From Now to 2020: MEMS

Although no one expects a molecular, self-replicating robot to be built in the near future, there has been slow but genuine, steady progress made in a variety of areas that continue to pique the curiosity of both skeptics and advocates alike. Rather than debate the merits of nanotechnology, physicists and chemists are actually building prototypes of these devices in their laboratories.

From now to 2020, the first generation of micromachines, called microelectromechanical systems (MEMS), are expected to find wide commercial application. MEMS are miniature sensors and motors about the size of a dust particle. Although they are still a far cry from true molecular machines, prototypes of MEMS are already entering the marketplace, creating a $2.2 billion industry which is expected to exceed $15 billion early in the next century.

What is driving the MEMS market are the same etching techniques that were first pioneered by the microchip industry. Instead of etching millions of transistors, scientists are now etching tiny sensors and motors onto silicon wafers. In addition, tiny X-ray beams are being used to etch polymers, which can be electroplated to create metallic molds.

Already, a motion detector no larger than a whisker of hair is being used in air bags. A hair-thin piece of silicon which can detect sudden decelerations of a car is now replacing earlier, clumsier air-bag motion detectors. Denso Corporation, a Toyota affiliate, has made one of the world's smallest motors, a micro-engine .7 millimeter in size, which can propel a micro-car at two inches per second.

MEMS could revolutionize a number of fields. For example, in medicine, MEMS might be able to:

- dramatically reduce the current cost of $20,000 laboratory spectrometers down to $10
- create a complete laboratory-on-a-chip, capable of full medical diagnostics and chemical analysis
- create micro-devices which can thread blood vessels, driven by a micro-car.

MEMS may also become a standard part of industry with:

- steel and other building materials containing millions of cheap pressure-sensitive MEMS embedded within them to detect stress, which could save the lives of thousands in case of an earthquake
- airplane wings with MEMS placed on the surface to reduce drag and increase efficiency

The military is also interested. They are studying MEMS called "surveillance dust," which can be sprayed over a battlefield. This dust, which will hover in the air for hours, can be equipped with infrared detectors, radios, even poison-gas detectors, and can be used to locate enemy positions.

One of the people backing MEMS technology is Intel co-founder Gordon Moore, who coined Moore's law. He says, "It took a long time for the transistor to have an impact. MEMS is a really intriguing technology, and I believe it will have a significant impact in the next century."

## 2020 to 2050: Building Molecular Machines

Beyond 2020, MEMS may be replaced by true molecular machines. Already, tantalizing results have been achieved in the laboratory involving the manipulation of structures no bigger than a few atoms across.

In 1989, two scientists at IBM were able to drag thirty-five individual xenon atoms along a nickel surface using a scanning tunneling microscope until they spelled "IBM" in letters consisting of single atoms. The scanning tunneling microscope involves dragging a tiny needle across the surface of an object. By measuring the slight electrical disturbances created by the atoms as the needle moves, scientists are able to detect the presence of individual atoms. (This is somewhat similar to the way blind people read Braille by running their fingers across bumps on a sheet of paper.) Electrons "tunnel" across the electrical barrier between the needle and the atoms, thereby creating a current which measures the positions of the atoms. But the scanning tunneling microscope can also be used to physically move atoms around one by one.

In 1996, another breakthrough was made when scientists at the IBM Research Laboratory in Zurich showed that they could use the needle of a scanning tunneling microscope to move individual molecules to create stable hexagonal rings of molecules at room temperature. (Previous work had manipulated molecules at very low temperatures.)

In late 1996, the physicists at the same lab topped their previous feat and made a workable abacus out of single atoms. Although thoroughly impractical, it points to the rapid progress being made in the field. The "beads" of the abacus are actually made of buckyballs (i.e., sixty carbon atoms arranged like the segments on the outside of a soccer ball, named after Buckminster Fuller). Each buckyball is then placed in grooves that are only one copper atom wide. The needle of the scanning tunneling microscope is then used to slide each buckyball down each groove, mimicking a real abacus. (One physicist compared this feat to using the Eiffel Tower to slide the beads of an abacus.)

One criticism of nanotechnology is that there is no equivalent of a wire by which to conduct electricity and connect the various parts of the molecular machine. This problem, however, is being solved by the introduction of an extraordinary state of matter called the "carbon nanotube." Recently, there has been a flurry of interest among chemists in these hollow cylinders made of carbon molecules which have 100 times the strength of steel yet only 1/60 of its weight. Nanotubes are so thin it would take 50,000 of them side by side to cover the width of a human hair.

These fantastic fibers are created in the same way one makes buckyballs, by heating ordinary carbon into a vapor and allowing it to condense in a vacuum or inert gas. The carbon atoms will rearrange themselves in buckyballs and nanotubes without any prompting. Ordinary buckyballs condense into soccer-like pentagons and hexagons. However, nanotubes condense into a series of hexagons in the shape of a cylinder.

These nanotubes are so tough that they are being called the "ultimate fiber." They are also conductors of electricity, which may give them enormous applications in computers. Although only grams of nanotubes have been manufactured at a time, chemists envision a time when tons of them are commercially available. Uses for these fantastic fibers abound. It may be possible, for example, to make a molecular transistor out of nanotubes. By inserting an ordinary sixty-atom buckyball inside a nanotube (which just fits inside one type of nanotube), the buckyball can travel down the nanotube, thereby changing its electrical properties, in effect acting as a molecular switch for the transmission of electricity. Even novel electrical devices with no counterpart in ordinary electronics can be constructed. For example, it may be possible to insert a metal atom inside a buckyball, which in turn is then inserted into a nanotube, creating the equivalent of an atom-wide metal wire, with novel electrical properties.

Other applications include manufacturing molecular wires to connect other molecules, or building a molecular "sensor" that can "feel" the molecular texture of a surface (which could give us silicon wafers with ultra-pure specifications).

Perhaps the most fantastic application which seems to obey the laws of physics is to build a "sky hook," a mythical hook which is anchored in thin air. If a satellite orbits roughly 20,000 miles from the earth, the orbit is geosynchronous—i.e., it is stationary in the sky because the rotation of the earth exactly matches the satellite's orbit around the planet. Since the satellite appears motionless when viewed from the ground, a cable can be dropped from the satellite back down to earth, so the cable would appear to hover in midair; it would rise in the sky until it disappeared past the clouds, seemingly defying the laws of gravity. In principle, one could ascend this sky hook and venture into outer space without any booster rockets, thereby revolutionizing space travel.

Unfortunately, a simple calculation shows that the stresses on such a cable would be so great that it would break, making a sky hook impractical. However, Daniel T. Colbert at Rice University in Houston envisions building tough molecular cables out of nanotubes. He calculates that nanotubes, unlike any other substance on earth, would be able to support their own weight and the stresses of connecting the earth to a geosynchronous satellite.

In the era 2020 to 2050, we will probably see increasingly sophisticated machines made by manipulating individual atoms. The counterparts of gears, cogs, levers, and wires on the molecular level could be built in this time period. These micro devices, however, are still a far cry from the self-replicating micromachines envisioned by the gurus of nanotechnology. If they are possible at all, it will be in the era after 2050.

Perhaps Feynman, always the practical joker, may have the last laugh. Nanotechnology seems to violate no known law of physics, yet nanotechnology has yet to produce one solid piece of science or engineering. The attitude of most physicists is that the proof lies in the actual construction of these fabled self-replicating micromachines. Until then, the jury is out.

Given the rapid progress in manipulating individual atoms, the first generation of simple molecular machines may very well be built within the next decade. But the full promise of nanotechnology (self-replicating molecular machines that can move molecules at will) remains purely speculative at this point.

## Room Temperature Superconductors: The Holy Grail

In *Back to the Future*, Doc Brown's hover car floated on air. In principle, such a car could levitate off the ground by using magnetism, which may become a tremendously powerful force in the twenty-first century. Maglev (magnetic levitation) trains with powerful magnets can be made to hover a few inches off their metal tracks, coasting on a cushion of air without friction, and thus reach enormous velocities without using much power. Friction, which accounts for just about all the waste in mechanical systems, could be reduced to almost zero.

The key to building cheap hover cars or maglev trains rests on another bizarre property of the quantum world: superconductivity. Superconducting supermagnets could generate colossal magnetic fields with little energy and realize some of these outlandish dreams.

One reason large magnets are inefficient is that all electrical appliances give off heat. Even superb conductors of electricity, like silver or copper, have tiny amounts of resistance to electric currents. To an engineer, this is frustrating, because the excess heat is a power drain on the circuit, as well as a cause of costly breakdowns and failures. PCs, for example, because they do not have any moving parts (other than the disk drive), should, theoretically, last forever. In fact, they do eventually break down because of overheating caused by electrical resistance.

Superconductivity may change all this. This curious property was first noticed in 1911, when Heike Kamerlingh Onnes in Leiden discovered that mercury lost all electrical resistance when cooled to about 4.2 degrees above absolute zero. This pivotal discovery earned Heike Onnes the Nobel Prize.

Today, a surprising number of substances have been found to be superconductors; in fact, the number runs into the thousands. The problem is that these superconductors need large quantities of expensive liquid helium to cool them down to near absolute zero. Thus, although we can create huge magnetic fields via superconducting materials by dipping them into liquid helium, the cost is often prohibitive for commercial use.

Not surprisingly, the Holy Grail of superconductivity research is finding a "room temperature superconductor," which requires no cooling whatsoever. If such a superconductor is ever discovered, it will change modern industry in unimaginable ways. Since electricity is the energy source which powers our machines, lights up our cities, toasts our bread, entertains us at night, and performs our computations, every aspect of our life would be intimately affected. Some physicists claim that it would create a "second industrial revolution." Here are but some of the changes it would bring about:

- Supercomputers, as we saw in Chapter 5, face the "point one" barrier as transistors become smaller and smaller, thereby generating tremendous amounts of heat. Supercomputers made of room temperature superconductors would generate less heat, opening up new kinds of computer architectures.
- Much of our electrical power is lost in its transmission across great distances. Room temperature superconductors could slash these electrical losses, saving us billions of dollars.
- Reducing losses in electrical appliances could eliminate the need for new fossil fuel power plants, help solve the energy crisis, and alleviate the greenhouse effect. The electrical power needs of modern society would plunge as the efficiency of our machines skyrocketed.
- Costly breakdowns of our electrical machines, caused by heating due to electrical resistance, would also drop precipitously. All this adds to tremendous efficiencies and savings for the consumer and industry.
- Room temperature superconductors would also make possible cheap but powerful magnetic fields that could result in the development of maglev trains, hover cars, and even new types of MRI machines and particle accelerators.
- Superconductors could make possible extremely sensitive magnetic sensing devices, called SQUIDs, which can measure infinitesimal variations in a magnetic field. They are already being used in hospitals to measure the magnetic field of the body.

For decades, physicists despaired of ever finding this mythical superconductor. The quest for room temperature superconductors has been compared to the futile search by the ancient alchemists for the "philosophers' stone" which would turn lead into gold.

A breakthrough came unexpectedly in 1986 when K. Alexander Müller and J. Georg Bednorz of the IBM Research Laboratory in Zurich announced that they had created a ceramic superconductor at a record temperature of 35 degrees above absolute zero. This announcement, the first major discovery in this field in seventy years, caught the world of physics by surprise and won them the Nobel Prize. No one had suspected that lanthanum barium copper oxide ceramics, which are normally insulators, could become superconductors. Soon, Maw-Kuen Wu and Paul Chu showed that yttrium barium copper oxide (YBCO) could become superconducting at the balmy temperature of 93 degrees above absolute zero (or roughly –180 degrees C).

A stampede soon erupted, as lab after lab around the world rushed to see who could discover a room temperature superconductor. Within weeks, new records began to be set, each time racheting up the temperature at which these copper ceramics became superconducting.

Today, the world's record (which is surely to be broken soon) is held by Andreas Schilling and his colleagues at the Swiss Federal Institute of Technology in Zurich, who showed that mercury barium calcium copper oxide can become superconducting at 134 degrees K (or –139 degrees C). Tantalizing, sporadic results in various labs (none of them reproducible) have been reported of materials that become superconductors at temperatures as high as 250 degrees K (or –23 degrees C). If these reports can be verified, it would put us within striking distance of room temperature (about 300 degrees K).

Although scientists are still considerably short of reaching room temperature, already this new class of superconductors is causing a quiet revolution in science. These copper ceramic substances can be cooled by liquid nitrogen, which costs only 10 cents per quart (making it cheaper than Kool-Aid), while conventional superconductors must be cooled by liquid helium, which costs $4 per quart. The advantage of using liquid nitrogen for these ceramics opens up a host of commercial products.

Nonetheless, the excitement around superconductors in the late 1980s has largely subsided, replaced by the sober realization that true room temperature superconductors may still be years and more likely decades away.

There are several problems to be solved before commercially viable superconductors are available.

The first, of course, is to raise the critical temperature of these superconductors to room temperature. This would eliminate expensive refrigeration, including the use of liquid nitrogen.

Second, because these materials are ceramics rather than metals, they are difficult to extrude into wires. Ceramics also tend to be brittle, while metals are strong, ductile, and can be bent and twisted into any number of shapes. These copper oxides have the fragility of chalk and consist of tiny, irregular grains, which impedes the flow of electricity.

Third, these superconductors tend to lose superconductivity when exposed to large magnetic fields, because vortices of magnetism tend to develop within them, which also impedes the flow of electricity.

But from a more fundamental point of view, the most frustrating aspect of these ceramic superconductors is that scientists don't know precisely how they work. These high temperature superconductors test the limits of what we know about the quantum theory of ceramics.

The explanation for the first generation of superconducting materials was laid out by John Bardeen, Leon Cooper, and J. Robert Schrieffer, work for which they won the Nobel Prize. (This was, in fact, the second time Bardeen won the Nobel Prize; he also won one for developing the transistor.) The BCS theory, as it is called, is based on the fact that electrons moving in a lattice can create small disturbances and ripples. A second electron, caught in this ripple, can form an electron pair with the first, called a Cooper pair, which overcomes the natural repulsion of the electrons. According to the quantum theory, these Cooper pairs can then move without resistance through the lattice.

However, the second generation of superconductors—the ceramics—apparently do not follow such a simple pattern. In fact, no one has so far successfully explained why these ceramics become superconductors. Nobel Laureate Philip Anderson of Princeton says that previous theories are "a catalog of failure and it is time we opened our minds to new ways of thinking."

In principle, the quantum mechanics of individual atoms is well understood from the Schrödinger wave equation. The properties of simple lattices can also be derived from this equation. But for more complicated materials like ceramic copper oxides, the Schrödinger equation becomes so complicated that even supercomputers cannot solve it.

One key to explaining these new ceramic superconductors is that their atomic structure is based on regular lattices. Imagine stacking planes on top of each other, with electrons moving freely within each plane. Even slight changes in the lattice structure will destroy the superconducting properties of the ceramic. Unfortunately, no one knows why this lattice structure makes these copper ceramics superconductors.

In spite of these difficulties, high temperature superconductors are slowly entering the marketplace in modest but important ways.

## Maglev Trains

It is impossible to tell when room temperature superconductors may be discovered. There is no Moore's law for superconductors which makes reasonable predictions possible. They may be discovered tomorrow, or not at all. However, given the remarkably steady advances made in the last decade, one can speculate that they could be discovered as early as the next fifteen to twenty years.

But several countries, even without room temperature superconductors, within the next ten years will be building maglev trains to connect their main cities. These conventional maglev trains, too, ride on a cushion of air, except that their magnetic coils consume large amounts of power.

Since traffic congestion in the air and on the roads has impeded the expansion of commerce, a maglev train that can travel 450 to 500 kilometers per hour is an ideal way to connect cities about 600 kilometers apart.

Although the technology of maglev trains was first developed thirty years ago in the United States by physicists at the Brookhaven National Laboratory in New York, progress in their development rapidly shifted to Japan and Europe. The Japanese have already developed a maglev train called the ML-500R, which in 1979 set a speed record for a maglev train of 517 kilometers per hour (on a 4.3-mile test track). They hope to have an operational maglev train operating between Tokyo and Osaka by 2005.

The Germans have also built a working prototype maglev train, the TR-07, which regularly attains speeds of 400 to 450 kilometers per hour. The TR-07 consists of two cars that run on a 20-mile test track. The German government plans to have a maglev train linking Berlin and Hamburg in operation by 2005, which it hopes will be the heart of a transportation initiative to link eastern and western Germany.

In the United States, there was brief interest by the government in a maglev train during the 1980s, but the proposal, called the National Maglev Initiative, died a quiet death in 1994 due to lack of corresponding interest by the private sector.

But if a room temperature superconductor could be found, then the economics of the maglev train could be turned upside down. The cost of creating magnets for maglev trains would plummet. Until then, one expects steady but not spectacular progress in the development of maglev trains.

## Fusion: Harnessing a Piece of the Sun

Cheap, inexhaustible, and limitless—these are the criteria for any new energy source for the twenty-first century. Within the next thirty years, fossil fuels will become increasingly scarce and prohibitively expensive. We may also see the unmistakable effects of global warming.

What are the alternative sources of inexhaustible power in the future? Physicists see three possibilities:

- fusion power
- breeder reactors
- solar energy

All three are intimately connected to the laws of quantum physics.

The promise of fusion power is that it may one day light up our cities with the same cosmic force that energizes our sun and the stars in our galaxy. The basic fuel for fusion machines is ordinary sea water (rather than banana peels) of which there is an ample supply.

But when might we see fusion plants rising on the horizon to light up our cities? Harold P. Furth, director of the famed Princeton Plasma Physics Laboratory from 1980 to 1990, predicts, "By the middle of the next century, our grandchildren may be enjoying the fruits of that vision." He estimates that we will have the first industrial plants for fusion energy within fifty years.

Although the prospects for fusion power have been overhyped for years, what is working in Furth's favor is that oil reserves (which provide about 40 percent of the world's energy) are likely to begin to run out early in the next century. There are about a trillion barrels of proven and probable reserves, of which 77 percent is located in OPEC countries, 65 percent in the Persian Gulf. But early in the next century, the cost of oil will rise as it becomes more difficult to reclaim and refine this oil. Barring new significant discoveries, the price of oil will likely begin to skyrocket around 2020. Around 2040, the price of oil will become prohibitive, with the potential to perhaps precipitate a worldwide economic crisis unless measures are taken soon to prevent it. Although the precise year when this will begin to take place is impossible to predict, it is inevitable that fossil fuels in the not too distant future will become too expensive to fuel industry.

At the same time that the cost of oil is rising, energy demand will continue to grow, driven in large part by the industrialization of large portions of the Third World. World energy consumption is expected to *triple* by 2040, from 10 to 30 trillion watts (even assuming large gains in energy efficiency).

What will replace these fossil fuel sources? The United States has enough coal reserves to last perhaps five hundred years, yet the environmental damage caused by acid rain, the greenhouse effect, and air pollution seem to rule out any significant role for coal in the future.

This is where fusion would come in. Since there is virtually unlimited deuterium (a form of hydrogen) in ordinary sea water, the physicists at Princeton estimate that we have about a million- to 10-million-year supply of fusion power!

Unfortunately, for all its promise, fusion power is an elusive goal—for a simple reason. Hydrogen nuclei repel each other because they all have positive charges. To overcome this electrostatic repulsion, scientists have to heat them up and slam them together until they get close enough to stick. In the 1930s, physicists realized that quantum forces would then take over and fuse hydrogen into helium, releasing fabulous amounts of energy in the process. The problem is that the temperature required to bring these hydrogen nuclei together is on the order of 10 to 100 million degrees, far higher than any temperature ever found naturally on earth. In a hydrogen bomb, this problem is solved by detonating an atomic bomb, which serves as the trigger which heats up the hydrogen nuclei to fantastic temperatures. This leads us to the question: how do you safely contain a piece of the sun on the earth? A temperature of 10 to 100 million degrees is enough to vaporize any known container.

Fortunately, two basic fusion designs have been found which overcome this problem: one based on the principles governing the sun and the other based on the hydrogen bomb.

When a star is born, the intense gravitational field compresses the hydrogen gas of the star until temperatures reach 10 to 100 million degrees at its core, which is hot enough to fuse the hydrogen atoms into helium. The fusion process causes the hydrogen to ignite and form a star. This is called gravitational confinement.

On the earth, we cannot use gravitational fields to confine plasmas, but we can use a "magnetic bottle." A variation of this design, called magnetic confinement, is used in the Tokamak fusion reactor, based on a Russian design created by Andrei Sakharov and Igor Tamm in the 1950s, which is now widely copied around the world. In the Princeton Tokamak, for example, a magnetic field confines hydrogen gas contained within a doughnut-shaped tube. The magnetic fields prevent the plasma from hitting the walls, which would vaporize upon contact with the hot gas.

Huge coils of wire are wound around the Tokamak tube, which creates a magnetic field that contains the hydrogen gas. An electrical current is then fed into the tube, which heats up the gas. In 1994, the Princeton Tokamak set a world record, generating 9 million watts of power in a brief burst of energy lasting a fraction of a second. (By contrast, a typical nuclear fission plant generates 1,000 megawatts of power almost continuously.)

The second fusion design is called inertial confinement, which is used both for the hydrogen bomb and for the laser fusion machines at the Livermore National Laboratory in California. In a hydrogen bomb, a small atomic bomb is used to generate a burst of high-intensity X-rays, which is then used to compress a mass of lithium deuteride. As the lithium deuteride heats up, the hydrogen atoms are fused into helium, releasing the heat of thermonuclear fusion.

The fusion reactor at Livermore uses a series of laser beams focused on a small pellet of lithium deuteride to create fusion. The laser beams vaporize the surface of the pellet, causing a shock wave which implodes the pellet, generating enormous pressures and temperatures sufficient to unleash fusion.

Both designs extract power from the release of energy by fusion by boiling water. The fusion of hydrogen to helium produces a large flux of neutrons which penetrates a "blanket" surrounding the reactor which contains tubes carrying water. As the neutrons heat up the blanket, the water within these tubes starts to boil. The water turns to steam, which is then shot past the blades of a turbine, making it spin, as in a hydroelectric or coal plant. (The spinning magnet pushes electrons in a nearby wire, thereby creating an AC current in that wire which eventually reaches the wall sockets in our homes.)

Currently, the United States spends about $370 million per year on fusion research. Japan and Germany each spend 40 percent more than that.

The goal of both the Tokamak and laser fusion is to reach "break-even"—i.e., the point at which the amount of energy released equals the amount of energy consumed.

At present, neither design has reached break-even. The problem with the Tokamak design is that the hot plasma gas circulating in the doughnut has its own magnetic field, which interacts with the main magnetic field of the Tokamak. Together, these magnetic fields magnify small eddy currents or imperfections in the field, which in turn cause leakage of the plasma. So far, none of the various designs has been able to keep the plasma stable enough for long enough time.

The precise requirement for fusion is called Lawson's criterion, which has not yet been satisfied. (Lawson's criterion says that a self-sustaining fusion reaction will occur when the product of particle density multiplied by the time of confinement exceeds $10^{14}$ seconds per cubic centimeter.) Every time we see the sun and the stars or view pictures of thermonuclear explosions, we see a version of Lawson's criterion being satisfied on a cosmic scale.

In 1997, the fusion program was dealt a setback by the closing of the Princeton Tokamak laboratory due to budget cuts.

The next big step in fusion will therefore be the International Thermonuclear Experimental Reactor (ITER), a joint effort of Russia, Japan, the European Union, and the United States. Instead of operating for brief fractions of a second, the ITER will sustain fusion for thousands of seconds.

Because we are making steady progress toward reaching Lawson's criterion and break-even, the physicists at the Princeton Plasma Lab have made rough estimates of when we might expect fusion power to energize our cities:

- by 2010: the creation of a 1,000-megawatt fusion ITER plant
- by 2025: demonstration of a fusion power plant
- by 2035: the first commercial fusion power plant
- by 2050: widespread use of commercial fusion plants

## Fusion Versus Fission

Public opinion polls have shown that the public has grave reservations about fission power, which derives its energy from the splitting of the uranium or plutonium atom. They have read about the disasters at Three Mile Island and Chernobyl, the unsolved problems with nuclear waste, the radiation experiments done on unsuspecting citizens in the early days of nuclear research, and the seventeen disintegrating nuclear weapons sites which will require up to $500 billion to clean up.

A fusion reactor would have several advantages over a fission reactor. First, a fusion reactor will not create tons of high-level nuclear waste per year, as uranium reactors based on fission do. The only waste from a fusion reactor might be the steel hull, which gradually becomes radioactive, and any radioactive hydrogen that might escape. This is an enormous advantage, since the nuclear waste from fission plants is mind-boggling. Each year, for example, a large 1,000-megawatt nuclear fission plant produces thirty tons of high-level waste, and there are over 100 of these commercial plants in the United States. Such nuclear waste will be dangerous to life for thousands to millions of years.

Second, fusion plants are not subject to meltdowns. In fission reactors, the nuclear core remains hot even months after the fission process has stopped. This can cause the core to melt down (as it almost did in Three Mile Island). Such an accident can cause a catastrophic release of radiation if the core continues to melt through its vessel. By contrast, if someone accidentally cuts off the magnetic field of a fusion plant and the superhot gas leaks out, the gas may partially melt the containment structure, but then the fusion process stops, as Lawson's criterion is no longer satisfied, and the accident is over.

Third, a fusion reactor cannot go supercritical, like an atomic bomb. In a fission reactor, small bursts of power (called transients) can spiral out of control if the neutron population grows too rapidly (as it did in Chernobyl).

Fusion reactors, however, do share one disadvantage with fission reactors. The neutron radiation that both reactors create is strong enough to weaken the metal containment, causing "brittle fracture." Tiny microfractures caused by neutrons bouncing atoms out of the metal lattice can build up, creating a potential catastrophic collapse of the metal containment. It's still premature to state with any finality the pros and cons of fusion reactors, since none exist at the moment. Until a workable demonstration fusion reactor is built, possibly around 2025, any decisive analysis of its potential problems are premature.

### Breeders and Terrorism

The two other inexhaustible potential sources of energy are breeder reactors and solar energy.

Of all the sources of power, breeder reactors are perhaps the most politically sensitive. In the 1950s, analysts at the Atomic Energy Commission originally projected that the country would need 1,000 conventional reactors and 1,000 breeder reactors by the year 2000. Today, the United States has only a little over 100 reactors. And it has no breeder reactors.

The great advantage of breeder reactors is that they produce or "breed" fissionable plutonium from waste uranium. But this is also the difficulty with breeder reactors: they operate on highly enriched fuels, which pose the problem of potential criticality accidents, as well as the risk of sabotage or theft.

Early in the nuclear age, breeder reactors seemed attractive because of their "Midas touch." They are able to transmute waste uranium 238 into fissionable plutonium 239. In a breeder, waste uranium 238 gradually absorbs a series of neutrons. After "cooking" in a reactor for several months, much of the waste uranium turns into neptunium and eventually plutonium 239.

But breeder reactors have been a great disappointment. The very first breeder in the United States, the Experimental Breeder Reactor EBR-1 (and the first to generate electrical power), melted down in 1955 in one of the first nuclear reactor accidents. About 40 to 50 percent of its uranium core was destroyed in the meltdown. The second major breeder was the commercial breeder reactor called Fermi-1, located just outside Detroit. It created a major emergency when it partly melted down in 1966, with 2 percent of its core melting when its sodium coolant was jammed by a piece of zirconium the size of a beer can. Plagued with problems, such as sodium explosions, the plant was eventually closed down. The successor to Fermi-1 was the Clinch River breeder reactor, which was eventually closed down by Congress in 1983.

The Japanese and the French, which are vigorously pursuing large-scale breeder reactors, have not been particularly successful either. Because the fuel has to be chemically reprocessed with volatile solvents to extract the plutonium, there is the danger of fires, several of which occurred at The Hague processing plant in France, which services the Superphoenix reactor. The Japanese, too, experienced a major sodium leak in 1996 at their Monju breeder reactor; the subsequent cover-up caused considerable embarrassment for the government. (Ironically, Monju means "wisdom" in Japanese.) In 1997, they had a major explosion at the Tokaimura plant.

One concern with the breeder is that it is a primary target for terrorists because of the quantities of plutonium it produces. Not only is plutonium one of the most toxic substances known to science; a mere 5 to 10 pounds of plutonium is needed to manufacture a small atomic bomb. As a result, there is considerable apprehension about the breeder reactor program.

Furthermore, the shipment of plutonium is enormously hazardous. In 1995, the shipment of 200 tons of plutonium from the reprocessing plants of France to Japan ignited international protest as it sailed in international waters, raising the possibility of accidents and terrorism.

Because nuclear reactors cannot be used in cars and trucks, nuclear power can make only a small dent in the overall energy needs. (Nuclear power, because it generates only electricity, mainly displaces coal-fired plants and has little effect on oil, which is used mainly for transportation and heating.) Given the current uranium glut in the world market, the safety concerns of fission reactors after Three Mile Island and Chernobyl, the skyrocketing cost of nuclear power plants, and rising consumer resistance to nuclear power, it may be wise to let the uranium stay in the ground.

## Here Comes the Sun

Perhaps the most promising solution to the energy problem of the future is solar energy. Solar cells (commonly called photovoltaic cells) use yet another principle from quantum physics. At the turn of the century, it was noticed that light falling on certain metals generates small electrical currents. It was Einstein who explained this process, reasoning that a particle of light, called the "photon," falling on the metal bounced an electron out of the atom. Einstein's equations explaining the photoelectric effect eventually won him the Nobel Prize in physics. (Einstein's relativity theory, apparently, was too outlandish to be considered by the conservative Swedish Nobel Committee.)

This photoelectric effect is the same process which makes television cameras work. Light falling on the "retina" of the video camera ejects electrons, which generate a current of electricity. Solar cells convert light from the sun into electricity in the same way.

In principle, solar energy is unlimited. Each year, the energy the earth receives is 15,000 times the amount of energy consumed by the entire population of this planet. And it's all free. William Hoagland of the National Renewable Energy Laboratory in Golden, Colorado, who estimates that by 2025 worldwide demand for electricity will rise by 265 percent, optimistically predicts that 60 percent of the world's electricity by that time will come from the sun. The cost of marketing solar cells, however, has always been prohibitive. On one hand, the efficiency of solar cells has only been on the order of 15 percent. (The current record for solar cell efficiency is 30 percent.) At a cost of 10 cents per kilowatt-hour, solar energy is only beginning to become competitive with coal and oil.

However, all this will change in the twenty-first century. The price of solar electricity is plummeting with new advances in technology and with the mass production of solar cells. In the 1980s, for example, the price for photovoltaic cells dropped by a factor of forty. Proponents of solar technology point out that handheld calculators cost $300 when they first came out. Today, with the mass production of chips, the price has fallen by a factor of almost one hundred.

Advocates of solar energy have always claimed that what is needed is a jump start to get solar power off the ground. They point out that nuclear energy in the United States was jump-started by the federal government, with up to $100 billion given to basic research and subsidies to businesses involved in the uranium fuel cycle.

It is unlikely that such a jump start for solar energy will come from the federal government in today's political climate. However, in 1995, two commercial companies, the Enron Company, the largest supplier of natural gas, and Amoco Corporation, which owns Solarex and produces solar cells, began a joint venture to supply an entire city of 100,000 with solar electricity. They claimed that the $150 million solar plant they are building will be able to supply electricity at 5 cents per kilowatt-hour, about 3 cents cheaper than fossil fuel plants.

"If they pull this off, it can revolutionize the whole industry. If they fail, it is going to set back the technology ten years," says Robert H. Williams of Princeton University.

Every year, the economics of solar power become more favorable, while the cost of fission power soars and the problems with burning fossil fuel grow. Combined with increased energy efficiency and alternative fuels (such as wind power, geothermal power, co-generation, and so on), solar technology inevitably will continue to grow and prosper in the twenty-first century, in spite of the foot dragging by the politicians.

## The Electric/Hybrid Car

At the beginning of the twentieth century, it was not clear whether the steam-, electric-, or gasoline-driven car would eventually prevail. In 1895 in Michigan, for example, all three types of cars were found in roughly equal numbers. Thomas Edison and Henry Ford had a friendly rivalry over which version of the automobile would win out, the electric- or gasoline-powered car.

However, twenty years later, gasoline clearly had become the winner. The reason why the gasoline engine has continued to outperform present-day attempts at an electric car is simple: ounce for ounce, gasoline packs *100 times* the energy concentration of electric batteries. This is why electric cars have lagged behind gasoline-powered cars. (A common lead acid battery, for example, contains only 25–40 watt-hours per kilogram of battery.)

As a consequence, the range of most electric cars is about 100 miles, or about one-third that of gasoline cars, and it requires three to ten hours to recharge them. Some car batteries, like nickel-metal hydride batteries, can give cars a range of 300 miles on a single charge—even farther than many gasoline-driven cars—but they don't provide rapid acceleration. (The world's record, set in 1996, is 373 miles without recharging.) Conversely, batteries which give good acceleration do not have long ranges.

Nonetheless, the internal combustion engine may be an early casualty in the twenty-first century. The average gasoline-powered car consumes 3,000 gallons of gas in its lifetime, spewing 35 tons of carbon into the air, contributing to pollution, acid rain, and global warming. Cars and other motor vehicles produce over half of the pollution in urban areas, and one-fourth of the greenhouse gases. Cars, in fact, account for fully half the oil consumed by the United States.

Perhaps the most attractive of the alternatives to the internal combustion engine in the early part of the twenty-first century is the electric/hybrid, which will use sophisticated onboard computers to juggle the various components. It is rapidly gaining popularity among car designers. Current designs are already available which can be used economically for urban travel, where most trips are over short distances. Within the next decade, electric/hybrid engines are expected to rival the performance of internal combustion engines and become a permanent part of the urban landscape. It could pave the way for the fully electric car.

The hybrid takes advantage of the many ways in which electric batteries, motors, flywheels, and small gas engines can be configured. In one design, an electric battery is used for stop-and-go urban traffic, where endurance is not important. When the car is idling, the car's energy is shifted to a rapidly spinning flywheel. In this way, even when the car is stopped at a light or in heavy traffic, a large amount of energy is being stored quietly through the action of the spinning wheel. (Rotors made of new composite materials have been able to attain rotation rates of 100,000 revolutions per second, which allow them to store large amounts of kinetic energy.) Then, as you hit the accelerator pedal, the flywheel is engaged and the energy is then transferred immediately to the car wheels, providing instant acceleration, the most difficult thing for batteries to provide. For long-distance travel, you would simply switch on the small internal combustion engine, designed to release significantly less hydrocarbons and carbon monoxide than conventional engines. An alternative design is to have the small internal combustion engine feed energy into an electric battery and an electric motor, which in turn runs the car.

The electric/hybrid could benefit from the use of high-tech composites, the same ones NASA is using to make its next generation of spacecraft. The Swiss Esro car, for example, which is made of these composite fibers, gets an astonishing 120 to 150 miles per gallon of gas. It weighs only 880 pounds and seats two people, making it ideal for urban use. If the basic design is incorporated into an electric/hybrid, it would further increase its range and efficiency.

By 2010, electric/hybrids could become even more attractive with yet another type of design based on a combination of ultracapacitors and fuel cells, both of which have been used successfully in the U.S. space program. In its simplest form, capacitors are nothing but two parallel plates that store electric charge. But recent advances in this technology have given us ultracapacitors made of carbon and liquid electrolytes which can rival the electrical capacity of standard batteries.

Similarly, the fuel cell is an advanced device which promises even greater efficiencies for the electric car. The fuel cell, unlike lead acid batteries, is completely pollution-free. Electric batteries need to be recharged, which in turn requires an electric power plant which may pollute the environment. Fuel cells, by contrast, generate their energy by combining hydrogen and oxygen, releasing energy and water in the process. Fuel cells are attractive since they have an efficiency rating of 40 percent, twice the rating of the internal combustion engine.

Unfortunately, fuel cells have problems of their own which must be solved before they can be mass-produced for the public. They are still costly, and hydrogen is potentially explosive and must be handled with caution. But with mass production and economies of scale, the economics of a solar/hydrogen/electric combination will become increasingly attractive about 2010. In fact, Daimler-Benz announced in 1996 that it expects to sell the first Mercedes with fuel cells by 2006.

With technical problems of electric cars rapidly disappearing, the main stumbling block to the electric/hybrid is resistance by the oil and automobile industries, which in 1996 waged a fierce campaign that succeeded in reversing the 1990 mandate by the state of California to adopt zero-emission vehicles (ZEVs). The quotas for zero-emission vehicles for 1998 and 2001 were eliminated, leaving only the provision that by 2003, 10 percent of the cars in California will be ZEVs. Nonetheless, if the deadline of 2003 is not further pushed back or modified by pressure from the automobile industry, it would force enormous numbers of electric/hybrids onto the highways.

Daniel Sperling, director of the Institute of Transportation Studies at the University of California at Davis, estimates that by 2010 the number of electric cars could balloon to the millions, especially as foreign competitors begin to market their version of the electric/hybrid.

## The Next Generation of Lasers

One of the great inventions driving the information revolution is the laser, which already has the ability to send billions of telephone calls along beams of light by using fiber optics. In addition to telecommunications, other industries that have been created or revolutionized by the laser include laser printing, laser CDs, laser surgery, and laser cutters for industry.

The next generation of lasers, however, may open up entirely new realms, as they become microscopic, as well as gargantuan. They may eventually light up our living rooms with 3-D television.

Microlasers, as we saw in Chapter 5, are the key link in optical computers, which have several decisive advantages over silicon-based computers. However, optical computers today are still quite large and relatively feeble in power, mainly because of the size of the lasers and S-seeds (optical transistors). This may well change early in the twenty-first century.

The current generation of lasers used in CDs and fiber optic communications is based on the semiconductor diode laser, one of the workhorses of the electronics industry. The problem is that these diode lasers are one hundred times larger than the micron-sized silicon transistors that are routinely made by the computer industry. (A typical diode laser may measure 250 microns across.) Optical transistors are at the level that silicon transistors were back in the 1950s.

But the same photolithography techniques that produced the microprocessors of the 1970s and 1980s are now being used to etch out the microlasers of tomorrow. The process begins by spraying molecules of aluminum arsenide or gallium arsenide onto a silicon wafer. These molecular sprays can create layers which are no more than one atom thick. A typical microlaser may have up to 500 of these individual layers. Then through photolithography, a light beam is used to etch grooves into these layers, creating a series of microlasers shaped much like a Coke can.

In 1989, the first dramatic demonstration of a microlaser was successfully performed by scientists based primarily at Bell Labs; they were able to manufacture *one million lasers* on a tiny chip smaller than a fingernail. About the size of bacteria, these microlasers were lined up, by the millions, in careful arrays resembling neatly arranged barrels. Scientists have been able to make microlasers about one to five microns across. They believe that the ultimate physical limit in size for a microlaser may be about a third of a micron. (This limitation is a result of the fact that the size of the microlaser must be larger than the wavelength of the laser light. If the microlaser is smaller than the wavelength, then the material will not lase properly. The wavelength of laser light used in these experiments was one micron. Because of the refraction of light, the wavelength inside gallium arsenide is effectively smaller than one micron, or about 0.3 micron, which is probably the ultimate limit for this type of microlaser.)

This is still larger than the point one barrier facing the silicon microprocessor, which is 0.1 micron in size. So microlasers do not solve the point one barrier. But because laser beams can effortlessly crisscross each other, the optical computer can be made cubical in shape, vastly increasing power, and because the heat generated by these microlasers is quite small, the heat from optical lasers is minimal, allowing continued miniaturization of desktop computers.

At the other end of the scale are giant laser devices such as the X-ray laser. The advantage of X-ray lasers is that a great deal more information can be packed on them. The higher the frequency, the shorter the digital pulses that can be carried along the laser beam. With such a short wavelength (little more than the size of an atom), these lasers can serve a wide array of industrial uses. X-ray lasers can etch fine lines onto silicon microprocessors, perhaps breaking the point one barrier.

Furthermore, because X-rays have short wavelengths, X-ray lasers may be able to illuminate the microscopic world of bacteria and viruses in greater detail, giving us the atomic structure of these living entities.

As pointed out in earlier chapters, however, there are problems with the X-ray lasers. X-rays are quite powerful and difficult to work with, and since they do not reflect very well, it is difficult to find mirrors that can concentrate their power.

Another problem is that laser light is usually amplified by being reflected back and forth between two parallel mirrors, until it finally leaks out one of the mirrors. Because X-rays are quite penetrating, X-ray lasers do not have this advantage.

One approach to the amplification problem is to devise an energy source so large that only a single pass is required. One solution is the nuclear-driven X-ray laser, which was originally to be the centerpiece of President Ronald Reagan's ill-conceived Star Wars program. The X-ray pulse from a hydrogen bomb is so intense that a single pass through copper rods is sufficient to create an X-ray laser beam. Unfortunately, short of detonating a hydrogen bomb, there are large technical problems with generating an X-ray laser for commercial use.

Another approach is to modify the existing giant lasers at the Livermore National Laboratory. Although this is the most practical solution to the X-ray laser problem, it is not very practical for commercial applications, due to the enormous size and cost of these lasers, which approach the size of football fields.

Another way lasers may impact on our lives is through their use in creating 3-D TV to replace the conventional two-dimensional TV. Such a technological development could eventually alter the way we view entertainment. Although holographic TV is still many decades away, the barriers are strictly technical rather than conceptual.

Holograms use the fact that two beams of laser light can collide on a transparent photographic plate and create an interference pattern. (One beam contains the image and the other is a reference beam in phase with the first beam.) The interference pattern, consisting of tiny whorls and spiderweb-like lines, is captured on the plate. When a second laser beam is then shot through the plate, the precise three-dimensional wave front of the original beam is reproduced.

If the information stored in the interference patterns can be read into a computer, then the computer might be able to generate the original wave front on a screen and hence reproduce the original 3-D image.

A 3-D TV set of the future might look like a large crystal ball, such that when viewing into the ball one sees three-dimensional images dancing inside, or it may be in the shape of a wall screen, where the viewer sits in one spot and watches the motion of people in the screen. It could even be constructed in the shape of a planetarium, where one can view images in 360 degrees.

At present, 3-D TV is not possible because of the enormous amount of memory necessary to store holographic pictures. The MIT Media Lab has created some prototypes of holographic images using computers, but they cheat a bit: the image is slightly distorted, such that if you move your head sideways, the image changes as expected, but if you move your head vertically, the image does not change.

Using present technology, a holographic TV set would be a monstrosity. Large banks of supercomputers would have to be used to generate each holographic frame, and then only at slow speeds with primitive images. And sending holographic images via ordinary cable or telephone wires would be impossible.

By the middle of the next century, however, 3-D TV may become a reality, as computers become powerful enough to manipulate the information stored on holographic images.

## Beyond 2100: Antimatter Engines

For the most part, the technologies I've profiled so far may well be realized in the period from 2020 to 2100. Beyond that date, new technologies will likely come into play. The most intriguing of these is the antimatter engine. Every fan of *Star Trek* knows that the *Enterprise* gets its energy from antimatter. But not every Trekker knows that Gene Roddenberry actually stole that idea from quantum physics.

In 1928, P. A. M. Dirac, one of the founders of the quantum theory, devised an equation describing electrons obeying Einstein's relativity theory. In doing so, he realized that Einstein's famed equation, $E = mc^2$, is actually not completely correct. Einstein had to take the square root of an equation to arrive at this formula. The correct equation was actually:

$$E = + \text{ or } - mc^2$$

This minus sign was ignored by Einstein in formulating the relativity theory. But Dirac found that his theory of the electron was consistent only if the minus sign was included. He was forced to postulate an entirely new state of matter, antimatter, that had remarkable properties.

On the surface, antimatter is almost indistinguishable from ordinary matter. You can form antiatoms out of antielectrons and antiprotons. Even antipeople and antiplanets are theoretically possible. They will look just like ordinary people and ordinary planets. But that's where the similarity ends. The charge of antimatter is the opposite of that of ordinary matter; it will annihilate into a burst of energy upon contact with ordinary matter. Anyone holding a piece of antimatter in their hands would immediately explode with the force of thousands of hydrogen bombs.

Antimatter, therefore, is a nasty substance to hold in your hand, but perhaps an ideal substance to fuel space travel. It would leave no waste products and would generate enormous thrust.

Scientists have been playing with minute quantities of antimatter in the laboratory for decades, analyzing it among the decay products of radioactive atoms. (When I was in high school, I photographed tracks of antimatter emitted from radioactive sodium 22 for a science fair project.) Scientists also track the antimatter created when powerful atom smashers generate a beam of protons which is then slammed into a target. Antielectrons and antiprotons are among the thousands of debris particles, which can then be extracted by powerful magnets.

But obtaining large quantities of antiatoms and antimolecules is economically and practically prohibitive. Only recently, in 1995, have physicists using the particle accelerator in CERN in Switzerland been able to obtain the first minute quantities of antihydrogen in the lab.

Beams of antimatter can be made by firing beams of ordinary matter through a target, which creates a sudden burst of particle debris, some of which is made of antimatter. Then a magnetic field is used to separate out the antimatter from the matter. The physicists at CERN took a beam of antiprotons circulating in their particle accelerator and then shot it through a jet of xenon gas. The collisions between the antiprotons and the xenon atoms created antielectrons, some of which are briefly captured by antiprotons to create antihydrogen. Unfortunately, the antihydrogen lasted only 40 billionths of a second, too short a time for any detailed analysis.

Now that antihydrogen is known to exist, the next step is to create stable collections of antihydrogen in containers. Already, antimatter "traps" have been constructed which can successfully contain particles of antielectrons and antiprotons. The antimatter is held in these traps via a combination of magnetic and electric fields. The trick is now to combine them to create stable antihydrogen atoms. Physicists at the Max Planck Institute for Quantum Optics in Garching, Germany, are experimenting with using carbon dioxide lasers to force the antielectrons and antiprotons together in the traps to form antihydrogen.

As one can see, the expense of creating antihydrogen is enormous. Creating antimolecules will take many decades into the future. And creating enough antimatter to use in an engine using known technology would bankrupt the United States.

A secondary problem with antimatter is: where do you put it? Any box containing antimatter would instantly explode. Nor can you place antiatoms in a magnetic field or magnetic bottle, since the antiatoms will be neutral in charge. For example, plastic (or antiplastic) will pass effortlessly through the strongest magnetic field.

There is no scientific reason for ruling out the use of antimatter as a fuel in the distant future, other than the severe economic constraints placed on its creation. (One hope is that perhaps in the future an antimatter meteorite may be found in outer space. In 1977, a huge "fountain" of antimatter was discovered near the center of our galaxy, so perhaps antimatter can one day be found naturally.) But because the cost is prohibitive, one would have to invent new cost-effective schemes to create large quantities of antimatter. As a result, antimatter engines are unlikely until far into the future.

## Defying the Known Laws of Physics

It's always dangerous to make predictions stating that certain things are impossible. Too many times, naysayers have lived to see these very things come to pass. Legions of naysayers who told the Wright brothers, James Watt, and Thomas Edison that the airplane, the steam engine, and the lightbulb were impossible lived to see these inventions change the course of history.

Because we have a reasonable grasp of the laws of nature, however, we can say that certain technologies are incompatible with the known laws of electrodynamics, quantum theory, relativity, etc. This doesn't mean that the following inventions are impossible, only that they are highly unlikely with our current understanding of the laws of nature.

Here are several futuristic technologies that have been much discussed but which are beyond our understanding and reach, and may possibly remain so forever:

### PORTABLE RAY GUNS

In *The War of the Worlds*, H. G. Wells introduced the concept of "heat rays" fired by Martian walking machines that devastated entire cities and reduced the human race to slavery.

Today, laser beams can be every bit as powerful: we can generate millions of watts of laser power which can blast through steel. In fact, the only limitation to the power that can be packed onto a laser beam seems to be the stability of the lasing material (which heats up, cracks, and becomes unstable at high power levels) and the energy source.

The problem, though, lies in creating a portable power pack that can be held in your hand. If you want the energy of a nuclear power plant to erupt from a ray gun, then there is a slight problem: you need to be connected to a nuclear power plant.

President Reagan faced this identical problem when he proposed the Stars Wars plan in 1983. He needed a portable power pack that could be placed on small satellites orbiting the earth. The only known portable source of power of that kind is a hydrogen bomb. When a hydrogen bomb detonates, its burst of X-rays can be diverted, as we have seen, along copper rods to generate intense X-ray laser beams. Indeed, Edward Teller's original Star Wars scheme was to place thousands of hydrogen bombs in each of his "Excalibur" X-ray lasers orbiting the earth.

However, it was later shown that even bomb-driven X-ray lasers do not have the energy or the ability to shoot down thousands of Russian warheads in a very short period of time.

A portable power pack for a ray gun, short of a hydrogen bomb, does not currently exist, and scientists don't know where to start to find one.

### FORCE FIELDS

Force fields—transparent, impenetrable walls made of pure energy—are a common feature in science fiction. To construct a possible force field, we must study the four fundamental forces that govern the universe: the electromagnetic, the gravitational, and the weak and strong nuclear forces. None of them is a likely candidate for a force field.

Electromagnetic fields are unlikely because certain objects are neutral under both electric and magnetic fields. They will fly right past an electromagnetic field. A piece of plastic, we noted earlier, can be sent through a powerful magnetic field without being deflected.

Gravitational fields are unlikely because they are attractive, not repulsive. They are also extremely weak. For example, if you comb your hair, the comb can lift tiny pieces of paper. Thus the gravitational pull of the planet can be neutralized by a simple comb!

The weak and strong nuclear forces are also unlikely, because they only act over atomic and nuclear distances. But force fields in science fiction novels work over distances of many feet to miles.

There are, however, several possible loopholes. One would be if there is a yet-unknown "fifth force." Several serious attempts have been made to find a fifth force that might, for example, act over a distance of several feet. There have been several reports that such a force had been found, but all were later shown to be spurious. Second, since our understanding of the weak and strong nuclear forces is still primitive, perhaps one day we could create force fields whose thickness is on the atomic or nuclear scale, but which extend over many feet or miles. Unfortunately, no one knows how to do this.

### TRANSPORTERS AND REPLICATORS

"Beam me up, Scotty," has become a popular expression to millions of *Star Trek* fans. Most viewers, when asked, think that although the *Enter-*

*prise*'s warp-drive engine is farfetched, the transporter may eventually be built. Actually, the situation is probably the reverse. At least we have a theory of space warps and wormholes from Einstein's equations and the quantum theory. Physicists, however, raise their hands in complete befuddlement when it comes to postulating a mechanism for a transporter. We don't have a clue.

The very idea of taking something apart, atom for atom, shooting it across space, and then reassembling it is beyond comprehension. Each of these three steps is beyond any physics known to science.

First, we cannot take people apart atom for atom because we cannot calculate the location of each atom. That information alone would exhaust all the computers on earth. Second, we cannot send atoms across space, even by radio. And third, we wouldn't know how to reassemble a human even if we knew the location of all of his or her atoms.

### INVISIBILITY

In the novel, *The Invisible Man*, H. G. Wells's hero becomes invisible during an accident. His body floats in the fourth dimension, just slightly off the three dimensions of our universe, and hence appears invisible to us, although he can see everything that happens in our three-dimensional world.

Unfortunately, there is no known way of making someone invisible. What makes an object visible or invisible is the structure of the atom's electron shells. For objects which are opaque, their electron shells simply absorb or scatter incoming light. For objects which are transparent, the electron shells absorb the incoming light but then scatter them, re-creating the original wave front. At present, we do not know how to manipulate the atomic shell structure of atoms to change their optical properties at will and make objects invisible.

Although these ideas are farfetched and probably violate basic laws of physics, there is one futuristic technology that is well within our understanding, and that is starships to probe the nearby stars.

# 14

# To Reach for the Stars

"There is no way back into the past. The choice is the Universe—or nothing."
—H. G. WELLS

"Two possibilities exist: either we are alone in the universe or we are not. Both are equally terrifying."
—ARTHUR C. CLARKE

LIKE A SEARING BOLT of lightning, the discovery of fossilized worm-like structures in a piece of Martian rock has riveted international attention on the Red Planet, galvanizing President Clinton to declare: "Today, Rock 84001 speaks to us across all those billions of years and millions of miles. . . . It speaks of the possibility of life. If this discovery is confirmed, it will surely be one of the most stunning insights into our universe that science has ever uncovered." With these words, in the summer of 1996 the President capsulized the thrill and excitement generated by the possible discovery of life on Mars. In 1997, scientists even speculated that life may exist on the moons of Jupiter.

Undoubtedly, we will witness a vast number of stunning discoveries and milestones in space in the twenty-first century as scientists expand the present boundaries of knowledge. We will see a series of mobile autonomous robots exploring the surface of Mars, and the successor to the Space Shuttle, the X-33 VentureStar, will soar into space to dock with the new Space Station Alpha being constructed cooperatively by several nations.

We will also have new types of telescopes able to detect earth-like planets outside of our solar system, which, if detected, would almost certainly spur scientists to design the first starships to explore the nearby stars for intelligent life.

Because the physical and engineering laws of rocketry are fairly well known, it is possible to make reasonable predictions about the future course of space exploration in the twenty-first and even the twenty-second century. In this chapter, I will outline how the space program will likely evolve in the period from now to 2020, for which NASA has already laid out some broad goals, from 2020 to 2050, when new propulsion systems make interplanetary travel commonplace, and from 2050 to the twenty-second century and beyond, when humanity contemplates the colonization of other planets.

Space colonization is not just idle speculation and wishful thinking, but a matter of the long-term survival of our species. The earth lies in the middle of a cosmic shooting gallery. On a time scale of millennia to millions of years, it is inevitable that a meteor or comet or other natural disaster will destroy most life on earth. This means that one day our species will have to find a new home in outer space. Seeking a new home in space is a practical matter of survival.

## From Now to 2020

The invasion of Mars has begun! When H. G. Wells wrote his classic *War of the Worlds* in 1898, he assumed that Martians would invade the earth. Actually, it is the inhabitants of the earth who are invading Mars. Ten NASA space probes are due to rendezvous with the Red Planet from 1997 to 2007, averaging one probe per year. This may eventually pave the way for a permanent robot base and even piloted missions to Mars late in the twenty-first century.

The pace of space missions to Mars will undoubtedly quicken if the findings by the scientists at the Johnson Space Center are confirmed concerning the existence of microbial life on Mars. As reported around the world, these scientists have been studying an unusual class of meteors which landed in Antarctica thousands of years ago. (Because Antarctica is almost completely white, finding meteors amid the barren icy terrain is quite easy.) These meteors have precisely the same mineral and gaseous content found on Mars, and as a result scientists are convinced that they originate from the Red Planet. Because Mars is half the size of the earth, and hence has a very weak gravitational pull, scientists theorize that meteor impacts on Mars might have blasted fragments of the Martian soil into deep space, where they drifted for millions of years before some of them landed on earth. Fourteen Martian meteors have now been recovered from Antarctica.

One meteor which caught their attention was a four-pound rock about the size of a large potato dubbed ALH84001. It was shot out of Mars 16 million years ago and drifted in space until it landed in the South Pole some 13,000 years ago. Detailed analysis of its contents have suggested the existence of microbial life on Mars 3.6 billion years ago. Photographs of these wormlike structures look almost identical to the remnants of early fossilized microbial life found on earth rocks which also date back 3 billion years.

This astonishing claim was further supported when the British announced a few months later that their Martian meteor also harbored signs of organic material consistent with life.

All this has sparked intense public interest in the first missions to Mars in two decades: the Pathfinder and the Mars Global Surveyor missions, which were launched in 1996 and reached Mars in 1997. They represent just the beginning of a concentrated effort by NASA to explore a neighboring planet.

Already, NASA is preparing a timetable for the exploration of the Red Planet. The first missions to Mars will survey the planet in greater detail than ever before, and will place on the planet the buglike two-foot-long Mars Rover (based on the insectoid robots of Rodney Brooks featured in Chapter 4), which can independently scout out the rough terrain without detailed commands from the earth. (Because radio transmissions take roughly ten minutes to reach Mars, it is too cumbersome to guide the Rover via remote control.)

Later Rovers will possess a small shovel to take soil samples, to hunt for traces of life forms, and to survey the composition of the soil. All this will culminate in the first robotic recovery of Martian rock by 2005. (Since the Martian gravity is weak, lifting off the surface of Mars for the return trip is not much of a problem.) One possibility being studied by NASA scientists is producing fuel propellant and oxygen directly from Martian resources, given the fact that the soil is probably rich in ice and the atmosphere has plenty of carbon dioxide. This would significantly reduce the cost of a retrieval mission and might be the key to setting up a permanent robotic base on Mars.

If it is confirmed that Mars did have microbial life, then the question is: how did this life form evolve on such an inhospitable planet?

## Mars: The Frozen Desert

From the Mariner and Viking probes, we know that Mars is a "frozen desert," a cold, harsh planet with subfreezing temperatures, bleak desert terrain, huge planetary storms, and a thin, unbreathable carbon dioxide atmosphere about 1 percent the density of the earth's atmosphere. (Anyone unfortunate enough to be caught on Mars without a space suit would suffocate, freeze, and eventually explode.)

But it wasn't always that way. Several billion years ago, Mars had abundant lakes, seas, and possibly oceans. We see ancient riverbeds and remnants of islands carved out by running water which once ran freely on the Martian surface, proving that the climate of Mars was radically different in the past. Carl Sagan noted: "Between 4.0 and 3.8 billion years ago, conditions on Mars . . . may have favored the emergence of life. The surface of Mars is covered with evidence of ancient rivers, lakes and perhaps even oceans more than 100 meters deep." (It was so lush and conducive to the formation of life that some scientists have speculated that perhaps life first began on Mars, rather than the earth, and Martian meteors seeded the earth with microbial life. "Who is to say that we are not all Martians?" asks Richard Zare of Stanford University.)

One serious proposal NASA is considering for sometime beyond 2010, is to set up some sort of permanent robotic base on Mars that would monitor conditions on the planet and mine the ground for useful chemicals, creating a self-sustaining base of operations.

In 1997, plans to send more interplanetary probes into space received another shot in the arm with the announcement that Europa, the ice-covered moon of Jupiter, may have conditions compatible with life. Although Europa is a harsh, frigid world, beneath its permanent icy crust there may be a vast ocean of liquid water heated by volcanic activity, radioactive decay, as well as gravitational energy from Jupiter's immense tidal forces. The Galileo probe currently circling around Jupiter came within 363 miles of Europa and took photographs of what appear to be red-colored seas with floating icebergs. In the same way that microbes live near volcanic vents on the earth's seabed, life may exist near volcanic vents on Europa's oceans. "I'm sure there's life there," says astronomer John Delany of the University of Washington.

Although proposals to send people to Mars and beyond evoke stirring memories of President John Kennedy's famous decision to land men on the moon, such a crash program would be very risky and yield little science for the money. At a minimum cost of $500 billion dollars, by some estimates, the piloted mission to Mars would be exorbitant in cost, wasteful of scarce resources, and extremely dangerous.

NASA has wisely decided not to repeat the same mistake made in the 1960s, when the space program was largely driven by the Cold War and collapsed after the politicians lost interest in the moon. It is difficult to chart the future of space travel because the driving force behind the space program has often been politics, rather than science, with politicians demanding that astronauts perform glamorous but largely ceremonial stunts in space which could be done by robots for a fraction of the cost. As one politician put it: "No Buck Rogers, no bucks."

As William Walter wrote in his book *Space Age:* "Our love affair with the moon seemed to have become merely a flirtation, a one-night stand inflamed by the passions of the Cold War." LBJ perhaps put the attitude of our national leaders best when he said grimly that he "didn't want to go to bed by the light of a Communist moon." Isaac Asimov, satirizing Washington's rapidly fading interest in space exploration after we reached the moon, was equally succinct: "We scored a touchdown. We won the game, now we can go home."

NASA's new missions to Mars reflect director Daniel Goldin's streamlined motto: "smaller, faster, cheaper, better." Instead of sending one extremely expensive space mission to Mars every twenty years (like the disastrous billion-dollar Mars Observer, which probably exploded in 1993 just as it approached Mars), the new NASA plan is to spread out the risk and cost, sending ten smaller but more sophisticated probes over the next ten years.

## The Space Station in 2002

For all his energy and vision, NASA director Goldin has, unfortunately, been saddled with a few white elephants left by his predecessors, like the Space Station Alpha and the Space Shuttle. The prime example is the Space Station, for which a true scientific mission has yet to be found.

When finally built in June 2002, the International Space Station Alpha will be a far cry from the sleek, breathtaking orbiting space port that Stanley Kubrick and Arthur C. Clarke envisioned in the movie *2001.* Alpha will weigh a paltry 443 tons, and, if you include its expansive solar panels, will be about the size of a football field, measuring 361 by 290 feet. It will look, as one critic put it, like a bathtub with wings, and will carry only a crew of six astronauts, working out of seven laboratories, orbiting about 200 miles above the earth.

About 67 launches will be required to hoist all the materials into space, including 22 Space Shuttle missions, with the rest of the launches made by the Russians. The total price tag could be $60 billion, or about *$1 billion per launch.* About $43 billion will come from the United States and the rest from foreign countries. The General Accounting Office, calling NASA's estimates far too optimistic, put the cost at a more realistic $93.9 billion.

Albert Wheelon, former CEO of Hughes Aircraft and a member of the 1993 presidential commission on the Space Station, said, "The science value is wildly exaggerated. . . . It's a job program plain and simple." The National Research Council, which includes some of the top space scientists, said it "cannot be supported on scientific grounds."

From a purely scientific point of view, the main criticism of Alpha is that it does very little science for the $100 billion price tag. Almost all the planned experiments on Alpha can be performed at a fraction of the cost via single rockets or smaller orbiting stations like the Russian Mir.

The Space Station Alpha is scheduled to be finished by 2002, costing perhaps $100 billion. (Courtesy National Aeronautics and Space Administration)

Originally, one of the scientific missions was to investigate "microgravity"—i.e., the manufacture of exotic materials and proteins in the weightlessness of outer space. But as physicist Allan Bromley, former President George Bush's science adviser, said, "microgravity is of microimportance."

The feeling of most scientists is summarized by space scientist James Van Allen, who said, "The shuttle and space station represent the opposite of everything Goldin says he wants. They are bigger, slower, more expensive, and worse."

In 1997, the Space Station was dealt two more blows. The National Research Council estimated that there was a 50 percent chance that the Space Station could suffer a catastrophic collision with a small meteor in space during its fifteen-year lifetime. Some of these meteors would be too small to detect by radar but would be big enough to rupture the outer hull. Also, the Russians, facing a disintegrating economy, may not be able to finish their contribution to the Space Station, the central Service Module. "From the day we started with the Russians, it has been the perils of Pauline," concluded Goldin. This set the program back as much as one year.

Although the Space Station may become obsolete even before it is built in 2002, at least NASA is ready to replace the Space Shuttle.

## X-33: Workhorse of the Twenty-first Century

The Space Shuttle has been called "the most effective device known to man for destroying dollar bills" by Congressman Dana Rohrabacher. Another relic of the Cold War, the Space Shuttle has been an albatross for NASA officials. With soaring costs and a feeble, embarrassing record of only eight launches per year, the Shuttle has become a black hole for tax dollars. The Shuttle can carry twenty-seven-ton loads, but at a staggering cost of $800 million per launch. It costs about $15,000 to send a pound of payload into space on the Shuttle, which is more than twice the price of gold (roughly $6,000 per pound).

As a result, the European Space Agency, with its nimble Arianne rocket, has seized two-thirds of the launch market, which was once the exclusive domain of the United States.

But this dismal track record may be turning around. In July 1996, the Clinton administration awarded $1 billion to Lockheed Martin to develop a radically new rocket design that would be cheap and efficient. Many space scientists have hailed this as the beginning of a new era of cheap and frequent space travel.

This sleek new design, the first in three decades, is the X-33 VentureStar, a reusable launch vehicle, a prototype of which will have its maiden voyage in March 1999 and is scheduled to complete fifteen test flights before 2000.

The key to the VentureStar concept is that it will be cheap and truly reusable, cutting costs by a factor of ten. This change in the economics of space travel could alter the way people view outer space. Space travel, once regarded as being impossibly expensive, may eventually become relatively commonplace. Astronomer John Lewis of the University of Arizona even envisions the day when a trip into outer space will cost only a bit more than a transatlantic flight, making space travel accessible to the public.

The X-33 has an unorthodox shape and resembles the *Millennium Falcon* featured in the movie *Star Wars.* Like the *Millennium Falcon,* it both soars into space and lands on a conventional airfield. There is no need for

The X-33 VentureStar, a reusable launch vehicle, scheduled to replace the Space Shuttle early in the next century, uses advanced composite resins, which are stronger but lighter than steel. (Courtesy National Aeronautics and Space Administration)

huge booster rockets that are wastefully discarded. The goal is to have thirty launches per year, about four times more than the current Space Shuttle.

A sixty-seven-foot, half-size prototype of the VentureStar is scheduled for its maiden flight in 1999. By 2006, the full-sized craft should be operational. By 2008, says Gene Austin, NASA's X-33 program manager, the RLV "will simply be cargo-and-crew delivery flights for space-station support and commercial launch capability." By 2012, it should completely take over from the current Space Shuttle. Eventually, the RLV will be handed over to industry.

## The Orient Express

The VentureStar may be joined early in the next century by yet another radically different launch vehicle: the aerospace plane. Dubbed the Orient Express by President Reagan because of its ability to reach Tokyo from New York in about an hour, the hypersonic plane is intended to take off and land like an ordinary jet, but soar in space like a rocket. As a result, it does not require heavy oxygen tanks or huge booster rockets, like conventional rockets, because it sucks in oxygen directly from the air, like a jet airplane, and cruises at Mach 2 or Mach 3 (1,500 to 2,200 mph). As it reaches the upper atmosphere, where the air is too thin to power its jet engines, the rocket engines turn on and the plane accelerates to Mach 23, soaring like a rocket through space into orbit.

As William Safire of the *New York Times* said: "This time, we are embarked on the development of a 4,000-mile-an-hour National Aerospace Plane (NASP) that will zip past the French-British Concorde the way a souped-up Corvette passes an antique tin lizzie."

The original NASP was a ten-year, $1.5-billion effort which was terminated in 1992. Now officially called the Advanced Space Transportation Technology Program, the building of the hypersonic jet will be supervised by NASA's Marshall Space Flight Center in Huntsville, Alabama. Testing of the engine will begin in 2000. The first flight of a small-scale system is planned for 2002. And a large-scale test of the system is expected by 2005.

In 1997, NASA awarded a $33.4-million contract to Microcraft Inc. to develop a new engine for the hypersonic plane. The project, called Hyper-X, will create four unmanned reusable vehicles which should reach Mach 5 to Mach 10 by 1998.

Ultimately, the goal of hypersonic launch vehicles is to reduce the cost of launching low-earth-orbit satellites by 95 percent by 2009. If these launch vehicles attain such a vast cost reduction, sending frequent payloads into space, even possibly commercial passengers, may become a reality.

This new generation of vehicles is made possible by revolutionary advances in materials technology, which have produced tough, resilient, and lightweight resins for the hull, replacing the clumsy (and potentially dangerous) ceramic tiles that make up the heat shield of the Space Shuttle. (During reentry into the atmosphere, when the Space Shuttle experiences blistering temperatures created by air friction, the loss of a few of these precious tiles could cause a catastrophic penetration of the hull. Engineers have been known to cross their fingers during reentry and pray silently, hoping that the tiles don't come loose.) By contrast, the RLVs use advanced lightweight graphite composites and aluminum-lithium resins which considerably reduce the weight of the rocket and make it more efficient.

These high-tech composites are five times lighter than steel, although they are much stronger. They cost $1.50 to $2.00 per pound (compared with 20 to 40 cents per pound for steel), but this price is sure to come down with mass production. These composites are so advanced that some engineers have called for passenger cars and trains to be made of the same space-age materials, which would increase the safety and efficiency of ground transportation. The composites are formed by making fibers spun from carbon, glass, and other materials and then fusing them into a matrix of plastic, ceramic, or metal. The creation of these revolutionary new materials, in turn, has been accelerated by the computer and quantum revolutions.

Supercomputers, modeling the airflow over the fuselage in virtual reality, have given us the ability to calculate the temperatures and strains experienced by spacecraft flying at hypersonic velocities without performing costly experiments. Because the mathematics of airflow and aerodynamics are well established, these supercomputers can give an accurate description of the hostile environment faced by the hull of the spacecraft as it enters the atmosphere at about 17,000 miles per hour.

## From 2020 to 2050

Beyond the year 2020, radically different types of rockets will be required to serve a new function: to carry out long-haul interplanetary missions in deep space, including servicing a robot base on the moon, probing the asteroid belt and comets, and even supplying a manned base on Mars. By then, missions to the planets will become routine. What is needed is a cheap and reliable means of transport.

Several competing designs have been proposed to power the rockets of the future, including the ion engine, the nuclear rocket, the rail gun, and the solar sail. Many of them suffer from major drawbacks. The nuclear rocket, for example, poses a dangerous health hazard if a meltdown occurred in space. The rail gun accelerates objects so rapidly that it will flatten most payloads. And the solar sail is quite difficult to build and maintain in space. Physicist Freeman Dyson has said, "I declare solar-electric [ion] propulsion to be the winner in space because it allows us to push as far in the directions of speed, efficiency, and economy as the laws of physics allow."

Like the chemical rocket, which was the workhorse of the twentieth century, the solar-electric ion engine will probably perform the yeoman's work late in the next century. The ion engine operates very much like the electron gun found in a TV set. It derives its energy from solar cells which generate electricity, which is then used to heat up and ionize a gas such as cesium or xenon. These charged ions are then pulled toward a charged plate, which is used to shoot them out the end of a gun.

The ion engine is almost the exact opposite of the Saturn rocket, which generated 9 million pounds of thrust over just a few minutes in its trip to the moon. The problem is that chemical rockets like the Saturn produce enormous power only for a brief period of time. The ion engine, by contrast, emits only a thin beam of ions and hence generates a modest amount of thrust, but it can maintain this thrust almost indefinitely.

The two engines are analogous to the tortoise and the hare. Chemical rockets, like the hare, are fine for rapidly blasting out of the earth's gravitational field, but they exhaust fuel so rapidly that they can be turned on for only a few minutes. Ion engines, like the turtle, are able to go long distances because their small but steady acceleration can be maintained for years.

Physics tells us that the important quantity is not the thrust, but the product of the thrust and the time of duration, which is called the "specific impulse." What you lose in thrust you can more than make up in duration, as the hare found out to his regret. A chemical rocket can typically attain specific impulses of about 500 seconds, while an ion engine can attain specific impulses of 10,000 seconds, with the maximum being approximately 400,000 seconds. (By convention, the unit for specific impulse is the second.)

The simplest design for long-haul interplanetary missions will use a combination of rockets. Chemical rockets will be needed to escape the earth's gravitational pull, but once in space, rockets will use ion engines to accelerate steadily to high velocities and coast to the outer planets and beyond.

Because the ion engine slowly builds up speed over a long period of time, it is ideally suited for long-haul missions around the solar system where time is not the main consideration. One can imagine hauling large cargo shipments between planets via ion engines. It might form the backbone of a cosmic Interstate Railway network in the heavens.

"A diversified system of solar-electric spacecraft would make the entire solar system about as accessible for commerce or for exploration as the surface of the earth was in the age of steamships," says Freeman Dyson.

At the Lewis Research Center in Cleveland, tests are currently being conducted on the Kuiper Express, a vehicle powered by an ion engine which will explore the comets located beyond the orbit of Neptune, in what is called the Kuiper Belt. The Kuiper Express uses an ion engine powered by solar cells which ionize xenon gas. Electrodes then pull the ions out of a gun, thereby creating thrust. The Kuiper Express gets its energy from two large solar panels which can extract light from the sun even in the deep space beyond Pluto, where the sun is not much brighter than the nearby stars.

## Extra-Solar Planets in Space

Later in the twenty-first century, interest will gradually shift away from our own solar system to the nearby stars. By 1997, scientists were ecstatic to have discovered thirteen extra-solar planets orbiting nearby stars in familiar constellations like Virgo, the Big Dipper, and Pegasus. Unfortunately, all of them are huge, Jupiter-like planets, and are probably uninhabited.

But beyond 2020, our instruments may become sensitive enough to detect tiny, earth-like planets circling nearby star systems, which could encourage scientists to reach for the stars. "The Holy Grail is to find an extrasolar planet that is capable of supporting life," says astronomer Alan P. Boss of the Carnegie Institution in Washington.

Just as the possibility of discovery life on Mars will drive much of space exploration in the early part of the twenty-first century, the possibility of finding earth-like planets outside our solar system will probably drive interstellar space exploration through the end of the next century.

We know from Newton's laws of motion that our own sun wobbles a bit because of the presence of giant planets like Jupiter and Saturn. Similarly, giant planets will tug at their nearby stars, causing them to oscillate. Since the orbiting planets do not emit any light of their own, our telescopes will only detect the wobbling of the star it orbits around.

Using these methods, astronomers have found a planet orbiting the star 47 Ursae Major, 200 trillion miles from the earth in the Big Dipper; the planet is twice the size of Jupiter. Another planet, with six times the mass of Jupiter, circles the star 70 Virginis in the constellation Virgo. Most of these planets are 20 to 40 light-years away, probably too far for our space probes even in the next century. But in June 1996, a remarkably close planet was discovered by the astronomers at the University of Pittsburgh. A Jupiter-sized planet only 8.1 light-years away, it orbits the star Lalande 21185, a small red star which is the fourth-nearest star to the earth. They also found evidence that there are two smaller planets in that solar system.

Although they did not find an earth-like planet, what is encouraging is that this star system is so close to the earth and seems to resemble our solar system. It is within striking distance of a future starship, giving added incentive to build rockets in the next century that can travel light-years into space. Its close proximity and its similarity to our own solar system prompted Alan Boss to say, "It's fantastic! This is the one we've been waiting for."

Unfortunately, because all these planets are larger than Jupiter, it is likely that they are gas giants made of hydrogen, so the chances of finding carbon-based life forms like ourselves are quite small. So far, the limitations of the technology prevent us from finding planets smaller than Jupiter.

## Finding Earth-like Planets in Space

Within ten years, the next generation of astronomical instruments may be able to find scores of planets as tiny as the earth, capable of harboring life as we know it. This could open up a new era of astronomy, potentially changing our conception of life in the universe, showing that the conditions for life in the universe are not as restrictive as we have up to now imagined.

This new class of instruments introduces a new concept in telescopes, the optical interference of light. Up to now, optical telescopes have been limited by the size of their mirrors. Telescopes are "light buckets"; the more light you can collect at night, the greater the resolution of the image. Ultimately, however, you hit the physical limits of machining mirrors. The 200-inch mirror at Mount Palomar, for example, was such an engineering feat that it held the world record for the biggest telescope mirror for about six decades. Modern telescopes like the twin Keck telescopes in Hawaii are larger than Mount Palomar and use novel designs, such as mirrors made of separately movable pieces of glass. But astronomers are gradually reaching the limit of what can be done with giant glass mirrors.

A new generation of interference telescopes uses a trick to increase their resolution. Because of recent advances in instrumentation, it is now possible to combine the light coming from two distinct telescopes separated by a large distance. (By allowing these light signals from two telescopes to collide precisely in the middle, the two wave fronts interact and produce an interference pattern. By carefully analyzing this interference pattern, one can obtain an image as if it were made by a supertelescope whose mirror size is equal to the separation distance of the two telescopes. Thus, instead of building mirrors which are miles across, which is physically impossible, one can use two smaller telescopes separated by miles to simulate this single giant telescope.)

Satellites will also greatly accelerate our search. In 2001, NASA will launch Kepler, a satellite so sensitive it should be able to detect up to 2,400 new planets, about 100 of which are expected to be earth-like. By 2007, Kepler will be joined by other satellites, like the Space Interferometry Mission and the Terrestrial Planet Finder. These satellites are so accurate they can, if placed on the earth, see an astronaut on the moon passing a flashlight from one hand to the other.

These earth-like planets may well contain the most precious quantity in the universe: liquid water, the "universal solvent." As far as we know, only liquid water has the capability of dissolving complex carbon-based molecules so they combine to form the precursors of life: proteins and nucleic acids.

## Creating a Garden of Eden in Space

When long-haul missions to the planets become commonplace after 2020, some in the scientific community will begin looking at creating colonies in space, but most of this discussion will be hypothetical. Although the cost of sending payloads into space will drop considerably by this time, it will still be far too expensive to send the kind of large payloads into space necessary to attempt to construct colonies in space. Furthermore, the hostile conditions in outer space, where people are continually threatened with cosmic rays, solar winds, small meteorites, and subfreezing temperatures, make life-support systems notoriously expensive.

This will not deter some thinkers, however, from laying out reasonable scientific hypotheses about the cost of constructing space colonies, such as setting up a colony on the moon or terraforming another planet (i.e., making them more earth-like in their climate). As the Russian visionary Konstantin Tsiolkovsky once said: "Earth is the Cradle of Mankind, but one cannot stay in the cradle forever."

The discovery of ice on the moon in 1996 in the southern polar region raises the long-term possibility of building a moon base. Previously, scientists discounted the possibility of ice on the moon because the scorching sunlight beating down on the surface of the moon is strong enough to vaporize any ice. However, the Clementine spacecraft detected ice in a crater which was perpetually in the shadows. The possibility of ice on the moon raises the possibility of creating a permanent moon base and also using the ice for rocket propellant, by breaking it up into hydrogen and oxygen.

Another challenge is to terraform Mars or Venus, our closest neighbors in space, a formidable task given their hostile atmospheres. Scientists who have seriously considered terraforming Venus say that it is out of the question. Temperatures on Venus soar to a blistering 900 degrees F (hotter than a baker's oven) because its carbon dioxide atmosphere has trapped enormous quantities of energy from the sun. It is the greenhouse effect gone amok. In addition, a spaceship would be crushed like an eggshell—Venus's dense atmosphere is a staggering ninety times that of the earth—and the spaceship would probably disintegrate because of the corrosive sulfuric acid in the clouds. "All proposals for terraforming Venus are still brute-force, inelegant, and absurdly expensive," Carl Sagan concluded.

Mars is a somewhat better bet. Moviegoers got a glimpse of this terraforming process in the movie *Total Recall*, in which Arnold Schwarzenegger is thrown into the desert of Mars without a space suit. As his blood is about to boil and his skin to rupture, an ancient alien terraforming device is activated and enormous quantities of water are released from the frozen soil to re-create the ancient oceans of Mars.

But to bring water back to Mars in real life would be a daunting task, something beyond anything conceivable for at least a century or more.

Some scientists, however, have speculated about creating lakes and seas on Mars by using comets, which are gigantic icebergs in space. In 1986, scientists sent a spacecraft to take close-up pictures of Halley's comet. By the twenty-second century, we should have considerable experience landing on comets. Since about 200 million comets probably lie within the Kuiper Belt in our solar system, scientists have advocated placing rocket thrusters on them to deflect their trajectories an infinitesimal amount, sufficient to alter their paths so they impact on Mars. As the comets burn up in the thin atmosphere, they would create steam clouds that would eventually produce vast rainstorms on the planet.

In order to raise the temperature of the planet, scientists have suggested the possibility of generating a mini-greenhouse effect on Mars, by deliberately injecting small amounts of chlorofluorocarbon (CFC) chemicals (the same ones which are now largely banned on the earth) and ammonia into the atmosphere. Chlorides, for example, could be mined from the salt beds left over from the ancient Martian oceans and seas.

To haul large quantities of these greenhouse gases from the earth would be prohibitively expensive, however, requiring several centuries' worth of interplanetary missions. A better idea a century or two into the future would be to create robot chemical stations on Mars which would manufacture these chemicals directly from the soil or atmosphere. Robots would mine the surface of Mars, build large chemical factories, and create chemical reactions which generate greenhouse gases.

As the mini-greenhouse effect raised temperatures on Mars, the ice frozen in the underground permafrost and ice caps would begin to melt, which could also provide large quantities of water. The process, however, is excruciatingly slow, and even if possible, could take another century before temperatures and pressures begin to reach those of the earth.

Mars is the only planet in the solar system which has even the remotest chance of being terraformed. Mercury is too hot and desolate, Jupiter and the other gas giants are made of hydrogen gas and are too cold and distant. One moon of Saturn, Titan, has a nitrogen/methane atmosphere, but it's simply too cold on Titan for anything other than robotic missions.

Astronomer John Lewis, however, believes that the asteroid belt might make a suitable habitat for space colonists. He believes that asteroids could be hollowed out to provide secure housing for perhaps millions of colonists. (Living inside an asteroid would provide protection from cosmic rays, solar winds, and meteorite bombardments.) They can also be mined for minerals to construct factories and cities, and since they also contain quantities of helium, colonizers could build fusion plants to energize their machines.

To Lewis, the colonization of outer space has little to do with romance and glamour. Sooner or later, he believes, humans will be forced to leave the earth as the population soars and resources are depleted; he views mining the moon, planets, and asteroids as strictly a matter of self-preservation.

If Mars cannot be terraformed, moon bases fail, and asteroids cannot be hollowed out, we will likely eventually have to leave the solar system to search for habitable planets at some distant point in the far future. Given the fact that earth-like planets will inevitably be found outside our solar system, there will be an increasing chorus of scientific voices calling for an effort to send interstellar probes to the nearby stars.

## Beyond 2050: To Build a Starship

Up to now, the laws of physics have been fairly straightforward in setting the framework for the exploration of deep space. The chief uncertainty has been politics. The specific impulse from a rocket necessary for long-haul interplanetary missions is only a few thousand seconds, which is well within the reach of ion engines.

The problems with starships, however, stretch the boundaries of known physics and the resources of the planet. A number of designs have been proposed, each with distinctive advantages and disadvantages. The laws of the physics of interstellar flight are well known, but it's notoriously difficult to build a starship which is economical and can attain sufficient speed to reach the stars.

The fundamental problems with starships are twofold. First, the distances separating us from the stars are truly staggering. Although it might take a light beam, traveling at about 186,000 miles per second, about a day to reach the outer planets of our solar system from the earth, it would take light four years to reach the nearest star, Alpha Centauri, and a hundred years to reach many of the familiar stars we see at night. A starship traveling at a small fraction of the speed of light may therefore take centuries to reach the nearby stars. Second, according to Einstein, nothing can go faster than light.

The specific impulse necessary to approach the speed of light is *30 million* seconds, which exceeds the capacity of all the rocket designs discussed so far by a huge margin. Nevertheless, some ambitious rocket designers have made proposals for interstellar travel. At the top of the list in terms of star travel are various fusion machines. Ordinary fusion machines, which are scaled-up varieties of the prototypes found on the earth, have specific impulses between 2,500 and 400,000 seconds. This is rather disappointing, since that number is still too small to reach the "magic million" figure necessary to approach a reasonable percentage of the speed of light.

The most intriguing fusion device, however, is the hypothetical ramjet fusion engine, which sucks in interstellar hydrogen as its fuel as it soars through space at nearly the velocity of light. It can be made remarkably lightweight, since it relies on extracting resources from its environment, rather than carrying fuel, similar to the way that conventional jet airplanes suck in air as an oxidizer.

In shape, the ramjet resembles a large funnel (called the ram scoop), which sucks in hydrogen molecules as it moves forward. The design was originally proposed in 1960 by Robert Bussard, who estimated that a

The fusion ramjet scoops up hydrogen from deep space in its funnel as fuel. It is one design that may eventually take us to the stars, if certain questions about the proton-proton fusion process can be solved. (Courtesy Robert O'Keefe)

starship weighing 1,000 tons could accelerate indefinitely at one g (or 32 feet per second squared). This is convenient, since the people in the starship would then be pushed to the floor and feel artificial gravity comparable to that on the earth. He estimated that the starship could gradually approach the speed of light within a year.

Cruising comfortably along at one g force in nearly earth-like conditions, the crew aboard the starship could reach the nearest star within five years. But since time slows down aboard the starship, according to Einstein's special theory of relativity, the crew could reach the Pleiades star cluster (M45), which is 400 light-years away, in as little as eleven years, by the clocks aboard the starship. After twenty-five shipboard years, such a ship could even reach the Great Andromeda Galaxy (although over 2 million years would have passed on the earth).

As attractive as the fusion ramjet is, it has fallen out of favor. The fundamental weakness of the proposal is that the ramjet relies on fusing protons extracted from deep space. The proton-proton fusion process is much more difficult to attain than the standard deuterium-tritium fusion process used in our prototype fusion machines on earth. Fusion technology is not yet developed enough to make any definitive statements about ramjet fusion power, especially the more difficult proton-proton fusion process. But because ramjets do not violate any known law of physics but simply fail on formidable technical grounds, there have been a series of modifications of the idea over the decades. With a better understanding of fusion over the decades, the ramjet may yet prove to be a viable possibility.

## Nuclear Pulsed Rocket

Perhaps the strangest device that has been seriously proposed for star travel is the nuclear pulsed rocket, which would use multiple hydrogen bomb detonations to push a starship forward. By sending out a series of mini-hydrogen bombs from the rear, the starship would be pushed forward by each nuclear shock wave.

The basic principle was proposed in 1946 by Stanislaw Ulam of Los Alamos (who originally designed the first H-bomb and proposed computerizing DNA analysis). He showed that if a steel plate were coated with a graphite layer and placed a certain distance from an atomic bomb, the blast wave would push the plate rather than destroy it. If shock absorbers were then attached to the plate, the assembly would then be propelled by the nuclear explosion.

The hypothetical nuclear pulsed rocket has been through various incarnations, such as Project Orion (1958–65) and Project Daedalus (1973–78). The maximum specific impulse attainable with the nuclear pulsed engine would be about a million seconds, sufficient to reach the nearby stars in several decades to centuries. The Daedalus probe, by one calculation, could use successive micro-explosions to gradually boost its way to 12 percent of the speed of light on a one-way, fifty-year mission to Barnard's star (5.9 light-years away).

There are a number of steep obstacles to this type of rocket, such as the intense X-ray radiation and heat released from the blast, which could endanger the crew and the structural integrity of the ship. And because it involves fine-tuning the physics of thermonuclear explosions, it requires advanced knowledge of nuclear testing, which flies in the face of the current drive to sign a Comprehensive Test Ban Treaty. Furthermore, this kind of technology can be used for warfare as well and could proliferate new nuclear dangers.

The risks of tinkering with this volatile technology probably outweigh the benefits. (Theodore Taylor, the nuclear bomb designer who led Project Orion for the General Dynamics Corporation in the 1960s, now calls for the complete abolition of nuclear weapons.)

## Photonic Engines and Sails

At the exponential rate at which physicists are developing lasers for a surprising number of purposes, the possibility of using lasers to energize a starship by the end of the twenty-first century is not beyond the bounds of physics. A design called the photonic engine is basically a powerful laser which uses light pressure to propel itself into space. Several variations have been proposed. One would use powerful ground-based lasers on the earth or the moon to propel a solar sail into space. Normally, solar sails which use the sun's light pressure have problems because of the weakness of sunlight in deep space. From Saturn or Neptune, the sun is not much brighter than an ordinary star. Not surprisingly, the solar sail must be huge, on the order of hundreds of miles across, to capture enough sunlight to glide in space. Also, it is difficult to maneuver such a spacecraft for the return voyage. Sailboats can tack in the water to change directions, but solar sails cannot (unless they use sophisticated, untested ways of tacking against the magnetic field of outer space). Most important, they are very slow, taking years to build up a steady velocity. And once they reach a high velocity, they have difficulty slowing down.

Many of these problems could be solved if we used a laser beam on the moon to push the sail. For the return trip, the solar sail could whip around the star and use this "slingshot effect" to propel its return voyage. (However, this means the ship could not stop and land on any planet. It must deploy small probes if it wants to gain information about extra-solar planets.)

The biggest problem with the solar sail/laser engine is the power requirement. One calculation showed that the power of the laser beam would have to be 1,000 times the current output of the earth. And the amount of time necessary to reach the stars may be a few hundred years, too long to guarantee the political stability of the laser beam!

Another variation would have a more modest sail accompanied by a ramjet of some sort. The combination of a laser and ramjet may help to solve some of the problems with power, specific impulse, and so on.

There are even more futuristic designs, such as an antimatter engine and warp engines. Such machines are centuries down the line, if it is possible to build them at all. As we cautioned earlier, the cost of an antimatter engine, could one be built, is truly prohibitive economically. The possibility of warp engines will be discussed in Chapter 16.

## Suspended Animation

Most of the designs outlined above require decades, if not centuries, for a ship to reach the nearby stars. Thus starships will probably have to be unpiloted, at least initially. However, if we are serious about sending humans into space, then we must resort to some type of suspended animation for these grueling extended voyages.

Suspended animation is more primitive than we are led to believe from the popular media, even though a number of celebrities have willed their bodies to be frozen in liquid nitrogen after death.

Unfortunately, there are severe technical problems with suspended animation. One is that ice crystals form inside the cells as the body is frozen. These ice crystals grow until they eventually rupture the cell walls. Anyone who is frozen will suffer irreparable damage to their vital organs. The thawing process is also harmful to tissue. As you raise the temperature to the melting point, ice crystals begin to fuse together, which squeezes, deforms, and even ruptures cells.

For example, at present it's difficult to keep kidneys or livers alive by freezing for more than three days, and hearts and lungs for more than half a day. That's why, beyond sperm and blood cells, human body parts are notoriously difficult to freeze.

Some scientists have tried to freeze tissue extremely rapidly to minimize the formation of these lethal ice crystals. This process is called vitrification. Although this rapid-freeze method does in fact retard the formation of ice crystals, yet another problem arises. The lipids that are found in the cell membrane, which are usually in liquid form, become a gel (much the way animal fat congeals when it cools). As a result, the cell membrane becomes leaky and the cells quickly die as their delicate chemical balance is disrupted.

Nature, however, has devised a number of clever mechanisms by which cold-blooded animals can survive a harsh, icy winter. Fish, for example, can swim in subfreezing Arctic waters and frogs can be frozen solid and still be thawed out alive.

Recent investigations have now unraveled the biological mechanisms which make this possible. Fish have evolved a way to produce proteins that act as an antifreeze, allowing them to swim in Arctic waters about two degrees Centigrade below freezing, which is sufficient for them to survive in those icy-cold waters. Similarly, frogs have developed two mechanisms that allow them to survive even after being frozen in a block of ice. First, frogs have antifreeze chemicals, such as glucose. More important, frogs have the ability to maintain high glucose levels within the cell; thus ice crystals never form on the inside of the cell, even if the frog is frozen solid.

Adapting these methods, scientists have been able to prolong the life of some mammalian organs by a few hours, but not on the order of weeks or years needed for spaceflight.

In short, suspended animation is yet an unproven technology.

### Beyond 2100: Our Place Among the Stars

The fate of humanity ultimately must lie in the stars. This is not wishful thinking on the part of hopeless visionaries; it is mandated by the laws of quantum physics. Eventually, physics tell us, the earth must die.

Since it is inevitable that the earth will be destroyed sometime in the future, the space program may ultimately be our only salvation as a species. At some point in the distant future, either we stay on the planet and die with it or we leave and migrate to the stars.

Carl Sagan has written that human life is too precious to be restricted to one planet. Just as animal species increase their survivability by dispersing and migrating to different regions, humanity must eventually explore other worlds, if only out of self-interest. It is our fate to reach for the stars.

The upper limit for the existence of the earth is about 5 billion years, when the sun exhausts its hydrogen fuel and mutates into a red giant star. At that time, the atmosphere of the sun will expand enormously until it reaches the orbit of Mars. On earth, the oceans will gradually boil, the mountains will melt, the sky will be on fire, and the earth will be burnt into a cinder.

The poets have long asked whether the earth will die in fire or ice. The laws of quantum physics dictate the answer: the earth will die in fire. But even before that ultimate time 5 billion years from now when the sun exhausts its fuel, humanity will face a series of environmental disasters which could threaten its existence, such as cosmic collisions, new ice ages, and supernova explosions.

### Cosmic Collisions

The earth lies within a cosmic shooting gallery filled with thousands of NEOs (Near Earth Objects) that could wipe out life on earth. Some scientists at the Jet Propulsion Lab at Caltech believe that 2,000 or more mountain-sized asteroids are lurking in space undetected. In 1991, NASA estimated that there are 1,000 to 4,000 asteroids that cross the earth's orbit which are greater than half a mile across and which could inflict enormous destruction on human civilization. The astronomers at the University of Arizona estimate that there are 500,000 near earth asteroids greater than a hundred meters across, and 100 million earth-crossing asteroids about ten meters across.

Surprisingly enough, every year, on average, there is an asteroid impact creating about 100 kilotons of explosive force. (Fortunately, these asteroids usually break up high in the atmosphere and rarely hit the earth's surface.)

In June 1996, a close call with an NEO took place. This time asteroid 1996JA1, about one-third of a mile across, came within 280,000 miles of the earth, or a bit farther than the moon. It would have hit the earth with the force of about 10,000 megatons of explosive power (greater than the combined U.S./Russian nuclear weapons stockpile).

There were several deeply unsettling facts concerning both the 1993 asteroid and 1996JA1. First, they were undetected, suddenly appearing almost out of nowhere. Second, they were discovered not by any government-sponsored monitoring organization (there is none) but by mere accident. (Two students at the University of Arizona stumbled across 1996JA1.)

An asteroid only a kilometer across would create cosmic havoc by impacting on the earth. Astronomer Tom Gehrels of the University of Arizona estimates it would have the energy of a million Hiroshima bombs. If it "hit on the West Coast," he adds, "the East Coast would go down in an earthquake; all your buildings in New York would collapse." The shock wave would flatten much of the United States. If it hit the oceans, the tidal wave it created could be a mile high, enough to flood most coastal cities on earth. On land, the dust and dirt of an asteroid impact sent into the atmosphere would cut off the sun and cause temperatures to plunge on earth.

The most recent giant impact took place in Siberia, on June 30, 1908, near the Tunguska River, when a meteor or comet about fifty yards across exploded in midair, flattening up to 1,000 square miles of forest, as if a giant hand came down from the sky. The tremors were recorded as far as London.

About 15,000 years ago, a meteor hit Arizona, carving out the famous Barringer Crater, creating a hole almost three-quarters of a mile across. It was caused by an iron meteor about the size of a ten-story building.

And 64.9 million years ago (according to radioactive dating) the dinosaurs may have been killed off by the comet or meteor that hit the Yucatán in Mexico, gouging out an enormous crater about 180 miles across, making it the largest object to hit the earth in the last billion years.

One conclusion from all this is that a future meteor or comet impact which could threaten human civilization is inevitable. Furthermore, on the basis of previous incidences, we can even give a rough estimate of the time scale on which to expect another collision. Extrapolating from Newton's laws of motion, there are 400 earth-crossing asteroids greater than one kilometer which definitely will hit the earth at some time in the future.

Within the next 300 years, we therefore expect to see another Tunguska-sized impact, which could wipe out an entire city. On the scale of thousands of years, we expect to see another Barringer type of impact, which can destroy a region. And on the scale of millions of years, we expect to see another impact that may threaten human existence.

Unfortunately, NEOs have a high "giggle factor." As a result, NASA has allocated only $1 million per year to identify these planet-killing objects. Most of the work locating these NEOs is performed by a handful of amateurs.

## To Die in Fire and Ice

Another ice age will certainly occur, most likely on the scale of 10,000 years or so. The last great ice age might have affected the evolution of our species, dividing *Homo sapiens* into different races about 100,000 years ago. A brief warming spell in this ice age 10,000 years ago made civilization possible. However, civilization may come to a halt when the brief warming spell we are living within ends. Large parts of North America may once again be submerged under up to a mile of ice, as they were in the last ice age. Unfortunately, no one knows what causes ice ages, but the most popular theory holds that they are caused by minute wobblings in the earth's rotation.

If we do not die in ice, we could die in fire. Supernovas occurring within a few light-years of the earth could bathe the planet in a lethal rain of X-rays, killing all life on earth. Supernovas within our own galaxy take place once every 500 years or so. By analyzing a supernova that erupted in 1987, physicists were able to confirm our theory about the energy generated by supernovas, which are caused when the fusion process within an aging star suddenly shuts off, creating a massive gravitational collapse. Fortunately, because we know a fair amount about stellar evolution, we will have plenty of warning if a nearby star is about to go supernova.

But assuming that some form of looming catastrophe forces us to leave our solar system, what might we find? Is anyone out there?

## Aliens from Space

Astronomer Frank Drake of the University of California at Santa Cruz made the first reasonable estimate of the number of planets harboring intelligent life within our own Milky Way galaxy, which contains roughly 200 billion stars. Making a series of reasonable assumptions (e.g., about the number of stars that are like our own, the number of stars that have planets, the number of planets that are earth-like, the number of earth-like planets with life, and so on) one comes up with an estimate that perhaps as many as 10,000 planets exist in our galaxy which would harbor intelligent life.

As a result, a great many scientists believe that the universe is teaming with intelligent life forms. What divides us is whether or not they have visited the earth.

During World War II, when the scientists at the Manhattan Project were worrying about the progress of the German advance in Europe, they would often ask questions about the universe during lunch. Once, the conversation turned to aliens from space. Nobel Laureate Enrico Fermi, who kept an open mind about these things, would interrupt and ask, "But where are they?"

Fermi's question bedevils us even today. Like most scientists, I believe there is intelligent life in outer space. It's simply arrogant to believe that we are the only intelligent life form among billions of earth-like planets in the universe. However, the SETI (Search for Extraterrestrial Intelligence) Project has so far detected no sign of any intelligent life.

Astronomers have scanned for radio and TV emissions within 100 light-years of the earth and have found no sign of any signal from space indicating intelligent life. Our planet has been radiating electromagnetic radiation in the form of radio and TV for the past fifty years; as a result, there is a sphere surrounding the earth, 50 light-years in radius, expanding at the speed of light. The expanding sphere contains a vast representative cross section of the cultural achievements of the planet earth. Any planet within 50 light-years of earth should be able to detect our signals. (Although once they decipher some of our programs, they may question whether or not there is intelligent life on earth.)

It's puzzling that we do not detect any alien emissions. This does not stop people (including scientists) from speculating about what other beings might look like.

## Will They Look Like Us?

In countless TV documentaries, eyewitness accounts, sensational books, tabloid headlines, films of alien "autopsies," interviews with abductees, a consistent picture emerges of the "standard alien": small, frail, pale, with big eyes and a big head.

But if you look at the rich diversity of life on just this planet, we see that nature has created millions of possible body types that are far more imaginative than the rather conservative designs offered in science fiction, most of which are small variations of the human body type.

(The myth that alien life in the universe must look like humans was in part stimulated by the aliens seen in the movies. Perhaps the Screen Actors Guild contract demands that all aliens be played by union members!)

Most exobiologists believe that there are only a few basic criteria for intelligent life. If one examines how our species became intelligent, for example, we need only look at our hands. Our thumbs, which were originally used to grab tree branches, were key to our manipulating the environment. Those apes who were driven from the forest 5 million years ago who could adapt their tree-swinging thumbs to grasp tools survived and flourished, while those who could not perished. In other words, Man did not create tools. Tools created Man.

Our stereoscopic eyes are the eyes of a hunter. In general, animals with eyes to the side of their face are less intelligent than animals with eyes to the front of their face. This is because animals with eyes to the side are prey, like rabbits and deer, and have to keep a vigilant eye for the presence of predators, whereas animals with eyes to the front, like wolves, tigers, cats, and lions, use their stereoscopic eyes to home in on the prey.

Finally, we are a social species, for whom communication and culture are crucial. Language enables us to accumulate culture and science across hundreds of generations, giving us the wisdom and insights of people we never met.

When viewed from the perspective of exobiology, we can now summarize the few criteria for intelligent life in space:

First, some form of eyes. (This does not mean the two eyes we find on our face. It could mean multiple eyes, or even an entirely new sensory organ for gathering information about the environment.)

Second, some form of hand to manipulate the environment. (This does not mean two hands. There can be multiple hands, or even tentacles.)

Third, some form of language to accumulate knowledge and culture.

Notice that these three criteria give a tremendous amount of latitude to construct new intelligent life forms. The body shape that Hollywood associates with intelligent beings (e.g., bilateral symmetry, apelike head, neck, torso, arms, and legs) has very little to do with these three criteria. Even on the earth, one can imagine other life forms evolving these three attributes and gradually becoming intelligent. They, in turn, would not resemble us in any way. For example, whales are relatively intelligent, but they breathe through a hole in the top of their head.

The last question that scientists ask is: what will alien civilizations look like? To soar across hundreds of light-years of space, they must be hundreds if not thousands of years ahead of us technologically. Scientists who search for extraterrestrial life have taken this problem very seriously. By using the laws of physics to project how alien civilizations may derive their energy thousands of years into the future, we can obtain a better picture of how sophisticated such civilizations must be. And by using physics to determine the nature of civilizations thousands of years ahead of us in technology, we can begin to see our own future.

# 15

# Toward a Planetary Civilization

"Destiny is not a matter of chance—
it is a matter of choice.
It is not a thing to be waited for—
it is a thing to be achieved."
—WILLIAM JENNINGS BRYAN

IN THE PAST, scientific revolutions, such as the introduction of gunpowder, machines, steam power, electricity, and the atomic bomb all changed civilization beyond recognition. A question we should all ask ourselves is: how will the biomolecular, computer, and quantum revolutions similarly reshape the twenty-first century?

The pace of scientific discovery is already accelerating into the next century. The biomolecular revolution will give us a complete genetic description of all living things, giving us the possibility of becoming choreographers of life on earth. The computer revolution will give us computer power that is virtually free and unlimited, eventually placing artificial intelligence within reach. And the quantum revolution will give us new materials, new energy sources, and perhaps the ability to create new forms of matter.

In view of this, what might our civilization look like several centuries into the future on the basis of such rapid progress?

Of course, no one has a crystal ball to foresee how the future of civilization will unfold. However, there is one field of science in which this question is the focus of investigation.

Astrophysicists have actively explored what types of civilizations may exist far in the distant future, perhaps centuries or millennia beyond ours. Astrophysicists use the laws of physics to propose speculative guidelines for the analysis of extraterrestrial civilizations, which may serve as a model to guide our own thinking about the evolution of our planet for the next several thousand years.

Since the universe is roughly 15 billion years old, it is possible that there are civilizations in the galaxy which are literally millions of years ahead of ours. And with some 200 billion stars within our own Milky Way galaxy and trillions of galaxies within the visible universe, it is a distinct possibility that there are thousands of civilizations in space unimaginably ahead of ours in their science and technology.

To focus the seemingly hopeless search for extraterrestrial intelligence in outer space, astrophysicists have searched for life in space by analyzing characteristic energy signatures in space that may serve as a guide. Russian astronomer Nikolai Kardashev introduced convenient categories, which he called Type I, II, and III civilizations, to classify extraterrestrial civilizations, based on the natural progression of energy consumption.

Based purely on physical considerations, any civilization in outer space will rely successively on three main sources of energy: their planet, their star, and their galaxy, corresponding to Type I, II, and III civilizations, respectively. The energy output of each civilization is roughly 10 billion times larger than the previous one. But even that staggering number can be bridged by any modestly expanding civilization.

Assume, for the moment, that our own world economy grows at a rather anemic rate of 1 percent per year, which is very conservative. Since economic growth is fueled by increased consumption of energy, we would find a corresponding growth in energy as well. Within a hundred to a few hundred years, our world will approach a planetary Type I civilization.

At such a growth rate, the transition from a planetary Type I civilization to a stellar Type II civilization will take longer, perhaps 2,500 years. A more realistic growth rate of 2 percent per year would reduce that figure to 1,200 years. And a 3 percent annual growth rate would reduce that even further to 800 years.

Eventually, the energy needs of a Type II civilization will outgrow even the energy output of its star. It will be forced to go to nearby star systems in search of resources and energy, eventually transforming it into a galactic civilization.

The transition from Type II to Type III will take much longer, since that civilization must master interstellar travel. But one can assume that within a hundred thousand to a few million years (depending on its progress in developing interstellar travel), a stellar Type II civilization will make the transition to a galactic Type III civilization.

Where does this put us? On this cosmic scale, we are a Type 0 civilization—we derive our energy from dead plants (e.g., fossil fuels). We are like infants, just beginning to contemplate the vast universe of possible civilizations. Our civilization is so new that even a hundred years ago we still got most of our energy from burning wood and coal, and any discussion of extraterrestrial energy sources would have been considered madness.

## Dangers Faced by Type 0 Civilizations

Of these three transitions, perhaps the most perilous one is the transition from a Type 0 to a Type I civilization. Like a child learning how to walk, it suddenly becomes aware of new life-threatening dangers in its quest to explore and master its world. The more it learns about the universe around it, the more it learns of potential dangers, such as ice ages, meteor and comet impacts, supernova explosions, and environmental threats, such as the collapse of its atmosphere or the proliferation of nuclear weapons.

Furthermore, a Type 0 civilization is like a spoiled child, unable to control its self-destructive temper tantrums and outbursts. Its immature history is still haunted by the brutal sectarian, fundamentalist, nationalist, and racial hatreds of the past millennia. A Type 0 civilization is still split along deep fracture lines created thousands of years in the past.

The main danger faced by a Type 0 civilization occurs after its discovery of the chemical elements of the periodic chart. Inevitably, any intelligent civilization in the galaxy will discover two things: element 92 (uranium) and a chemical industry. With the discovery of uranium comes the possibility of annihilating themselves with nuclear weapons. With the creation of a chemical industry comes the possibility of polluting their environment with toxins and destroying their life-giving atmosphere.

Given the fact that astrophysicists do not see evidence of life in nearby star systems, even though Drake's equations predict the existence of thousands of intelligent civilizations in our galaxy, it is possible that our galaxy is filled with the ruins of Type 0 civilizations which either settled old grudges and jealousies via element 92 or else uncontrollably polluted their planet.

If these twin global disasters can be averted, then inevitably their science will rise to unlock the secret of life, artificial intelligence, and the atom, as they stumble upon the biomolecular, computer, and quantum revolutions, which will pave the way for their society to rise to the level of a planetary civilization. The computer revolution will link all their peoples with a powerful global telecommunications and economic network;

the biomolecular revolution will give them the knowledge to cure disease and feed their expanding population; and the quantum revolution will give them the power and materials to build a planetary society.

## Type I: A Planetary Civilization

By the time a civilization has reached Type I status, it has achieved a rare political stability. A Type I civilization is necessarily a planetary one. Only a planetary civilization can truly make the decisions that affect the planetary flow of energy and resources. A Type I civilization, for example, will derive much of its energy from planetary sources—i.e., from the oceans, the atmosphere, and from deep within its planet. It will modify its weather and mine its oceans, using planetary resources that are only a dream today.

Consider the modification of the weather. A simple hurricane can unleash more energy than a hundred hydrogen bombs. The manipulation of the weather today is still a distant possibility. But since the weather in one area intimately affects the weather in another, weather control or modification is possible only if we have cooperation on the part of many nations. Similarly, if one nation releases large quantities of greenhouse-producing or ozone-depleting gases, the entire planet is affected by it. Thus, to control the weather or eliminate planetary environmental threats, a Type I civilization necessarily must function with a high degree of cooperation among its peoples. This is what it means to be a planetary civilization.

A Type I civilization's energy consumption, compared with ours, is so large that, viewed from outer space, the planet will appear like a bright Christmas tree ornament. By contrast, our Type 0 planet, when photographed from outer space, has only faint filaments and dim patches of light corresponding to the large cities of the United States (mainly between Boston and Washington), Europe, and Japan (mainly around Tokyo).

Type I civilizations are still, however, quite vulnerable to astronomical and environmental catastrophes. Because of the difficulties of terraforming planets and the enormous distances separating them from the nearby stars, a civilization that has reached Type I status may spend many centuries inhabiting a single planet, which poses a risk to its long-term survival. It may send tiny exploratory parties to the planets and even nearby stars, setting up small outposts, but sustaining large, permanent colonies will strain its resources.

As time goes on, a Type I civilization will develop a planetary communication system, a planetary culture, and a planetary economy. There will be instantaneous communication linking society, which will tend to gradually erase long-standing cultural and national barriers which sometimes lead to war. The divisions and scars that afflict a Type 0 civilization will fade into history with the abundant material wealth and energy resources of a Type I society.

Assuming that the laws of biological evolution on a Type I planet are similar to ours, one can also conclude that their evolution will cease. Evolution tends to accelerate when there are isolated pockets of individuals and harsh environmental conditions. Within a small colony or tribe, small genetic differences due to inbreeding are gradually magnified, creating genetic "drift" within the same species. (In ancient times on the earth, for example, one could expect to marry someone from the same or neighboring tribe, with a total breeding population numbering less than a hundred. Today, breeding populations are usually in the millions.) In general, the larger the breeding population, the slower the rate of evolution.

Because a Type I civilization will no longer have isolated breeding populations, there will be a gradual mixing of peoples which will terminate their evolution as a species.

## Type II Civilizations: Invulnerable to Any Natural Disaster

By the time a civilization has reached Type II status, however, it will become immortal, enduring throughout the life of the universe. Nothing known in nature can physically destroy a Type II civilization. A Type II civilization has the ability to fend off scores of astronomical or ecological disasters by means of the power of its technology. Potentially disastrous meteor or comet impacts can be prevented by deflecting away any cosmic debris in space which threatens to hit its planet. On a scale of millennia, ice ages can be averted by modifying the weather—e.g., by controlling the jet stream near its polar caps or perhaps making micro-adjustments to the planet's spin

Because the planet's engines produce large amounts of heat, it requires a highly sophisticated waste management and recycling system. However, with centuries of experience in managing and recycling its wastes, it will not face catastrophes caused by the collapse of its environment.

Perhaps the greatest danger faced by a Type II civilization is posed by an eruption of a nearby supernova, whose sudden burst of deadly X-rays could fry nearby planets. But by monitoring its nearby stars, a Type II civilization will have centuries in which to build space arks capable of carrying its peoples to colonies on nearby solar systems if they detect that one of its nearby stars is dying.

A Type II civilization, by definition consuming 10 billion times more energy than a Type I civilization, will have exhausted planetary resources. Its energy requirements will be so large that it will derive energy directly from its own sun. It does this not by passively getting sunlight from solar collectors, but by actively sending giant spaceships to the sun to direct the sun's energy back to the home planet. (The United Federation of Planets featured on *Star Trek* is on the verge of attaining Type II status. They have had planetary government for centuries and are just at the stage where they can ignite dying stars.) Princeton physicist Freeman Dyson has speculated that a Type II civilization could build a huge sphere around its sun, thereby capturing all its energy and also sealing itself off from the rest of the universe.

Finding a Type II civilization in space may be a bit difficult since it may choose to conceal its radio and TV emissions. However, it cannot violate the Second Law of Thermodynamics. Specifically, there is no known way to prevent its machines from generating copious quantities of waste heat, which should be clearly visible by means of infrared detectors on the earth. As seen from outer space, a Type II civilization will have the infrared energy output of a small star. Since they are immortal, Type II civilizations may be quite common. Dyson has even advocated building special infrared detectors in hopes of spotting nearby Type II civilizations.

Type II civilizations would also produce large quantities of radio and TV signals. Scientists of the SETI Project, however, have conducted intense searches for electromagnetic signals from many stars, one frequency at a time, and have found nothing. Ironically, however, our galaxy may be teeming with Type II civilizations that have avoided detection by our radio telescopes—perhaps because, instead of broadcasting on one frequency, which is terribly inefficient, they have adopted the much more efficient method of scrambling their messages across the entire radio band and then unscrambling them at the receiving end. If we were to listen in on such scrambled messages, we would hear only gibberish, indistinguishable from noise. Thus, it is possible that the radio waves we detect in space may be full of Type II transmissions that we have not yet been able to recognize.

## Type III Civilizations: Conquering the Galaxy

The transition from a Type II to a Type III civilization will take more time, since its evolution depends on mastering interstellar travel, an extraordinarily difficult task. But if such civilizations have starships that can attain a substantial fraction of the speed of light, then colonizing other portions of the galaxy may well be possible.

Although Hollywood glamorizes heroic captains leading courageous teams of explorers to seek out extraterrestrial life and suitable planets to live on, this is perhaps the most inefficient way to explore the galaxy. The simplest way for a Type III civilization to map out promising star systems that can be colonized is to send thousands of "Von Neumann probes" into space, small robotic probes that land on the moons of distant star systems and build robotic self-replicating factories. Using the methane, ores, and other chemicals extracted from the atmosphere and soil, the robotic factories would be able to build thousands of replicas of itself, which will then blast off into deep space in search of even more star systems. The process can repeat itself endlessly, with each cycle multiplying the number of Von Neumann probes by a factor of thousands.

In this way, millions of star systems could be analyzed in the shortest possible time. (Von Neumann probes, in fact, were the basis for the monoliths seen in the movies *2001* and *2010.)* Freeman Dyson even envisions such lightweight probes as the product of bioengineering and artificial intelligence, capable of "eating" the methane on distant moons. (The probe would land on moons, rather than planets, because it is easier to escape the weaker gravitational field of a moon. From the vantage point of a moon, these probes can detect whether there are any signs of intelligent life on the planets.) He calls these probes "astro-chickens": small, compact, genetically engineered creatures capable of spaceflight which can thrive and sustain themselves on the hostile environment of faraway moons and send messages back to the home planet.

The ultimate merger of artificial intelligence and biotechnology may eventually produce an ideal Von Neumann probe. Such an advanced probe would be a living creature in every sense of the word, able to repair damage to itself, find "food" in the frozen surfaces of distant moons, and also produce thousands of "children" to continue the exploration of the galaxy. It would carry out all the functions of a living creature. It would also have a high degree of artificial intelligence, able to carry out its primary mission (to explore other star systems) and make independent decisions on its own consistent with its overall mission. It would also have emotions to help it function in outer space. It would feel "pain" and hence avoid danger, experience "pleasure" as it refueled on a distant moon, feel "maternal" toward its young progeny, and feel "joy" and a sense of accomplishment from carrying out its primary mission.

If a Type III civilization sends such probes at half the speed of light, it could then wait as signals came flooding in concerning interesting star systems. Within a thousand years, every star system within 500 light-years could be mapped by these Von Neumann probes. Within a hundred thousand years, it could explore all the stars in fully half of its galaxy. Because these Von Neumann probes are so efficient, a Type III civilization can very quickly determine which star systems are suitable for colonization.

There has been speculation by some scientists about whether a Type III civilization exists within our own galaxy. Being immortal, such a civilization may already have explored large portions of our galaxy and left behind Von Neumann probes, as in *2001.* Another theory holds that a Type III civilization, being thousands of years ahead of us in technology, may simply be uninterested in us. After all, when we see an anthill, do we bend down and offer the ants trinkets, medicine, knowledge, and science? On the contrary, some might have the urge to step on a few of them.

Even more ambitious would be for an advanced civilization to harness the "Planck energy," the energy necessary to tear the fabric of space and time. Although this fabulous energy scale seems hopelessly beyond the capabilities of our Type 0 civilization, it is well within the scope of a mature Type I or higher civilization, which according to our previous assumption possesses roughly 100 billion to a billion trillion times the energy output of our Type 0 civilization.

For a civilization with such a cosmic energy output, it may be possible to open up holes in space (assuming that these wormholes do not violate the laws of quantum physics). This may provide perhaps the most efficient way of reaching out to the stars to create a galactic civilization, using dimensional windows rather than clumsy starships to explore unseen worlds.

## Toward a Planetary Civilization

On earth we are still a Type 0 civilization: we are still hopelessly fractured into bickering, jealous nations and deeply split along racial, religious, and national lines. Mining the oceans or manipulating the weather is out of the question when we can barely send feeble space probes to nearby planets and can't even take care of our own food and energy needs. At present, the world is experiencing two conflicting trends. It is both becoming increasingly fragmented, as civil and ethnic wars and national interests dominate many parts of the world, and becoming increasingly unified, with new levels of cooperation between nations on a global scale and the emergence of common trading partnerships, such as the European Union.

To see which trend will ultimately dominate, think ahead to the world a hundred years from now. With some Asian nations achieving spectacular annual growth rates of 10 percent, it is not unrealistic to assume that the world growth rate for the next century may average a bit below 5 percent, as the Third World becomes increasingly industrialized. At that rate, in a century the gross world product and world energy consumption of the planet will grow by a factor of 130 times.

The economic, technical, and scientific achievements of a century from now may dwarf anything which is conceivable at present by a factor of over a hundred. Entire regions of the world, many of which are pockets of wretched poverty today, will be industrialized by that time. Much of this wealth, of course, will not be distributed evenly, but the passions and hatreds that fired up the nationalism and sectarianism of the past may gradually subside as people become wealthier and have a larger stake in the system. It is hard for firebrands to light the torch of separatism and fragmentation when the people are well fed and content. As one wag once noted: "There is no such thing as a fat nationalist."

By the late twenty-first century, there will also be enormous social, political, and economic pressures to forge a planetary civilization generated by the global economy. Of course, there will always be ruling elites trying to jealously protect their influence and power. For many decades beyond the end of the twenty-first century, they may try to resist the global trends that are creating a Type I civilization on the earth. However, every decade their power will diminish because of the enormous social and economic forces unleashed by these scientific revolutions.

## Planetary Collapse

One of the forces driving us toward a planetary civilization is the fear of planetary collapse, exemplified by the possible disintegration of the ozone layer, which has overcome governmental inertia and national rivalries and galvanized the United Nations. Ozone is a thin life-protecting layer fifteen miles above the earth's surface which absorbs harmful ultraviolet radiation. The discovery in 1982 that there was a huge hole in the ozone layer opening up over the South Pole, about the size of the United States, caused great international concern. Satellite data confirmed that ozone levels were also dropping ominously over the Northern Hemisphere by almost 1 percent per year in some areas. If ozone depletion is not reversed, there could be 60 million more skin cancers by 2075, not to mention the withering of important food crops and the deaths of animals crucial to the food chain.

When scientists finally proved that chlorofluorocarbons (CFCs), commonly used as a refrigerant, posed a clear and immediate danger to the ozone layer, thirty-one nations rapidly banded together and signed a historic agreement in Montreal in 1987 to begin phasing out our CFCs by the year 2000.

The Montreal Protocols on ozone depletion and the historic UNsponsored Earth Summit in 1992 in Rio de Janeiro were watershed events focusing international attention on the issues of biodiversity, pollution, overpopulation, etc. At present, there are now over 170 international treaties in force guarding different aspects of the environment.

The threat of atmospheric collapse, however, does not always generate planetary cooperation. On the contrary, although scientists almost universally believe that the greenhouse effect could raise global temperatures to dangerous levels in the next century, nations have dragged their feet on this question. The threat of global warming was clearly laid out in the UN's 1995 Intergovernmental Panel on Climate Change (IPCC), representing the authority of 2,000 scientists around the world. The report presented a grim tale of planetary collapse in the next century if carbon dioxide levels continue to rise: one-third to one-half of the world's mountain glaciers could melt, one-third of all ecosystems could be radically disrupted, sea levels could rise 15 to 90 centimeters by 2100, 92 million people in coastal areas like Bangladesh would be at risk, millions could die as malaria and other deadly tropical diseases spread, starvation could be widespread as growing areas turn into dust bowls and desert.

Because global warming is driven by fossil fuel consumption, and since many nations are heavily dependent on oil and coal, carbon dioxide levels (already the highest in 150,000 years) will continue their steady rise into the next century. In the decades ahead, when global warming begins to visibly disrupt the world's weather and ecology, the reluctant nations of the world may finally become frightened enough to take action, including levying a "carbon tax" or phasing out oil and coal burning. The threat of planetary collapse will inevitably forge international cooperation, even if done reluctantly.

## Population Explosion vs. Diminishing Resources

One of the most pressing long-range global problems, both environmentally and socially, is the human population explosion, which puts a tremendous strain on the planet's resources. It took several million years to reach a world population of a billion people, which happened about 1830. The next billion was added only a century later. The population doubled again by 1975 to 4 billion. In the twenty years that followed, the world population has soared to 5.7 billion people. Every year, we add 90 million more people to the planet. One twentieth of all the humans who have ever walked the earth are alive today.

This unprecedented population explosion places enormous stress on the food supply, the ecosystem, and biodiversity. According to the World Watch Institute, in 1997 worldwide harvesting of fish peaked at 100 million tons per year. Similarly, world grain production is peaking at around 1.7 billion tons per year. Meanwhile, the total area of deforestation is equal to the size of the continental United States. This means fewer croplands for growing food and the extinction of entire species of plants and animals. Some biologists estimate that we might lose a million species by the end of the century, and as many as a quarter of all species on the earth by the middle of the twenty-first century.

Biologist Robert W. Kates emphasizes that historically there have actually been three waves of population explosions, all of them coinciding with the introduction of new technology and science. The first population explosion began about a million years ago when humans discovered tool making, triggering an increase in world population from a few hundred thousand individuals to 5 million. The second revolution, which started about 10,000 years ago, came with the discovery of agriculture and the domestication of animals and plants. This time, the population grew a hundredfold, to about 500 million. The third population explosion started several hundred years ago with the industrial revolution.

The question is: can the world continue to feed its people when the population is still galloping ahead at a rapid pace?

This tremendous rise in population may one day come to an end. The UN estimates that the world's population will gradually slow down, reaching 6 billion in 1999, 7 billion in 2011, 8 billion in 2025, 9 billion in 2041, and 10 billion in 2071. It may even ultimately level off at around 12 billion in the twenty-second century.

The reason for this is that every industrialized nation has stabilized its population. In fact, Japan and Germany are even experiencing negative population growth. Every industrialized country experiences a rapid increase in population as modern medicine and sanitation reduce the death rate, then a stabilization of the population growth as it industrializes. At present, thirty nations (representing 820 million people) have reached a stable population.

As biologist Kates says: "Development is the best contraceptive." This is because economic development, he notes, "lessens the need or desire for more children because more children survive, which decreases the need for child labor and increases the need for educated children. Development also cuts the time available for childbearing and rearing and creates more opportunity for women to gain an education and find salaried work. Finally, it improves access to birth control technology." (Ironically, the elderly are the fastest-growing sector in the industrialized world, while the young are the fastest-growing sector in the Third World.)

Clearly our Type 0 civilization faces enormous environmental dangers into the next century. Nations that have historically resisted cooperating with each other may be forced to address these global issues and work cooperatively.

## Reversing the Great Diaspora

Culture has been both a gift and a curse to humanity. For 99 percent of human existence, we lived in small primitive, nomadic tribes that could economically support perhaps no more than fifty or so individuals. (Studies have shown that when a tribe expands beyond roughly this number, it cannot support and feed all the additional members and it will split up.) What held these tribes together was culture—the rituals, customs, and language that provided protection and support on the part of friends and relatives. The heroic tales and myths told around ancient campfires cemented the bonds within the tribe. This also gave us the current diversity of humanity and the rich mosaic of thousands of languages, religions, and customs of today.

But culture was also a curse. Many of these mythologies enforced an "us" versus "them" ideology that caused fierce rivalries and tribal wars between these nomadic cultures.

About 100,000 years ago, soon after modern *Homo sapiens* emerged in Africa, the Great Diaspora began, when these small wandering tribes began to spread out from Africa, probably due to changing climatic conditions. Perhaps no more than a few thousand individuals ventured north, eventually settling in the Middle East and Southern Europe. About 50,000 years ago, a second split sent a splinter group into Asia and eventually even the Americas. But because of the genetic isolation created by this Great Diaspora, humanity, adapting to the harsh environmental conditions, began to separate into the races we see today. Thus, not only were these tribes separated by culture, they were now separated by race as well.

In the next century, however, the current scientific revolutions are unleashing forces which, for the first time in 100,000 years, are beginning to erode the forces maintaining the Great Diaspora. The ancient, centrifugal tendencies which enforced the Great Diaspora are gradually evaporating. We see this tendency toward a planetary civilization on several fronts: the rise of a global economy, the decline of nations, the rise of an international middle class, the development of a common global language, and the rise of a planetary culture.

## The End of the Era of Nations?

The greatest obstacle to a planetary civilization is the obvious fact that political power resides with jealous nations. Clearly, we live in the era of nations. Furthermore, the reign of nations will continue for most of the twenty-first century. However, we sometimes forget that nations are a relatively new phenomenon on the historical stage, riding mainly on the coattails of the industrial revolution and the rise of capitalism, and that nations are not an eternal concept.

Before the industrial revolution, power rested mainly with feudal principalities. The power of nations was quite minimal, often reflecting more the political ambitions of monarchies and the imagination of mapmakers than the power of a functioning political unit. Alvin Toffler writes: "Even the greatest of emperors typically ruled over a patchwork of tiny, locally governed communities." Before the industrial revolution, writer S. E. Finer traveled from town to town within the same "nation" and observed that he had to change laws as frequently as he changed horses. "Kings and princes held power in bits and blobs," he concluded. Kings and princes could rhapsodize about great national destinies and fortunes, but the laws and customs in each town were largely controlled locally.

Germany did not exist in its modern form until the late nineteenth century, when Otto von Bismarck, the "Iron Chancellor," carved out the modern German state in 1871 from about 350 quarreling German principalities and Prussia. The feuding city-states of Italy, immortalized by Machiavelli in the devilishly candid political tract *The Prince*, put aside their centuries-old bickering only in 1870.

Similarly, the current nations of the Third World are a recent phenomenon. One of the reasons why there is so much strife and turmoil in the Middle East and Africa is that they were carved out by the Great Powers, especially Britain and France. More often than not, the Great Powers sliced up the various regions of the world according to a divide-and-conquer strategy, which made them easier to govern from London or Paris. Unfortunately, these artificial political boundaries, which were often chosen to increase the bickering between ethnic groups, are now the source of tremendous political instability in those parts of the world.

But although we are still in the thick of the era of nations, it is also possible to see how this era will end. Commercial bonds are becoming global in nature. National boundaries are giving way to economic forces, much the way that feudal principalities gave way to nations with the coming of the industrial revolution.

Kenichi Ohmae, author of *The End of the Nation State*, a former senior partner of McKinsey & Company and a consultant for international financiers, writes: "Traditional nation states have become unnatural, even impossible, business units in a global economy." He foresees that the twilight of nations is coming because of the enormous economic pressures being brought to bear by the expanding global economy.

"Nation states are political organisms, and in their economic blood streams cholesterol steadily builds up. Over time, arteries harden and the organism's vitality decays," he writes.

His views are not unique, but are echoed by many other political writers. French political theorist Denis de Rougement says, "The nation state, which regards itself as absolutely sovereign, is obviously too small to play a real role at the global level. . . . Not one of our 28 European states can any longer by itself assure its military defense and prosperity, its technological resources . . ."

Toffler adds: "We are moving towards a world system composed of units densely interrelated like the neurons in a brain rather than organized like the departments of a bureaucracy," he writes. Others, like Harvard economist and diplomat John Kenneth Galbraith, see the potential rise of a world government of some sort, replacing the anemic United Nations of today.

As John Lennon said in his song "Imagine," perhaps it's not hard to imagine a world without nations.

But in addition to the rise of a global economy and the weakening of nations, there is another, equally powerful force that is pushing for stability and a planetary civilization, and this is the rise of the international middle class.

## The Rise of the Middle Class

Throughout most of recorded history, tiny political elites have ruled, often brutally, over a large mass of impoverished people. Only the elites had the education, the knowledge, the wealth, and the military might to hold on to power.

It is a truism that political ruling elites act mainly to perpetuate their own political power, and a planetary civilization does not meet that criterion. Although the present elites and rulers will resist this tendency toward unification with all their formidable power, there is another engine that is driving the world in the direction of unification, and this is a relatively new but perhaps potentially profound power: the international middle class. With the rise of an international middle class, the power of these ruling elites is being diluted.

Most of the human race lives in the Third World, which is finally undergoing massive industrialization, 300 years after Europe. And just as the industrialization of Europe eventually toppled the monarchies and empires of the old order, the industrialization of the Third World is creating a new middle class that will be an engine of social change.

The middle class, being selfish like all other classes, has a stake in preserving harmony and promoting international trade and the free flow of information. Moreover, armed with fax machines, the Internet, satellite dishes, and cellular phones, the international middle class is able to mount formidable political campaigns as well. When people taste a bit of affluence, they want more.

Ohmae makes the point that the growing middle class of the world has rising expectations and wants. He feels that when people achieve an income of roughly $5,000 per capita, or $20,000 for a family of four, there is a subtle but far-reaching psychological change. When they no longer have to ask where their next meal comes from, or what diseases might kill the next family member, people begin to demand consumer goods, which in turn introduces them to the universe of luxury items flowing in from industrialized countries. More important, with a certain level of affluence and stability, "people will inevitably start to look around them and ask why they cannot have what others have. Equally important, they will start to ask why they were not able to have it in the past."

## De Facto Planetary Language and Culture

At present, there are approximately 6,000 languages spoken on the planet, reflecting the deep historic divisions created by the Great Diaspora. Within the next century, however, 90 percent or more of these languages could disappear, according to Michael E. Krauss, director of the University of Alaska's Native Language Center. He believes only 250 to 600 languages will survive.

Of these, English has already emerged as the lingua franca for business and science. Not only are most business and scientific meetings, conferences, and transactions conducted in English, but it is also the language of the Internet, which is unifying at least 30 million computer users. Already, any ambitious person wishing to participate in the global economy or the sciences must speak English, which serves as a common denominator for all global human activities.

As columnist William Safire says: "English will be the world's first second language in A.D. 2100." He estimates that a billion people speak English in the world today. That number will climb rapidly in the next century.

Ironically, the computer revolution will both help to preserve and to destroy the ancient languages of the Great Diaspora. On the one hand, the rise of English as the dominant planetary language is accelerating due to the establishment of global telecommunications, international travel, business, and science. But the computer revolution will also help to preserve many of the smaller endangered languages, some of which are spoken by only a handful of aging elders. Many of these languages will survive in the form of tape and dictionaries which are stored on computer.

This common language is also facilitating the rise of a common planetary culture. One of the first people in this century who saw that the seeds for a planetary culture were germinating because of the telecommunications revolution was Aldous Huxley, who wrote in *Those Barren Leaves:* "Cheap printing, wireless telephones, trains, motor cars, gramophones and all the rest are making it possible to consolidate tribes, not of a few thousands, but of millions."

This trend is accelerating due to the Internet. Bill Gates claims, "The information highway is going to break down boundaries and may promote a world culture or at least a sharing of cultural activities and values. . . . I think people want a sense of belonging to many communities including a world community."

News and telecommunications are being unified with CNN, Sky Television, and numerous other news channels, which are beamed into the most isolated parts of the world. Even harsh theocracies, like the one in Iran, cannot fully stop the proliferation of satellite dishes, which are growing like weeds in the Third World, and which bring new ideas and subversive scenes of affluence in other lands.

This trend toward a common planetary civilization may not be aesthetically pleasing to everyone. For example, among the United States' great exports are movies and rock and roll. Films starring Arnold Schwarzenegger are immensely popular because they are action-oriented and easily understood by different cultures. Rock and roll has found a ready-made international audience among the increasingly affluent and rebellious teenagers around the world.

Although not everyone may find a planetary culture to their taste, it may have the salutary effect of breaking down cultural barriers, as we move toward a Type I civilization.

# 16

# Masters of Space and Time

"The hardest thing to understand is why we can understand anything at all."

—ALBERT EINSTEIN

"The Bible tells you how to go to heaven, not how the heavens go."

—Pope JOHN PAUL II

MATTER. LIFE. THE MIND.

As we've seen, these three pillars of modern science are no longer shrouded in mystery, for the basic laws of the quantum theory, DNA, and computers have been discovered in the twentieth century. In the twenty-first century, however, we will learn how to manipulate these three almost at will, making the transition from being observers of the dance of nature to becoming active choreographers. We will also witness the intense cross-pollination among these three, which will typify science of the twenty-first century.

But our description of the future of science is still incomplete without a fourth ingredient that makes up our understanding of the universe: space-time. Recently, there has been intense activity by physicists in explaining the secret of space and time. Ultimately, what we learn may answer some of our deepest questions about the fabric of space and time, such as whether space can be torn, whether time can be reversed, and how the universe was born and will eventually die. The study of space-time may ultimately answer one of the most intriguing questions about the future: the final destiny of all intelligent life in the universe.

## The Fourth Pillar: Space-Time

In the movie *Star Trek: First Contact* and in Isaac Asimov's *Foundation* trilogy, the discovery of warp drive marks a watershed in galactic history, the "coming of age" of humanity, marking the transition from planetary isolation and ignorance to joining the galactic fraternity of planets. In such a grand vision, we are truly children of the stars, by birth and by destiny, ready to take our rightful place among the civilizations dwelling on far-flung star systems.

Is warp drive possible? Is it compatible with the laws of physics, or is it just a figment of the imagination of science fiction writers?

Interestingly enough, the answer to these questions takes us to the outer limits of physical knowledge. While achieving warp drive may not be possible in the twenty-first century, it is not out of the question in terms of our understanding of physics, especially for Type I or Type II civilizations.

In the previous chapters, we have seen how reasonable predictions could be made about the future of science because the basic laws of the quantum theory, computers, and DNA have largely been discovered. However, to construct a time machine or develop warp drive, we have to push our knowledge of physics to the very limit. Ultimately, we need a "theory of everything" to explain whether it is possible to bend time into a pretzel or punch a hole in space. The physics of warp drives will take us on a strange but fascinating odyssey through curved space, parallel universes, and the tenth dimension.

Developing a warp drive may be consistent with the laws of physics *if* one can open up a "wormhole" or hole in space. To visualize a wormhole, take a sheet of paper and mark two points, A and B. Usually, we are told that the shortest distance between two points is a straight line. But that is only true in two dimensions (i.e., on a flat sheet of paper). If we bend the sheet of paper so that points A and B touch and drill a hole connecting A and B, then the shortest distance between A and B is actually a wormhole.

Similarly, we can take two parallel sheets of paper and place one on top of the other. We can mark point A on one sheet and B on the other. Then we can join points A and B by drilling a hole through both of them, thereby connecting the two parallel sheets via a wormhole. An ant that has been crawling on one sheet of paper may accidentally fall through the wormhole and wind up on an entirely new sheet. Suddenly, the confused ant finds that an entirely new universe has opened up because it went through the hole.

Mathematicians call these strange configurations "multiply connected spaces," which give rise to all sorts of delicious paradoxes that violate our common-sense notions about space. Although multiply connected spaces sound bizarre, they are perfectly logical if our universe is connected via wormholes.

Wormholes were first introduced to the public over a century ago in a book written by an Oxford mathematician. Perhaps realizing that adults might frown on the idea of multiply connected spaces, he wrote the book under a pseudonym and wrote it for children. His name was Charles Dodgson, his pseudonym was Lewis Carroll, and the book was *Through the Looking Glass.*

The Looking Glass is in fact a wormhole. On one side is the bucolic countryside of Oxford, England. On the other side is Wonderland. By entering the Looking Glass, one smoothly leaves one universe and enters another via the wormhole. Like Siamese twins joined at the hip, Oxford and Wonderland are joined together by the Looking Glass.

Using such a wormhole, we can, in theory, leap across light-years of space and go "faster than the speed of light" without violating relativity. Notice that our velocity is quite small as we enter the Looking Glass. At no point is our body exceeding the speed of light. But the net effect is to go much faster than the speed of light if we measure the absolute distance that we have traveled.

Unfortunately, Lewis Carroll was a mathematician, not a physicist, and hence did not know if his creation was possible. All this changed, however, when Einstein wrote down his general theory of relativity in 1915. Included within general relativity is the possibility of building wormholes and even time machines.

General relativity is based on the idea that space is curved, and that the "forces" we see around us, like gravity, are actually an illusion created by the bending of space and time.

For example, if we place a heavy rock on a bed, the rock will sink into the mattress. If we shoot a small marble along the surface of the bed, then it will execute a curved path around the rock. From a distance, it looks as if the rock has exerted a mysterious "force" on the marble, moving it into an orbit. Similarly, Newton believed that a mysterious "force" called gravity is acting on the earth. (Newton himself understood the problem with this picture, since nothing is touching the earth, yet it moves. In his writings, Newton was deeply bothered by the fact that the earth could move without anything touching it and considered this to be a major blemish in his theory.)

What is really happening is that the marble is being pushed—i.e., by the bedsheet. Similarly, what is pushing the earth in its orbit around the sun is space itself. Thus, Einstein was led to believe that gravity was determined by the geometry of space-time—i.e., that gravity was an illusion caused by the bending of the fabric of space-time.

In other words, the reason why we can keep our feet firmly planted on the ground, rather than being flung into outer space, is that the earth is warping the four-dimensional space-time continuum around our bodies.

(More precisely, general relativity is based on the equivalence principle, i.e., that the laws of physics are locally the same in either a gravitating or an accelerating frame. Since light bends in an accelerating frame, it must also bend inside a gravitating frame as well. But since light sweeps out the geometry of space as it moves, according to the laws of optics, the geometry of space must also be curved due to gravity. Thus, gravity can be viewed as the result of the curvature of space.)

## Black Holes and Wormholes

But perhaps the most interesting distortion of space and time is found in a black hole. By definition, a black hole is an object so heavy that light itself cannot escape its enormous gravitational pull. Since light speed is the ultimate velocity, this means that nothing can escape a black hole once it has fallen in, including light. About a dozen black holes have now been seen in outer space, detected either by the Hubble Space Telescope or by ground-based radio telescopes. These black holes are found in the center of massive galaxies roughly 50 million light-years from earth, including the galaxies M-87 and NGC-4258.

Since black holes are invisible by definition, astrophysicists identify them in space using indirect means. Our instruments reveal a vast cosmic hurricane, an angry maelstrom of hot gases swirling about a tiny nucleus. Instruments from the space telescope have clocked winds circulating near the black hole traveling at a million miles per hour. At the very center lies a tiny dot of light, about a light-year across, which weighs as much as a million stars. Within this center lies the invisible spinning black hole.

By using Newton's laws of motion, one can calculate the rough mass of the spinning object, and hence its escape velocity. We find that its escape velocity equals the speed of light, so not even light can escape its gravitational field.

But perhaps the most revealing picture of black holes comes from Einstein's theory of curved space. In our previous analogy, if the rock becomes heavier and heavier, eventually it will sink farther into the bedsheet, so that the fabric will resemble the neck of a long funnel. If the neck becomes long enough, as in a black hole, it will eventually make contact with another funnel coming from a parallel universe, which is another universe on the other side of the black hole. This may open up a white hole in the other universe. The final configuration looks like two parallel sheets connected in the middle by a cylindrical tunnel.

This bridge connecting two parallel universes is called the Einstein-Rosen bridge. Einstein himself was not worried by the bridge, because anyone foolish enough to fall into a black hole would be crushed at the center, where the curvature and gravitational pull became infinite. Thus, when Einstein died, he believed that communication between the two universes was impossible.

But in 1963 mathematician Roy Kerr found the first realistic description of a rotating black hole. Instead of collapsing to a point, as in a stationary black hole, it collapsed to a rapidly rotating ring of neutrons. The fact that the black hole is spinning is crucial: the centrifugal force keeps the ring from collapsing into a point.

Kerr proved that anyone falling through the ring would not die, as commonly thought, but would actually fall through the ring into another, parallel universe. Touching the spinning ring is suicide, but falling through the ring is not. In other words, the wormhole is a spinning ring of neutrons, corresponding to the frame of the Looking Glass. Putting your hand through the Looking Glass is analogous to putting your hand through the rotating Kerr black hole, which is now a wormhole linking your universe to a parallel universe.

Since that time, literally hundreds of wormhole configurations have been discovered by physicists. In fact, it is now a relatively simple matter to embed a wormhole into a physically relevant universe.

## Time Travel

Not only was Einstein aware of the strange behavior of the Einstein-Rosen bridge, he also realized that his equations allowed for time travel. Because space and time are so intricately related, any wormhole that connects two distant regions of space can also connect two time eras.

To understand time travel, consider first that Newton thought that time was like an arrow. Once fired, it traveled in a straight line, never deviating from its path. Time never strayed in its uniform march throughout the heavens. One second on the earth equaled one second on the moon or Mars.

However, Einstein introduced the idea that time was more like a river. It meanders through the universe, speeding up and slowing as it encounters the gravitational field of a passing star or planet. One second on the earth is different from one second on the moon or Mars. (In fact, a clock on the moon beats slightly faster than a clock on earth.)

The new wrinkle on all of this which is generating intense interest is that the river of time can have whirlpools that close in on themselves or can fork into two rivers.

In 1949, for example, mathematician Kurt Gödel, Einstein's colleague at the Institute for Advanced Study at Princeton, showed that if the universe were filled with a rotating fluid or gas, then anyone walking in such a universe could eventually come back to the original spot, but displaced backward in time. Time travel in the Gödel universe would be a fact of life.

Einstein was deeply troubled by the Einstein-Rosen bridge and the Gödel time machine, for it meant that there may be a flaw in his theory of gravity. Finally, he concluded that both could be eliminated on physical grounds—i.e., anyone falling into the Einstein-Rosen bridge would be killed, and the universe does not rotate, it expands, as in the Big Bang theory. Mathematically, wormholes and time machines were perfectly consistent. But physically, they were impossible.

However, after Einstein's death, so many solutions of Einstein's equations have been discovered that allow for time machines and wormholes that physicists are now taking them seriously. In addition to the rotating universe of Gödel and the spinning black hole of Kerr, other configurations that allow for time travel include an infinite rotating cylinder, colliding cosmic strings, and negative energy.

Time machines, of course, pose all sorts of delicate issues involving cause and effect—i.e., time paradoxes. For example, if a hunter goes back in time to hunt dinosaurs and accidentally steps on a rodentlike creature who happens to be the direct ancestor of all humans, does the hunter disappear? If you go back in time and shoot your parents before you are born, your existence is an impossibility.

Another paradox occurs when you fulfill your past. Let's say that you are a young inventor struggling to build a time machine. Suddenly, an elderly man appears before you and offers the secret of time travel. He gives you the blueprints for a time machine on one condition: that when you become old, you will go back in time and give yourself the secret of time travel. Then the question is: where did the secret of time travel come from?

The answer to all of these paradoxes ultimately may lie in the quantum theory.

**Problems with Wormholes and Time Machines**

Although time machines and wormholes are allowed by Einstein's theory, this does not mean that they can be built. Several major hurdles would have to be crossed to build such a device.

First, the energy scale at which these space-time anomalies can occur is far beyond anything attainable on earth. The amount of energy is on the order of the Planck energy, or $10^{19}$ billion electron volts, roughly a quadrillion times the energy of the now-canceled Superconducting Supercollider. In other words, wormholes and time machines might be built by advanced Type I or more likely Type II civilizations, which can manipulate energy billions of times larger than what we can generate today.

(Thinking about this, I could imagine how Newton must have felt three centuries ago. He could calculate how fast you had to leap to reach the moon. One had to attain an escape velocity of 25,000 miles per hour. But what kind of vehicles did Newton have back in the 1600s? Horses and carriages. Such a velocity must have seemed beyond imagination. The situation is similar today. We physicists can calculate that all these distortions of space and time occur if you attain the Planck energy. But what do we have today? "Horses and carriages" called hydrogen bombs and rockets, far too puny to reach the Planck energy.)

Another possibility is to use "negative matter" (which is different from antimatter). This strange form of matter has never been seen. If enough negative matter could be concentrated in one place, then conceivably one might be able to open up a hole in space. Traditionally, negative energy and negative mass were thought to be physically possible. But recently the quantum theory has shown that negative energy is, in fact, possible. The quantum theory states that if we take two parallel uncharged metal plates separated by a space, the vacuum between them is not empty, but is actually frothing with virtual electron-antielectron annihilations. The net effect of all this quantum activity in the vacuum is to create the "Casimir effect"—i.e., a net attraction between these uncharged plates. Such an attraction has been experimentally measured. If one can somehow magnify the Casimir effect, then one can conceivably create a crude time machine.

In one proposal, a wormhole could connect two sets of Casimir plates. If someone were to fall between one set of Casimir plates, he would be instantly transported to the other set. If the plates were displaced in space, then the system could be used as a warp drive system. If the plates were displaced in time, then the system would act as a time machine.

But the last hurdle faced by these theories is perhaps the most important: they may not be physically stable. It is believed by some physicists that quantum forces acting on the wormhole may destabilize it, so that the opening closes up. Or the radiation coming from the wormhole as we enter it may be so great that it either kills us or closes up the wormhole. The problem is that Einstein's equations become useless at the instant when we enter the wormhole. Quantum effects overwhelm gravity.

To resolve this delicate question of quantum corrections to the wormhole takes us to an entirely new realm. Ultimately, solving the problem of warp drive, time machines, and quantum gravity may involve solving the "theory of everything." So in order to determine whether wormholes are really stable, and to resolve the paradoxes of time machines, one must factor in the quantum theory. This requires an understanding of the four fundamental forces.

## Toward a Theory of Everything

One of the great achievements of modern science has been the identification of the four fundamental forces of nature: the gravitational force (which holds the solar system and galaxy together), the electromagnetic force (which includes light, radar, radio, TV, microwaves, and so on), the weak nuclear force (which governs the radioactive decay of the elements), and the strong nuclear force (which makes the sun and the stars shine throughout the universe).

Our understanding of the basic equations underlying each of these forces has made possible many of the predictions in this book. In fact, the equations describing each of the four forces could be written in small script on just one of these pages. Amazingly enough, all physical knowledge at a fundamental level can be derived from this one sheet of paper.

But the crowning achievement of the past 2,000 years of science would be a "theory of everything," which would summarize these four forces in a single, coherent equation perhaps no more than one inch long. Einstein spent the last thirty years of his life futilely searching for this fabled theory. He was the first to point the way toward unification, but was ultimately unsuccessful.

Not only would a theory of everything be philosophically and aesthetically pleasing, finally tying up all the loose ends in physics into a unified whole, but it would also be of immense help in resolving some of the thorniest problems in physics, such as whether wormholes exist, whether time machines are possible, what happens at the center of black holes, and where the Big Bang came from. As cosmologist Steven Hawking has stated, this theory would give us the ability to "read the mind of God."

A unified theory would also clarify the energy scale and the dynamics at which one might be able to manipulate the power of space and time. A mature Type I or Type II civilization, commanding energy on a scale billions of times greater than that found on earth, might be on the threshold of becoming masters of space and time. We can only dream of punching holes in space or leaping into the tenth dimension. For an advanced civilization in space, these might be commonplace.

A theory of everything would also answer the time paradoxes mentioned earlier. If the quantum fluctuations around a wormhole can be controlled and stabilized, then it may be possible to go back in time and change the past. However, at that point another quantum universe opens up, and time "forks" into two rivers, each one leading to a new universe. For example, if we go back in time to save Abraham Lincoln at the Ford Theater, then in one universe Lincoln is saved and the direction of time is altered. However, the universe you came from is unchanged. Your past cannot be altered. You have merely saved the life of a quantum double of Lincoln in a quantum parallel universe. In this way, all the paradoxes found in science fiction can be answered by the quantum theory.

## Two Polar Opposites

Einstein once said, "Nature shows us only the tail of the lion. But I do not doubt that the lion belongs to it even though he cannot at once reveal himself because of his enormous size."

By the tail of the lion, Einstein meant the universe as we perceive it. The lion itself was the fabled unified field theory, which we puny humans cannot yet see in all its majesty.

At present, the tail of the lion is represented by two theories, the quantum theory (which describes the electromagnetic, weak, and strong forces) and general relativity (which describes gravity). Some of the greatest minds in twentieth-century physics, such as Einstein, Werner Heisenberg, and Wolfgang Pauli, have struggled to create a unified field theory, and have failed. These two theories are based on entirely different assumptions, different equations, and different physical pictures.

For the past fifty years there has been a cold war between general relativity and quantum theory; each has developed independently of the other, and each has had unparalleled success as long as it has stayed within its domain. However, the two theories must necessarily collide at the instant of the Big Bang, when gravitational forces and temperatures were so great that even particles would have been ripped apart, or at the center of black holes. At these energies, Einstein's theory becomes useless and the quantum theory takes over. One can calculate the temperature at which quantum effects overwhelm general relativity: it is $10^{38}$ degrees

Kelvin, which is a trillion trillion times hotter than the center of a hydrogen bomb explosion.

In fact, these two theories—the quantum theory and general relativity—seem to be the exact opposite of each other. But it is hard to believe that, at the most fundamental level, nature has created a universe where the left hand and the right hand do not coordinate with each other.

General relativity, for example, gives us a compelling description of gravity and the macrocosm, the world of galaxies, black holes, and expanding universes. It gives us a beautiful theory of gravity based on smooth, curved surfaces. "Forces" are created by the distortion of space and time. If an object feels a "force," it is only because it is moving in the curved space surrounding the object.

General relativity, however, has gaping holes in it. The theory breaks down at the center of a black hole, or at the instant of the Big Bang, when the curvature of space-time became infinite.

Similarly, the quantum theory gives us the most complete description of the microcosm, the ghostlike world of subatomic particles. The quantum theory is based on the idea that a "force" is created by the exchange of tiny discrete packets of energy, called "quanta." The quantum theory replaces the beautiful geometric picture of general relativity with its opposite: tiny packets of energy. For example, the quantum of light is called the photon. A quantum of the weak nuclear force is called the W-boson. And the quantum of the strong force is called the gluon. Each force has its own distinct quantum.

The most advanced form of the quantum theory is called the Standard Model, based on a bizarre, motley assortment of particles with strange names. (Back in the 1950s, so many subatomic particles were being found in our atom smashers that physicists were drowning in an ocean of particles. In frustration, J. Robert Oppenheimer, who directed the atomic bomb project, declared that the next Nobel Prize should go to the physicist who does *not* discover a new particle that year!)

Today, the Standard Model has been able to prune down the number of subatomic particles to the quarks, W- and Z-bosons, gluons, Higgs particles, electrons, and neutrinos. The Standard Model is perhaps the most successful physical theory of all time, reproducing nature to within one part in a billion.

But like general relativity, the Standard Model also has gaping holes. The theory contains nineteen numbers (e.g., the masses of the quarks, electrons, neutrinos, and the strength of their interactions) which are totally arbitrary. It gives no explanation as to why in nature there are three "generations" or carbon copies of the quarks, giving us a threefold redundancy which is deeply disturbing but at present impossible to explain. It is also one of the ugliest theories ever proposed. Its particles seem to have no rhyme or reason. It can be compared to taping together an aardvark, a whale, and a giraffe and calling it nature's most elegant and graceful creation, the end product of millions of years of evolution. Worse, it says nothing about gravity.

Unfortunately, all attempts to merge these two theories have failed. A naive quantum theory of gravity, for example, tries to break up gravity into tiny packets of gravitational energy, called "gravitons." But when these gravitons bump into each other, the theory blows up and produces infinities. So the naive merger of these two theories becomes useless. This problem of these infinities has defeated all efforts for the past fifty years.

## Superstrings

Many leading physicists are convinced that a theory of everything exists. Nobel Laureate Steven Weinberg compares this situation to the discovery of the North Pole. Nineteenth-century navigators knew that their compass needles, which were always pointing north, were converging on a single spot. No matter where they sailed on the earth, their needles would indicate the existence of a mythical spot called the North Pole. However, it wasn't until early in this century that the North Pole was finally reached. Similarly, Weinberg concludes: "If history is any guide at all, it seems to me to suggest that there *is* a final theory."

So far, the only theory that can remove these infinities is the ten-dimensional superstring theory, a theory which has dazzled the world of physics and astounded the world of mathematics with its elegant geometry. Both Weinberg and Nobel Laureate Murray Gell-Mann, who originated the quark model, agree on this point. Gell-Mann has said, "We now have, in superstring theory, a brilliant candidate for a unified theory of all the elementary particles, including the graviton, along with their interactions."

Einstein once said that all great theories are based on simple physical pictures. In fact, if a theory has no underlying physical picture, then it is probably worthless. Fortunately, superstring theory has an elegant physical picture which underlies its magical powers.

First, strings can vibrate much like violin strings. No one claims that the notes of a violin string, like A, B, or C sharp, are fundamental. Everyone knows that what is fundamental is the violin string itself. Likewise, the superstring can also have notes or resonances. Each vibration is equivalent to a subatomic particle in the zoo of particles we see around us, such as the quarks, the electrons, the neutrinos, etc. So the superstring can give a simple description of why we have such a messy collection of subatomic particles. In fact, there should be an infinite number of these particles, just as there are an infinite number of vibrations on a violin string.

If we had a super-microscope, we might be able to see that the electron, which appears to be a point particle, is actually a tiny vibrating string. When the string vibrates in different modes, it becomes a different particle. In this picture, the laws of physics are nothing but the harmonies of the superstring. The universe is nothing but a symphony of vibrating strings. (This, in a sense, fulfills the original dream of the ancient Greek Pythagoreans, who were the first to understand the laws of the harmony of strings. They suspected that the entire universe might be understood via the laws of harmony, but until now, no one knew how this could be done.)

Second, the superstring theory contains Einstein's theory of gravity. As the string moves in space and time, it forces the surrounding continuum to curve, exactly as Einstein predicted. So the superstring theory effortlessly reproduces all the predictions of general relativity, such as black holes and the expanding universe.

## The Tenth Dimension

One of the conceptually beautiful (and controversial) aspects of the theory is that it is formulated in ten-dimensional space-time. In fact, it is the only theory known to science that actually selects out its own dimension of space and time.

Defining a theory in hyperspace is a convenient way in which to absorb more and more forces. Back in 1919, it was recognized by mathematician Theordr Kaluza that if we add a fifth dimension to Einstein's four-dimensional theory of gravity, the fifth component of gravity reproduces the electromagnetic force of Maxwell. In this way, we see that vibrations of the unseen fifth dimension reproduce the properties of light. Similarly, by adding more and more dimensions, we can reproduce higher forces, such as the weak and strong nuclear forces.

To see how adding higher dimensions unify the fundamental forces, think of how the ancient Romans used to wage their wars. In ancient times, communicating between forces fighting battles on different fronts was a chaotic and messy affair, with messages carried by runners. That is why the Romans would always go into hyperspace—i.e., the third dimension—by seizing a hilltop. From the vantage point of the hill, one could see several battlefields down below as a single, coherent, and unified picture. Similarly, from the vantage point of the tenth dimension, one can look down below and see the four fundamental forces laid out as a single superforce.

Recently, there has been a flurry of activity in superstring research on the eleventh dimension. Edward Witten of the Institute for Advanced Study at Princeton and Paul Townsend of Cambridge University have shown that many of the mysteries of superstring theory become transparent if we formulate the theory in eleven dimensions. Witten has called it "M-theory." (Because the properties of M-theory are not completely understood by physicists, "M" can stand for "mystery," "magic," or "membrane," take your pick.)

At present, the fundamental problem with string theory is that millions of solutions of the theory have been found, but none which precisely matches the known spectrum of quarks, gluons, neutrinos, etc. Many people have despaired of ever finding *all* the solutions of string theory to find our universe. However, M-theory possesses a new type of symmetry, called "duality," which allows us to find solutions of string theory which were previously inaccessible (these are the so-called "nonperturbative" solutions). Our physical universe may be one of these nonperturbative solutions.

In fact, by reducing M-theory to lower dimensions, physicists have now found almost the complete set of possible universes in eight dimensions, and physicists are now rapidly completing an analysis of the possible six-dimensional universes as well. The next problem, which may still take years of hard effort, is to find all the four-dimensional universes and see if our universe is among them.

One astonishing aspect of M-theory is that it is formulated in eleven dimensions, and hence allows for the existence of other exotic objects called membranes. It now seems that strings coexist with various types of membranes in hyperspace.

All these recent discoveries have generated enormous interest in string theory.

## What Happened Before the Big Bang?

Not only would a quantum theory of gravity resolve what happens at the center of a black hole; it would also resolve what happened before the Big Bang.

At present, there is conclusive evidence that a cataclysmic explosion occurred roughly 15 billion years ago which sent the galaxies in the universe hurtling in all directions. Decades ago, physicist George Gamow and his colleagues predicted that the "echo" or afterglow of the Big Bang should be filling up the universe even today, radiating at a temperature just above absolute zero. It wasn't until 1992, however, that the Cosmic Background Explorer (COBE) satellite finally picked up this "echo" of the Big Bang. Physicists were elated to find that hundreds of data points perfectly matched the prediction of the theory. The COBE satellite detected the presence of a background microwave radiation, with a temperature of 3 degrees above absolute zero, which fills up the entire universe.

Although the Big Bang theory is on solid experimental grounds, the frustrating feature of Einstein's theory is that it says nothing about what happened before the Big Bang or why there was this cosmic explosion. In fact, Einstein's theory says that the universe was originally a pinpoint that had infinite density, which is physically impossible.

Infinite singularities are not allowed in nature, so ultimately a quantum theory of gravity should give us a clue as to where the Big Bang came from.

The superstring theory, being a completely finite theory, gives us deeper insight into the era before the Big Bang. The theory states that at the instant of creation, the universe was actually an infinitesimal ten-dimensional bubble. But this bubble (somewhat like a soap bubble) split into six- and four-dimensional bubbles. The six-dimensional universe suddenly collapsed, thereby expanding the four-dimensional universe into the standard Big Bang.

Furthermore, this excitement about quantizing gravity is fueling a new branch of physics called "quantum cosmology," which tries to apply the quantum theory to the universe at large. At first, quantum cosmology sounds like a contradiction in terms. The "quantum" deals with the very small, while "cosmology" deals with the very large, the universe itself. However, at the instant of creation, the universe was very small, so quantum effects dominated that early moment in time.

Quantum cosmology is based on the simple idea that we should treat the universe as a quantum object, in the same way that we treat the electron as a quantum object. In the quantum theory, we treat the electron as existing in several energy states at the same time. The electron is free to move between different orbits or energy states. This, in turn, gives us modern chemistry. Thus, according to Heisenberg's Uncertainty Principle, you never know precisely where the electron is. The electron thus exists in several "parallel states" simultaneously.

Now consider the universe to be similar to an electron. If we quantize the universe, the universe must now exist simultaneously in several "parallel universes." Once we quantize the universe, we are necessarily led to believe that the universe can exist in parallel quantum states. When applied to a universe, it gives us the "multiverse."

## The Multiverse

According to this startling new picture, in the beginning was Nothing. No space. No time. No matter or energy. But there was the quantum principle, which states that there must be uncertainty, so even Nothing became unstable, and tiny particles of Something began to form.

By analogy, think of boiling water, which is a purely quantum mechanical effect. The bubbles, which seem to come out of nowhere, suddenly expand and fill up the water. Similarly, in this picture Nothing begins to boil. Tiny bubbles began to form and then expand rapidly. Since each bubble represents an entire universe, we use the term "multiverse" to describe the infinite ensemble of universes. According to this picture, our universe was one of these bubbles, and the expansion is called the Big Bang.

At first, creating bubbles of Something in an ocean of Nothing might seem to violate the conservation of matter and energy. However, this is an illusion, because the matter-energy content of the universe is positive, but the gravitational energy is negative. In fact, the sum of the two is zero, so it takes no net energy to create a universe out of Nothing!

Different physicists have given their own spin to this picture.

Cosmologist Stephen Hawking believes that our universe is perhaps the most likely of all these infinite universes. In his picture, we coexist with an infinite sea of other bubbles (which he calls baby universes), but our universe is special. It is the most stable, and its probability of existing is the largest. He believes that all these baby universes are connected to each other by an infinite network of thin wormholes. (In fact, by adding up the contribution of these wormholes, he can present arguments why our present universe is so stable.) These wormholes are extremely small, so we do not have to worry about falling into one of them and finding ourselves in a parallel universe.

Steven Weinberg finds the idea of the multiverse an appealing one: "I find this an attractive picture and [it's] certainly worth thinking about very seriously. An important implication is that there wasn't a beginning; that there were increasingly large big bangs, so that the [multiverse] goes on forever—one doesn't have to grapple with the question of it before the bang. The [multiverse] has just been here all along. I find that a very satisfying picture."

(Parenthetically, the theory of the multiverse seems able to unite the Judeo-Christian account of Genesis, which starts with a definite beginning, and the Buddhist theory of Nirvana, which starts with a timeless universe. In this picture, we have Genesis taking place continually in Nirvana.)

Weinberg, however, believes that many of these parallel universes are dead—i.e., the proton is not stable in these universes, so that there is no DNA or stable matter. Although each bubble may represent a viable universe, most of them are probably uninteresting, consisting of a sea of electrons and neutrinos without any stable matter.

One advantage of this simple picture is that it offers a solution to one of the strangest experimental aspects of our universe.

For example, it is known that the physical constants of the universe lie within an extremely narrow band. If these constants (such as the masses and couplings of various subatomic particles) deviated slightly, then chaos would result and life would be impossible: the proton would decay, nuclei would be unstable, DNA could not form, and carbon-based life on earth would never happen.

This is not a trivial statement. So far, every important physical constant that has been tested is found to lie in this narrow region which is compatible with life. This is called the Anthropic Principle, the idea that the physical constants of the universe are such they make life possible. Some scientists have argued that this is purely a coincidence, which is difficult to believe. Other scientists have argued that it indicates the existence of a cosmic Providence, which chose this universe to have these physical constants so that life and consciousness would arise. But a new interpretation arises within the context of the multiverse.

If there are in fact an infinite number of universes, then in other universes the physical constants are, indeed, different. As Weinberg states, these universes are probably dead seas of electrons and neutrinos. But by chance there are universes in which the fundamental physical constants do make possible stable DNA. Our universe happens to be one of these, which explains why we are here to discuss the matter in the first place! In other words, the multiverse idea explains simply why the Anthropic Principle must hold.

## The Distant Future: The Fate of the Universe

No discussion of the future is complete without discussing the ultimate future, the fate of the universe itself. Using the laws of physics, we can narrow down the potential futures that lie ahead of us in the next hundred or so billion years.

Our bubble has been expanding for about 15 billion years, but scientists are not sure how long this expansion can last. It is not clear if the universe will ultimately die in fire or in ice.

If the density of the universe is above a certain critical point, this may create gravity powerful enough to reverse the cosmic expansion. The current red shift of starlight that we see in the heavens will gradually become a blue shift as gravity halts the expansion of the galaxies and even reverses their direction. As the contraction takes place, temperatures in the universe gradually start to soar. Billions more years into the future, the oceans will boil, the planets will melt, and the stars and galaxies will be compressed into a gigantic primordial atom. In this scenario, eventually the universe will collapse to a Big Crunch, and the universe dies in fire.

On the other hand, if there is not enough matter, then the universe may expand forever, until it gradually gets colder and colder, the inevitable consequence of the Second Law of Thermodynamics. In this scenario, eventually the universe will consist of dead stars and black holes, as temperatures drop to near absolute zero. Trillions upon trillions of years into the future, even the black holes will evaporate and the universe will decay into a gas of electrons and neutrinos. In this scenario, called the Big Chill or entropy death, the universe dies in ice.

At present, scientists are not sure which scenario is the correct one. The amount of visible matter in the universe is not enough to reverse the expansion, so astrophysicists have long thought that the universe will expand forever. Recently, however, astronomers have been convinced that perhaps 90 percent of the matter in the universe is in the form of nonluminous "dark matter." This mysterious dark matter, which no one has ever seen, has mass but is invisible. According to this new picture, dark matter surrounds the galaxies and prevents them from flying apart as they rotate. Since we don't know precisely how much dark matter there is in the universe, we cannot say for sure if there is enough to reverse the cosmic expansion.

Either way, however, the universe will eventually die, and all intelligent life with it. Nothing, it seems, can escape the death of the universe itself, including Type III civilizations. Either Type III civilizations in the universe will be incinerated as their machines fail to prevent temperatures from soaring to infinity, or they will gradually freeze as their machines grind to a halt and temperatures plunge to zero. Although Type III civilizations can muster the energy of a galaxy, this is still not enough energy to reverse the death of the universe.

So in either scenario, it seems as if the universe must die, and all intelligent life with it. Such an end seems like the ultimate existential absurdity—that intelligent life struggles over millions of years to rise from the swamp and reach for the stars, only to be snuffed out when the universe itself dies.

But there is a loophole to this dismal picture. There is the possibility that civilizations in space will eventually reach Type IV status, with the power to manipulate at will the fourth pillar of science, the space-time continuum. A Type IV civilization would be able to widen the microscopic wormholes that constantly connect the various universes, allowing them to pass between universes. If they have mastered the enormous power necessary to create these large wormholes between universes, then they might be able to tunnel their way through the wormhole and escape the death of the universe.

If so, then the theory of everything, which at first seems rather useless and devoid of practical application, may ultimately provide the salvation for intelligent life in the universe.

## Conclusion

When Isaac Newton walked along the beach, picking up seashells, he did not realize that the vast ocean of undiscovered truth that lay before him would contain such scientific wonders. He probably could not foresee the day when science would unravel the secret of life, the atom, and the mind.

Today, that ocean has yielded many of its secrets. Now a new ocean has opened up. As we have seen, it is a wondrous ocean of scientific possibilities and applications. Perhaps in our lifetime, we will see many of these marvels of science unfold before us. For we are no longer passive observers to the dance of nature; we are in the process of becoming active choreographers. With the basic laws of the quantum, DNA, and computers discovered, we are now embarking upon a much greater journey, one that ultimately promises to take us to the stars. As our understanding of the fourth pillar, space-time, increases, this opens up the possibility in the far future of being able to become masters of space and time.

Barring some natural catastrophe, war, or environmental collapse, we are on our way toward attaining, in the next century or two, the planetary power of a Type I civilization that will make us a truly planetary society. And what will make this possible is the power of these three revolutions. Ultimately, they will make it possible to fulfill our destiny and take our place among the stars. The harnessing of these scientific revolutions is the first step toward making the universe truly our backyard.

# Notes

CHAPTER 1 **Choreographers of Matter, Life, and Intelligence**

**p. 4 "Clearly, we are on the threshold . . ."** This can be seen, for example, by analyzing the number of pages published every year by scientific journals.

**p. 5 "(One is reminded . . ."** David Wallechinsky, *The People's Almanac Presents the Complete Idiosyncratic Compendium of the Twentieth Century*, Little, Brown, Boston, 1995; also *Parade* magazine, Sept. 10, 1995, p. 16.

**p. 8 "Second, that subatomic particles . . ."** The third postulate of the quantum theory states that the square of the absolute value of the Schrödinger wave function measures the probability of finding a particle at that particular point in space and time. Thus, Newtonian determinism, which claimed that all events could be described with infinite precision, is replaced by probabilities and waves. This in turn leads to the Heisenberg Uncertainty Principle—i.e., that you cannot accurately know both the position and the velocity of a particle at the same time.

**p. 9 "John Horgan, in his book . . ."** John Horgan, *The End of Science*, John Wiley, New York, 1996, p. 6. Interview.

**p. 10 " 'Arthur [is] an intelligent alien . . .' "** Sheldon Glashow and Leon Lederman, "The SSC: A Machine for the Nineties," *Physics Today*, March 1985, p. 332.

**p. 13 "Thurow writes . . ."** Lester C. Thurow, *The Future of Capitalism*, William Morrow, 1996, New York, p. 279.

**p. 13 "Already, the commodity prices . . ."** Ibid., p. 67.

**p. 13 " 'Today, knowledge and skills . . .' "** Ibid., p. 68.

**p. 13 "That list included . . ."** Ibid., p. 67.

**p. 17 "Russian astronomer Nikolai Kardashev . . ."** Freeman Dyson, *Disturbing the Universe*, Harper & Row, New York, 1979, p. 212.

**p. 18 "(To give a sense of scale . . ."** On the show, there is only one truly galactic civilization, called the Borg, which is probably Type III, and hence feared by all Type II civilizations. There is also a mysterious, almost godlike race of superbeings, called the Q, who can manipulate space, time, matter, and energy at will. This mythical race would probably qualify for an entirely new type of classification, perhaps a Type IV status.

## CHAPTER 2 The Invisible Computer

**p. 24 " 'ubiquitous computing' "** Interview with Mark Weiser. Weiser Web Page.

**p. 24 "As Weiser says, 'Disappearance . . .' "** *The Computer in the 21st Century,* Scientific American Books, New York, 1995, p. 78.

**p. 25 "falling cost of microchips . . ."** Ibid.

**pp. 25–26 "Ubiquitous computing was conceived . . ."** Ibid.

**p. 26 " 'There was a certain degree of mysticism . . .' "** Steven Lubar, *Info Culture,* (Boston, Houghton Mifflin, 1993), p. 336.

**p. 26 "To test their ideas . . ."** Ibid., pp. 368–69.

**p. 27 "(Ironically enough . . ."** Interview with Weiser.

**p. 27 "As Bill Gates admitted . . ."** Bill Gates, *The Road Ahead,* Viking, New York, 1995, pp. 273–74.

**p. 29 "Thomas George, general manager . . ."** *Forbes ASAP,* Feb. 2, 1996, p. 60.

**p. 29 "musical greeting cards . . ."** Paul Saffo, *The International Design Magazine,* Jan.–Feb. 1995, p. 74.

**p. 30 "The Motorola Power PC 620 . . ."** *Scientific American,* Jan. 1996, p. 62.

**p. 30 "That limit is the result . . ."** More precisely, the resolution of a light beam is given by the wavelength divided by the diameter of the aperture of the objective lens. This is Rayleigh's law.

**pp. 30–31 "Paul Saffo, director of the Institute . . ."** Interview with Paul Saffo, Feb. 14, 1996.

**p. 31 "Saffo says, 'If a meteor . . .' "** Ibid.

**p. 32 "Perhaps 100 tabs . . ."** *The Computer in the 21st Century,* pp. 78–80.

**p. 32 "The larger, foot-size *pads* . . ."** Ibid., pp. 85–87.

**p. 32 "At home, such boards . . ."** Ibid., pp. 87–88.

**p. 33 "The director of what . . ."** Interview with Neil Gershenfeld, July 26, 1996.

**p. 34 "As a result, the 'smart table' . . ."** Interview.

**p. 34 "Gershenfeld calls . . ."** Interview.

**p. 34 "Gershenfeld points out . . ."** Interview.

**p. 35 "In the future, shoes will think . . ."** Media Lab Web Page.

**p. 35 "These crude beginnings . . ."** Media Lab Web Page.

**p. 35 "Steve Mann of the Media Lab . . ."** *Business Week,* June 24, 1996, p. 119.

**p. 36 " 'Imagine a house that always . . .' "** *Scientific American,* April 1996, p. 68.

**p. 36 "If the computer scans a stranger's face . . ."** pp. 71–73.

**p. 37 "Therefore when the computer focuses only . . ."** pp. 72, 74.

**p. 37 "As James Gleick . . ."** *New York Times Magazine,* June 16, 1996, p. 28.

**p. 37 " 'People today do not put $5 billion . . .' "** Ibid.

**p. 37 "According to Carol H. Fancher . . ."** *Scientific American,* Aug. 1996, p. 43.

**p. 38 "Sales of the automobile industry . . ."** *The Economist,* June 22, 1996, p. 3.

**p. 38 " 'We'll see vehicles . . .' "** *Wall Street Journal Supplement,* July 1996, p. 8.

**p. 39 "They make it possible to determine your location . . ."** *New York Times,* March 5, 1996, p. D1.

**p. 39 " 'The commercial industry . . .' "** *New York Times,* March 5, 1996, p. D1.

**p. 39 "Prototypes of such highways . . ."** England, for example, already has the Trafficmaster system. *The Economist,* June 22, 1996, p. 16.

**p. 39 "By December 2001 . . ."** *International Herald Tribune,* June 29–30, 1996, p. 2.

**p. 40 "invisible computer . . ."** Interview with Weiser.

**p. 41 " 'Computer simulation is our only hope . . .' "** *Science News,* April 15, 1995, p. 235.

## CHAPTER 3 The Intelligent Planet

**p. 43 " 'Some are more aware of this . . .' "** *Fortune,* July 9, 1996, p. 46.

**p. 43 " 'Is it a fact . . .' "** *The Computer in the 21st Century,* Scientific American Books, New York, 1995, pp. 4–5.

**p. 45 "Already, young computer nerds . . ."** When Jerry Yang, twenty-seven, and David Filo, thirty, went public with their company, Yahoo!, in 1996, they found themselves $132 million richer on the day the company first traded publicly, even though it had not posted a dime of profit. *(USA Weekend,* May 10–12, 1996, p. 4.)

**p. 45 "(Clark reached billionaire status . . ."** *Business Week,* July 15, 1996, p. 63.

**p. 46 " 'The poor fellow's eyes . . .' "** Michio Kaku and Daniel Axelrod, *To Win a Nuclear War,* South End Press, Boston, 1987, p. 200.

**p. 47 "The Pentagon realized it could also . . ."** The Pentagon called this "first strike capability." It did not mean that it could execute a first strike. It meant that it could credibly threaten an adversary with a first strike.

**p. 47 "Because most cities would no longer exist . . ."** This idea was the product of Paul Baran, an immigrant from Eastern Europe, who was trying to make telecommunications invulnerable in case of a nuclear war. Other computer experts, who were interested in purely civilian applications, also contributed to the formation of ARPANET. *(New York Times,* Aug. 21, 1996, New York Times Web Page. See also Katie Hafter and Mathew Lyon, *Where Wizards Stay Up Late,* Simon & Schuster, New York, 1996.)

**p. 48 " 'There wasn't a photographer present . . .' "** *Newsweek,* Aug. 8, 1996, p. 57.

**p. 48 "By 1994, more than 45,000 . . ."** Nicholas Negroponte, *Being Digital,* Alfred A. Knopf, New York, 1995, p. 181.

**p. 49 "Today, the Internet is growing . . ."** Ibid., p. 182.

**p. 49 "It is truly the 'mother . . .' "** *Washington Post Magazine,* Aug. 4, 1996, p. 28.

**p. 49 "Certainly the potential . . ."** *Wall Street Journal,* June 17, 1996, p. R28.

**p. 49 "Cerf says, 'I'm not at all shy . . .' "** Ibid.

**p. 49 "In 1996, one could access . . ."** *New York Times,* June 21, 1996, p. D6; *Wall Street Journal,* Aug. 23, 1996, p. B1.

**p. 49 "(In 1996, for example, the switching station . . ."** *Wall Street Journal,* Aug. 23, 1996, p. B1.

**p. 50 "Vice President Al Gore . . ."** *The Computer in the 21st Century,* p. 156.

**p. 50 "This was one of the great . . ."** Steven Lubar, *Info Culture,* Houghton Mifflin, Boston, 1993, p. 134.

**p. 50 "Perhaps the most consistent critic . . ."** Interview with Clifford Stoll.

**p. 50 " 'They're irrelevant to cooking . . .' "** Ibid. Also Clifford Stoll, *Silicon Snake Oil*, Doubleday, New York, 1995, p. 10.

**p. 51 "But by 1980 . . ."** Lubar, *Info Culture*, p. 236.

**p. 51 "Larry Tesler, chief scientist . . ."** Another visionary on that original team was Alan Kay, whose fertile imagination spearheaded many of these innovations.

**p. 51 "The bottom line . . ."** Interview with Larry Tesler.

**p. 51 "Tesler rattles . . ."** Ibid.

**p. 51 " 'It's a match made in Maui' . . ."** *Wall Street Journal*, June 17, 1996, p. R6. The airlines, realizing that up to 20 percent of their operating expenses are for so-called distribution fees (commissions to travel agents and the computer reservation system), are anxious to put their reservations system on the Internet, especially to sell unsold seats.

**p. 51 " 'Look out, Merrill Lynch' . . ."** *Wall Street Journal*, June 17, 1996, p. R8.

**p. 51 "On-line bookstores will be . . ."** Amazon.com Inc, an on-line bookstore, offers a staggering number of books: one million titles. Jeffrey Bezos, its CEO, says, "You can't do that with a physical store, and you can't do that with a paper catalog, which would be the size of seven Manhattan phone books." *Wall Street Journal*, June 6, 1996, p. R6.

**p. 52 "As much as 15 percent of the $400 billion . . ."** *Wall Street Journal*, June 17, 1996, p. R10.

**p. 52 " 'And the sheer size of some . . .' "** Ibid., p. R6.

**p. 52 "Already, Technology/Clothing . . ."** *New York Times*, Feb. 19, 1996, p. C3.

**p. 52 "Microsoft CEO Bill Gates . . ."** *Fortune*, July 9, 1996, p. 46.

**p. 53 "(With an ISDN link . . ."** ISDN stands for integrated services digital network. A baud is a bit per second, and is named after the French inventor Emile Baudot. A 144 kilobaud ISDN signal can carry two 64 kilobaud signals.

**p. 54 "Annual installation of fiber optic . . ."** *New York Times*, Nov. 4, 1994, p. D5.

**p. 56 "Already, six major companies . . ."** *Time*, Aug. 12, 1996, p. 43.

**p. 57 "With the falling cost . . ."** *New York Times*, May 20, 1996, p. D7.

**p. 57 " 'For the first time . . .' "** Ibid.

**p. 57 "The market for LCDs . . ."** *New York Times*, Feb. 10, 1997, p. D5.

**p. 58 "At present, there are . . ."** Interview with David Nahamoo, June 1996.

**p. 59 " 'It's easier to simply build . . .' "** Interview with Pattie Maes.

**p. 59 "(She won the bet.)"** Ibid.

**p. 60 " 'In my vision . . .' "** *Time*, March 25, 1996, p. 58.

**p. 60 "She adds: 'Each succeeding . . .' "** *Scientific American*, Sept. 1995, pp. 85–86. Interview with Pattie Maes.

**p. 60 " 'Rather than manipulating a keyboard . . .' "** Ibid., p. 85.

**p. 61 "With 32 microprocessors . . ."** *Discover*, June 1996, p. 48.

**p. 61 " 'I was stunned . . .' "** *Time*, March 25, 1996, p. 55.

**p. 61 "But Douglas Hofstadter . . ."** *Washington Post*, Feb. 19, 1996, p. A11. Interview, May 27, 1997.

**p. 62 "(However, as computer expert Douglas Lenat . . ."** *Scientific American*, Sept. 1995, p. 81.

**p. 62 "The reason why expert systems . . ."** Dennis Shasha and Cathy Lazere, *Out of Their Minds*, Springer-Verlag, New York, 1995, p. 226.

**p. 64 " 'No one in 2015 would dream . . .' "** Daniel Crevier, *AI*, Basic Books, New York, 1993. p. 242. Interview, May 27, 1997.

**p. 64 "Lenat's goal . . ."** Ibid., p. 240.

**p. 65 " 'Napoleon died . . .' "** Ibid., p. 241.

**p. 65 "Lenat and his staff . . ."** David H. Freedman, *Brainmakers*, Simon & Schuster, 1993, p. 56.

**p. 65 "Lenat sums up his philosophy . . ."** *Scientific American*, Dec. 1991, p. 134.

**p. 65 "Maes, for example, thinks . . ."** Interview with Pattie Maes.

**p. 66 "After all, he reminds us . . ."** Crevier, *AI*, p. 243.

## CHAPTER 4 Machines That Think

**p. 71 "Unlike his namesake . . ."** Interview with Michael Wessler, MIT AI Lab, July 10, 1996.

**p. 71 "It spends most of the day . . ."** David H. Freedman, *Brainmakers*, Simon & Schuster, New York, 1994, p. 15.

**p. 72 "A simple feedback mechanism . . ."** Interview with Rodney Brooks, July 10, 1996.

**p. 72 " 'Insects must organize . . .' "** Freedman, *Brainmakers*, p. 24.

**p. 72 "Compared with the tiny . . ."** Interview with Rodney Brooks.

**p. 73 "NASA was so impressed . . ."** Interview with Donna Shirley, Aug. 14, 1996. Rodney Brooks and Anita M. Flynn, "Fast, Cheap and Out of Control," *Journal of the British Interplanetary Society*, vol. 42 (1989), pp. 478–85.

**p. 73 "As part of the Mars Pathfinder . . ."** NASA Web Page; Mars Pathfinder Web Page; interview with Donna Shirley.

**p. 73 "Five similar robots . . ."** Interview with Rodney Brooks.

**p. 73 "Brooks's papers, with provocative titles . . ."** Rodney A. Brooks, "Intelligence Without Reason," *Proceedings of the 1991 International Joint Conference on Artificial Intelligence*, 1991, pp. 569–95. Rodney A. Brooks, "Elephants Don't Play Chess," in *Designing Autonomous Agents*, MIT Press, Cambridge, Mass., 1990, pp. 3–15.

**p. 74 "Their philosophy was based . . ."** A Turing machine was introduced back in the 1930s by British mathematician Alan Turing, who wanted to isolate the essence of computing machines. Digital computers, no matter how complex, are Turing machines, which are sometimes called "universal computing machines." A Turing machine consists of an infinitely long binary input/output tape, a processor, and a program. The processor reads the input tape and can manipulate it to make an output tape based on instructions written on the program. The processor can perform only four operations: it can replace a 1 with a 0 and vice versa, and it can move one step forward or backward. Remarkably, these four operations are all that is necessary to simulate any modern digital computer.

**p. 75 "His machines, he says . . ."** *Scientific American*, Dec. 1991, p. 130.

**p. 75 "I said, "Well, I won't . . .' "** Crevier, *AI*, Basic Books, New York, 1993, p. 7.

**p. 75 "Marvin Minsky, co-founder . . ."** Freedman, *Brainmakers*, p. 29.

**p. 75 " 'Chess-playing programs don't . . .' "** Ibid.

**p. 76 " 'They're making things . . .' "** Ibid., p. 30.

**p. 76 " 'Complete agreement is a sign . . .' "** Ibid.

**p. 76 " 'Fully intelligent machines . . .' "** Hans Moravec, *Mind Children*, Harvard University Press, Cambridge, Mass., 1988, p. 20. Interview, Jan. 14, 1997.

**p. 77 "Eventually, medical robots . . ."** Ibid., p. 36.

**p. 77 "It resembles a large . . ."** Ibid., p. 37.

**p. 77 "As long as it conducted . . ."** George Harrar, *Radical Robots*, Simon & Schuster, New York, 1990, p. 36.

**p. 79 "Virasoro, however, is one of many . . ."** Interview with Miguel Virasoro, June 1992.

**p. 80 "By 2040, even desktop computers . . ."** For example, Daniel Crevier has given an estimate of the year in which supercomputers will overtake the human brain in terms of processing power. In the best case, the year is 2009. In the worst case, it is 2042. (Crevier, *AI*, p. 303.) Interview with Crevier.

**p. 80 "Creating faster and faster computers . . ."** We should point out that although the brain does appear to be a neural net, there are some scientists who argue a technical point: that the circuitry of neural nets can, in principle, be reproduced by a Turing machine. Thus, by indirect reasoning, the brain can still be considered to be a very complex Turing machine. The point here, however, is that such an approach is a very clumsy one and is not a practical one for brain researchers. A more practical way is to start directly with neural nets.

**p. 81 "To Virasoro, the brain is . . ."** As we mentioned earlier, we will include neural nets as part of AI, although some people separate it off as a special discipline.

**p. 81 "Like a proud father . . ."** Sejnowski is now a professor at the University of California at San Diego.

**p. 81 " 'It babbles.' "** William F. Allman, *Apprentices of Wonder*, Bantam Books, New York, 1988, p. 2.

**p. 82 "Each time a neural network . . ."** For example, the weights may be the resistances along the wires of the neural net. The point is that after each correct guess, the weights of that circuit are reinforced, so this particular circuit can be repeated once again. After an incorrect guess, the circuit is deemphasized, making this circuit less likely to be repeated in the future.

**p. 82 " 'Now it has discovered spaces . . .' "** Allman, *Apprentices of Wonder*, p. 2.

**p. 82 "As physicist Heinz Pagels . . ."** Heinz R. Pagels, *The Dreams of Reason*, Bantam Books, New York, 1988, p. 140.

**p. 82 " 'Some features are remnants . . .' "** Allman, *Apprentices of Wonder*, p. 179.

**p. 83 " 'Some of them, like Hopfield . . .' "** Pagels, *The Dreams of Reason*, p. 130.

**p. 84 " 'That was really important . . .' "** Allman, *Apprentices of Wonder*, p. 81.

**p. 84 "This in turn helped to . . ."** Hopfield's discovery was not the only one to help stimulate the revival of neural networks. Another discovery was "back propagation," which was an improvement on the way in which neurons interact within a neural net.

**p. 86 "He notes, 'If you want . . .' "** Allman, *Apprentices of Wonder*, p. 99.

**p. 86 " 'But it would be like . . .' "** *New York Newsday*, Jan. 22, 1991, p. 65.

**p. 87 "Using transistors . . ."** Allman, *Apprentices of Wonder*, p. 146.

**p. 87 " 'That's why I did . . .' "** Ibid., p. 147.

**p. 87 "In fact, he found that . . ."** Peter Coveney and Roger Highfield, *Frontiers of Complexity*, Ballantine Books, New York, p. 262.

**p. 87 "Rodney Brooks . . ."** Interview with Rodney Brooks.

**p. 88 " 'It's a paraplegic . . .' "** Ibid.

**p. 90 "Brooks even toyed . . ."** *Time*, March 25, 1996, p. 57.

**p. 90 "By finally merging . . ."** Charles Sheffield, Marcelo Alonso, and Morton A. Kaplan, *The World of 2044*, Paragon House, St. Paul, 1994, pp. 33–34.

**p. 91 "As Minsky has said . . ."** Marvin Minsky, *The Society of Mind*, Simon & Schuster, New York, 1985, p. 94.

**p. 91 " 'You can interpret that . . .' "** Crevier, *AI*, p. 266. Interview.

**p. 92 " 'We'll reimburse you!' "** Ibid., p. 267.

**p. 94 " 'I'd say yes, it is . . .' "** *Discover*, June 1996, p. 50.

**p. 94 "To Marvin Minsky . . ."** *Scientific American*, Nov. 1993, p. 38.

**p. 95 "McGinn claims this feat . . ."** *Time*, March 25, 1996, p. 53.

**p. 95 "Roger Penrose, the noted Oxford . . ."** However, his "proof" for the impossibility of thinking machines was not a rigorous proof in any sense of the word, but rather sophisticated and largely intuitive arguments drawn from analogies in mathematics and physics, especially Gödel's deep theorem on the incompleteness of arithmetic and Heisenberg's Uncertainty Principle.

In one variation of his arguments, Penrose cites the fact that certain numbers cannot be computed in a finite amount of time by Turing machines. But since humans can solve problems that Turing machines cannot, therefore one might conclude that humans cannot be machines. (The rebuttal to this is that our brains are neural nets, not Turing machines, and hence Penrose's arguments do not apply.)

**p. 96 " 'I was secretly pleased . . .' "** Pagels, *The Dreams of Reason*, p. 240.

**p. 96 "But as Turing stressed . . ."** Turing believed that computers would one day become so advanced that they would become operationally indistinguishable from humans. To prove this, he proposed the famous "Turing test." Let us place a computer in one black box and a human in another; we are then allowed to ask any question of either box. The test is: can we distinguish the human from the computer? When Turing first proposed his famous test, computers were so primitive that the test was never tried. However, because of the explosive growth in computer power, the Turing test was actually conducted on a PC. Although the PC ultimately failed the test, it was still sophisticated enough to fool some of the human judges.

## CHAPTER 5 Beyond Silicon: Cyborgs and the Ultimate Computer

**p. 99 "The iron laws of . . ."** Moore himself once spoofed his own law, showing that it predicts that by the year 2040 the semiconductor industry will surpass the entire world's gross domestic product.

**p. 100 "Supercomputers, like the Cray T90 . . ."** FLOP stands for "floating point operation," which represents the multiplication of two decimal numbers. "Giga" means a billion. "Terra" means a trillion. And "peta" means a quadrillion.

**p. 100 "In the previous chapter . . ."** *USA Today*, July 26, 1996, p. B1.

**p. 100 "In comparison, it is believed . . ."** Hans Moravec, *Mind Children*, Harvard University Press, Cambridge, Mass., 1988, p. 60. Interview.

**p. 102 "Other scientists have proposed . . ."** *Scientific American*, March 1994, p. 62.

**p. 102 "IBM, for example . . ."** *Scientific American*, Feb. 1995, p. 94.

**p. 102 " 'I won't quote anybody . . .' "** Ibid., p. 91.

**p. 103 "It eliminated wires . . ."** *New York Times*, Jan. 30, 1990, p. D8.

**p. 103 " 'This work is very significant . . .' "** Ibid.

**p. 104 "But a holographic memory . . ."** *Scientific American*, Nov. 1995, p. 70.

**p. 105 "For example, one can set . . ."** Note that we are losing some information here. In principle, one can economically generate far more numbers with four letters than with two. But so far DNA computers have not fully exploited this advantage.

**p. 105 " 'I have never seen . . .' "** *New York Times*, April 11, 1995, p. C10.

**p. 106 "DNA computers . . ."** Ibid.

**p. 106 "Lipton thinks . . ."** Ibid.

**p. 106 "Dan Boneh of Princeton . . ."** Ibid.

**p. 108 "Gary Frazier of Texas Instruments . . ."** The *Dallas Morning News*, Aug. 3, 1992, p. 5F.

**p. 109 "The essential point is this . . ."** This does not mean that quantum computers can calculate instantly. They also take time performing calculations, but for these difficult problems they process problems in what is called "polynomial time," i.e., the amount of time does not rise exponentially, but only as a power, which is a tremendous advantage over Turing machines.

**p. 111 " 'Now, I'm much more optimistic . . .' "** *Discover*, Oct. 1995, *Discover* magazine Web Page.

**p. 111 " 'In the not so distant future . . .' "** Ibid.

**p. 113 "These neurons sprouted . . ."** *Science News*, April 6, 1996, p. 223.

**p. 115 "Ralph Merkle of Xerox PARC . . ."** Ralph Merkle Home Page, Xerox PARC.

**p. 116 "Instead of natural selection . . ."** This might not be as silly as it first sounds. First, Minsky points out, the human brain does not know all that much. Thomas K. Landauer of Bellcore, for example, calculates that the average person learns at the rate of two bits per second. Minsky then reasons: "If one could maintain that rate for 12 hours every day for 100 years, the total could be about three billion bits—less than what we can currently store on a regular five-inch compact disc." *(Scientific American*, Oct. 1994, p. 113.)

Second, Minsky believes that, given enough time, technology will be able to transfer the mind, as envisioned by Moravec. He also believes that in the field of "nanotechnology" (to be discussed in Chapter 13), a team of tiny machines much smaller than a human hair may one day be constructed whose purpose is to connect a pile of silicon neurons in precisely the same way as the brain. He writes: "If we had a million construction machines that could build 1,000 parts per second, our task would take mere minutes." (Ibid.) Interview with Moravec.

**p. 116 " 'Our job is to see . . .' "** Ibid.

## CHAPTER 6 Second Thoughts: Will Humans Become Obsolete?

**p. 120 "The first prototype . . ."** *Scientific American*, Oct. 1992, p. 50.

**p. 120 " 'We now have demonstrated . . .' "** *Science News*, Feb. 10, 1996, p. 92.

**p. 122 " 'It's going to hit . . .' "** *Wall Street Journal*, June 17, 1996, p. R26.

**p. 122 " 'I would be running as far . . .' "** *Newsweek*, Sept. 2, 1996, p. 63.

**p. 122 " 'But it's going to hurt . . .' "** *Wall Street Journal*, June 17, 1996, p. R26.

**p. 123 " 'There's no limit to the demand . . .' "** Ibid.

**p. 125 "While stockbrokerage houses . . ."** Ibid.

**p. 125 " 'The future—mine, anyway . . .' "** *Washington Post*, June 21, 1996, p. A23.

**p. 125 " 'Then came the ignition . . .' "** Ibid.

**p. 127 "In France, 28.7 percent."** Hamish McRae, *The World in 2020*, Harvard Business School Press, Cambridge, Mass., 1994, p. 11.

**p. 127 "Michael Vlahos, a senior fellow . . ."** "The War After Byte City," Progress and Freedom Foundation, Web address: http://www.pff.org/pff/bigchange/wrbtcity.html. *Village Voice*, Feb. 6, 1996, p. 31.

**p. 127 "Frank Owen of *The Village Voice* . . ."** Ibid.

**p. 127 "As Barbara Ehrenreich observed . . ."** Ibid.

**p. 127 "He envisions vastly strengthening . . ."** Jeremy Rifkin, "A Radically Different World," *Forbes ASAP*, Dec. 2, 1996, p. 66. Interview, May 12, 1997.

**p. 129 "Furthermore, today 50 percent fewer Ph.D.'s . . ."** Lester C. Thurow, *The Future of Capitalism*, William Morrow, New York, 1996, p. 286.

**p. 129 " 'Put bluntly . . .' "** Ibid., p. 284.

**p. 129 " 'When technology and ideology . . .' "** Ibid., p. 326.

**p. 130 " 'Are we creating the next species . . .' "** Daniel Crevier, *AI*, Basic Books, New York, 1993, p. 341. Interview.

**p. 131 " 'Machines merely as clever . . .' "** Hans Moravec, *Mind Children*, Harvard University Press, Cambridge, Mass., 1988, p. 100. Interview.

**p. 132 " 'Had the ship's captain known . . .' "** Crevier, *AI*, p. 316.

**p. 133 "As Daniel Crevier has noted . . ."** Crevier, *AI*, p. 318. Interview.

## CHAPTER 7 Personal DNA Codes

**p. 139 "On his shoulders rests . . ."** Interview with Francis Collins, May 7, 1996.

**p. 140 " 'Success is nothing more . . .' "** Ibid.

**p. 140 " 'I feel I've been preparing . . .' "** Jeff Lyon and Peter Gorner, *Altered Fates*, W. W. Norton, New York, 1995, p. 359.

**p. 140 " 'Behavior patterns . . .' "** Interview with Francis Collins.

**p. 140 " 'If humanity begins to view itself . . .' "** Ibid.

**p. 140 "The task Collins . . ."** The precise number of genes in the human genome is not known. Some geneticists believe that the number could be less than 100,000, perhaps as low as 60,000.

**p. 141 "Eric Lander . . ."** Lyon and Gorner, *altered Fates*, p. 535.

**p. 141 "For example, the human brain . . ."** *New York Times*, Sept. 28, 1995, p. A24.

**p. 141 "By mid-1994, the list . . ."** Sir John Kendrew, *The Encyclopedia of Molecular Biology*, Blackwell Science, Cambridge, Eng., 1994, p. 489.

**p. 141 "By late 1996 . . ."** *New York Times*, Oct. 25, 1996, p. A18. *Science*, Oct. 1996.

**p. 141 "Given the astonishing advances . . ."** *New York Times*, Sept. 28, 1995, p. A24.

**p. 142 "Similarly, because DNA . . ."** Interview with Walter Gilbert, Dec. 30, 1996.

**p. 143 "*By the year 2000*, Gilbert claims . . ."** Lyons and Gorner, *Altered Fates*, p. 532. Interview.

**p. 143 "*By 2010*, the genetic profiles . . ."** In principle, the number of genetic diseases could be much larger than this, because almost any mutation in a gene can conceivably cause a genetic disease. The number 5,000 refers to the approximate number of genetic diseases that have been recorded by doctors. Interview.

**p. 143 " 'It is reasonably likely that . . .' "** *Scientific American*, Sept. 1995, p. 140.

**p. 143 "Gilbert claims, 'You'll be able . . .' "** Lyon and Gorner, *Altered Fates*, p. 532. Interview.

**p. 143 "The next century, Gilbert . . ."** Daniel Kevles and Leroy Hood, eds., *The Code of Codes*, Harvard University Press, Cambridge, Mass., 1992, p. 96.

**p. 144 " 'The possession of a genetic map . . .' "** Interview with Walter Gilbert. Kevles and Hood, *The Code of Codes*, p. 94.

**p. 145 " 'Medicine is basically going to change . . .' "** *Time* Special Issue, Fall 1996, p. 25.

**p. 146 "As a student, Collins was repelled . . ."** Interview with Francis Collins.

**p. 146 " 'The ability to describe . . .' "** Ibid.

**p. 146 "Biologist Stephen Jay Gould . . ."** Michael P. Murphy and Luke A. J. O'Neill, *What Is Life? The Next Fifty Years*, Cambridge University Press, Cambridge, Eng., 1995, p. 25.

**p. 146 "He recalled . . ."** Walter Moore, *Schrödinger: Life and Thought*, Cambridge University Press, Cambridge, Eng., 1989, p. 403.

**p. 146 "At Cambridge University . . ."** Watson recalled, "A major factor in [Crick's] leaving physics and developing an interest in biology had been the reading in 1946 of *What Is Life?* by the noted theoretical physicist Erwin Schrödinger." (James Watson, *The Double Helix*, Penguin Books, New York, 1968, p. 18.)

**p. 147 "Using this result, Watson and Crick . . ."** Watson, *The Double Helix*.

**p. 147 "The DNA, in turn, consists . . ."** These stand for the nucleic acids: A = adenine, C = cytosine, T = thymine, G = guanine.

**p. 149 " 'Whereas what was going on . . .' "** Interview with Francis Collins. *Time*, Jan. 17, 1994, p. 55.

**p. 150 "The counterpart of a magnifying lens . . ."** A modern way to magnify strands of DNA is to use the PCR (polymerase chain reaction) technique. By heating a DNA sample, we can separate the double strands. When we add DNA polymerase and cool the sample, each strand makes a copy of itself. Starting with one DNA molecule, we now have two. By alternately heating and cooling the sample several times, we can therefore increase the number of DNA molecules exponentially. Attaining increases by factors of millions and billions is not difficult with this technique.

**p. 150 "Thus we will know the genetic . . ."** This assumes that DNA molecules do not mutate too quickly over time. In the case of viruses, the mutation rate is so rapid that it may not be possible to completely reproduce their family tree.

**p. 151 " 'Gilbert's cost projections provoked . . .' "** Robert Cook-Deegan, *The Gene Wars*, W. W. Norton, New York, 1994, p. 111.

**p. 151 "Now it costs less than fifty cents . . ."** Interview with Francis Collins.

**p. 151 "For example, below is a table . . ."** Christopher Wills, *Exons, Introns, and Talking Genes*, Basic Books, New York, 1991, p. 273; Kevles and Hood, *The Code of Codes*, p. 137.

**p. 151 "This virus was chosen because . . ."** Since that time, many viruses have been decoded. In the early 1980s, the lambda virus (which infects *E. coli* bacteria) was decoded and all its 48,514 bases revealed, which in turn code for 50 different proteins.

**p. 151 "If spelled out in terms . . ."** Wills, *Exons, Introns, and Talking Genes*, p. 42; Kevles and Hood, *The Code of Codes*, p. 65.

**p. 152 "Smallpox is known to have . . ."** Enzo Russo and David Cove, *Genetic Engineering*, W. H. Freeman, New York, 1995, p. 54.

**p. 152 "One of the longest viruses . . ."** Thomas Lee, *The Human Genome Project*, Plenum Press, New York, 1991, p. 170; Wills, *Exons, Introns, and Talking Genes*, p. 317.

**p. 152 "The cell, *Hemophilus influenzae* . . ."** *New York Times*, Aug. 1, 1995, p. C1. The same team even announced that a second cell had also been sequenced: *Mycoplasma genitalium*, with 580,067 base pairs. The techniques are so advanced that the sequencing of this second bacteria was done in just three months.

**p. 152 "In early 1996, even this herculean feat . . ."** "Progress from the National Center for Human Genome Research," NIH, Bethesda, Md., April 24, 1996.

**p. 152 "In 1997, scientists at . . ."** *Science News*, Feb. 8, 1997, p. 84.

**p. 152 "Human sequencing has resulted . . ."** One unexpected result was the discovery that 95 to 98 percent of the information in our DNA is apparently nonsense, for which there is no obvious use. However, the remaining 2 to 5 percent is essential. The genes involved code for the vital protein molecules which are the workhorses of our body, making up the enzymes, tissues, and cell components that make our body work.

**p. 153 "This is reflected in the fact . . ."** "Progress from the National Center for Human Genome Research," April 24, 1996.

**p. 153 "We find that we share . . ."** Steve Jones, Robert Martin, David Pilbeam, and Sarah Bunney, eds., *The Cambridge Encyclopedia of Human Evolution*, Cambridge University Press, Cambridge, Eng., 1992, p. 310.

**p. 153 "From this, some scientists . . ."** Because the DNA sequence contained in the mitochondria of the cell is inherited only from the mother, it changes extremely slowly over many generations, and hence can be used to give us a "genetic clock" of the rate at which mutations occur in the human genome. *The Cambridge Encyclopedia of Human Evolution*, p. 396.

**p. 154 "In the future, not only . . ."** Thomas F. Lee, *Gene Future*, Plenum Press, New York, 1993, p. 67.

**p. 154 "In 1997, DNA evidence . . ."** *New York Times*, Feb. 5, 1996, p. A12; *Washington Post*, Jan. 5, 1997, p. A1. In 1993, DNA testing confirmed that certain bones discovered in Russia are, in fact, the remains of the last Czar, Nicholas II, and his family, killed by the Bolsheviks during the Russian Revolution. The Czar's DNA was compared with the DNA from Queen Elizabeth's husband, Prince Philip (whose grandmother was Czarina Alexandra's sister). *(Discover*, Jan. 1994, p. 90.) DNA testing by two groups, one in England and the other in Germany, has also shown that Anna Anderson Manahan, the woman claiming to be "Anastasia," who surfaced after the 1917 Russian Revolution, was actually an impostor. ("Research in the News," NIH Web Page, http://www.nih.gov.)

**p. 154 "This is, says Poinar . . ."** *New York Times*, Jan. 30, 1995, p. C10.

**p. 155 "The gene *PLA2* . . ."** *Science News*, Aug. 3, 1996, p. 77.

**p. 155 "They have reversed the verdict . . ."** *New York Times*, June 14, 1996, p. A12. Analysis of these cases has shown that most of these wrongly accused men were poor and could not afford expensive lawyers to counter misidentification by the police and the victims.

In an astonishing forty-eight cases, prisoners who were on death row since 1973 were released after DNA and other new evidence was obtained. In one dramatic case, on June 14, 1996, four young Chicago youths, two sentenced to death row, were shown to be innocent of a brutal rape and murder after spending eighteen years in jail. "The DNA is why we are here today," said Jeffrey Undangen, lawyer for one of them. They went from shackles to freedom overnight. (*New York Times*, June 15, 1996, p. 6.)

Nationwide, over five thousand DNA tests are conducted routinely, half in the FBI Washington laboratory, and the number is rapidly climbing. (*Washington Post*, Aug. 14, 1996, p. C3.)

**p. 155 "Of those, twenty-six states . . ."** *New York Times*, June 14, 1996, p. A12.

**p. 155 "For example . . ."** See note on the polymerase chain reaction, above.

**p. 156 " 'That day has arrived.' "** *New York Times*, June 11, 1996, p. C1.

**p. 156 "By 1992, a single machine could . . ."** Kevles and Hood, *The Code of Codes*, p. 146.

**p. 156 "Ulam, like Schrödinger, Delbrück . . ."** Cook-Deegan, *The Gene Wars*, pp. 283–85.

**p. 156 "By 1997, GenBank . . ."** Wills, *Exons, Introns, and Talking Genes*, pp. 51, 90.

**p. 157 "Robert Waterson, a mathematician . . ."** *New York Times*, June 11, 1996, p. C12.

**p. 157 " 'We figure that for us to . . .' "** Ibid.

**p. 157 " 'But from a computer science . . .' "** *New York Times*, June 11, 1996, p. C12.

**p. 157 "He and his colleagues noticed . . ."** Cook-Deegan, *The Gene Wars*, p. 293.

**p. 157 "Robert Cook-Deegan of the . . ."** Ibid.

**p. 158 " 'A new breed of scientist . . .' "** Ibid., pp. 293–94.

**p. 158 "The bio chip is a microchip . . ."** Wills, *Exons, Introns, and Talking Genes*, p. 97.

**p. 158 "A primitive bio chip already exists . . ."** Kevles and Hood, *The Code of Codes*, p. 147.

**p. 158 "This new diagnostic tool . . ."** *New York Times*, Aug. 18, 1996, sec. 3, p. 1.

**p. 159 " 'We have actually produced a prototype . . .' "** *Scientific American*, Sept. 1996, p. 42.

**p. 159 "(The traditional process . . ."** Ibid.

**p. 159 "As Walter Gilbert says . . ."** Kevles and Hood, *The Code of Codes*, p. 93.

**p. 160 "Yeast has proven . . ."** "Progress from the National Center for Human Genome Research," April 24, 1996.

## CHAPTER 8 Conquering Cancer—Fixing Our Genes

**p. 162 " 'She keeps us all in the game.' "** *Washington Post*, June 11, 1996, Health section, p. 11.

**p. 163 "Chemotherapy is out . . ."** Ibid.

**p. 163 " 'This is a desperate field . . .' "** Ibid.

**p. 163 "But 1996 also saw new hope . . ."** *Washington Post*, Aug. 29, 1996, p. A9.

**p. 164 " 'What an incredible time to . . .' "** Jeff Lyon and Peter Gorner, *Altered Fates*, W. W. Norton, New York, 1995, p. 24.

**p. 164 "Anderson predicts that . . ."** *Time* Special Issue, Fall 1996, p. 28.

**p. 164 "Leroy Hood . . ."** Lyon and Gorner, *Altered Fates*, p. 35.

**p. 164 "He says, 'You can sit . . .' "** Ibid., p. 28.

**p. 164 "Science 'is something . . .' "** Ibid., p. 37.

**p. 164 "By 1996, this number . . ."** Gene therapy experiments have involved SCID, cystic fibrosis, Gaucher's disease, familial hypercholesterolemia, hemophilia, purine nucleoside phosphorylase deficiency, alpha-1 antitrypsin deficiency, Fanconi's anemia, Hunter's syndrome, chronic granulomatous disease, rheumatoid arthritis, peripheral vascular disease, AIDS, and cancer (melanoma, renal cell, ovarian, neuroblastoma, brain, head and neck, lung, liver, breast, colon, prostate, mesothelioma, leukemia, lymphoma, multiple myeloma). See *Scientific American*, Sept. 1995, p. 128.

**p. 164 "About 30 diseases . . ."** *Time* Special Issue, Fall 1996, p. 29.

**p. 164 "Gene therapy experiments . . ."** *Science News*, Dec. 23 and 30, 1995, p. 428.

**p. 165 "Abigail Salyers and Dixie Whitt . . ."** Abigail Salyers and Dixie D. Whitt, *Bacterial Pathogenesis*, ASM Press, Washington, D.C., 1994, p. 100.

**p. 165 "As Sherwin B. Nuland . . ."** *Time* Special Issue, Fall 1996, p. 12.

**p. 165 "Cancer, one of the most dreaded . . ."** Enzo Russo and David Cove, *Genetic Engineering*, W. H. Freeman, New York, 1995, p. 4.

**p. 166 " 'The pieces of the puzzle . . .' "** Robert A. Weinberg, *Racing to the Beginning of the Road*, Random House, New York, 1996, p. 256.

**p. 166 "As Dennis Salmon . . ."** *Time*, April 25, 1994, p. 56.

**p. 166 "And by comparing . . ."** Russo and Cove, *Genetic Engineering*, p. 123.

**pp. 166–67 "In other words, a cell . . ."** A significant breakthrough came in 1975, when J. Michael Bishop and Harold E. Varmus of the University of California at San Francisco found that human cells contain mutations within their oncogenes which can make them become cancerous. That pathbreaking work won them the Nobel Prize.

**p. 167 "These oncogenes include the gene . . ."** Thomas F. Lee, *The Human Genome Project*, Plenum Press, New York, 1991, p. 198.

**p. 167 "*P-53* is so essential . . ."** *Time*, April 25, 1994, p. 60.

**p. 167 "Understanding *p-53* has . . ."** Although *p-53* was discovered in 1979 by Arnold J. Levine of Princeton and David Lane of the Molecular Research Council in Cambridge, only in the last six years have scientists realized how potent and how central this gene is to cancer.

**p. 168 "By analyzing the ways . . ."** *New York Times*, April 23, 1991, p. C9.

**p. 169 "All this changed in 1996, when scientists proved . . ."** Mikhail F. Denissenko, et al., "Preferential Formation of Benzo[a]pyrene Aducts at Lung Cancer Mutational Hotspots," in *P53*, *Science*, Oct. 18, 1996, p. 430.

**p. 169 "It may also help solve . . ."** One of the most dreaded cancers is breast cancer, which strikes 180,000 women in the United States each year, killing 46,000. Isolating the cause of breast cancer is also frustrating, since it is linked to a number of factors, such as high-fat diets, delayed childbirth, early menstruation, late menopause, family history, radiation and chemicals, etc. But what might link all these risk factors is genetics.

(The fact that the incidence of breast cancer grows directly with the number of menstrual cycles may partly help to explain its resurgence in modern times. Some scientists believe that because women today are menstruating earlier than in the last century and have fewer children, they have appreciably more menstrual cycles and hence possibly more breast cancers as a consequence. According to Boyd Eaton of Emory University, American women experience 3.5 times more menstrual cycles than our ancestors 10,000 years ago, and hence have 3.5 times the number of breast cancers.) *(Discover*, Oct. 1995.)

In 1990, an international team of scientists announced that they captured the first breast cancer gene in the laboratory, later christened *BRCA1*, found on chromosome 17. In high-risk families, having the *BRCA1* gene gives a woman a 85 percent chance of developing breast cancer. *(Science News*, Dec. 9, 1995, p. 395.)

Then, in 1995, a second breast cancer gene, *BRCA2*, was found on chromosome 13. Scientists at the Institute of Cancer Research in Sutton, England, have emphasized that 90 percent of all inherited cases of breast cancer can be attributed to one of these two genes.

Scientists emphasize that the *BRCA2* is also very long, containing perhaps 12,000 base pairs, and that only 7,000 base pairs have so far been identified.

Already, *BRCA1* and *BRCA2* are implicated in causing breast cancer among a specific group: Ashkenazi Jews, mainly from Eastern Europe, who make up about 90 percent of the 6 million Jews in the United States. Together, *BRCA1* and *BRCA2* may account for 25 percent of all early-onset cases of breast cancer in that group. *BRCA1* is quite large, and 125 different mutations have now been found. *BRCA2* may have even more.

**p. 170 "But if a cell contains . . ."** *Scientific American*, Feb. 1996, p. 95.

**p. 170 "Close examination of the telomeres . . ."** Ibid., p. 92.

**p. 170 "Cancer cells, because they can . . ."** *Time*, April 25, 1994, p. 58.

**p. 171 "As Richard Klausner . . ."** *Time* Special Issue, Fall 1996.

**p. 172 "But Lloyd Old . . ."** *Scientific American*, Sept. 1996, p. 138.

**p. 172 "For example, the protein . . ."** Ibid., p. 148.

**p. 172 "With modern techniques . . ."** Ibid., p. 141.

**p. 172 "The strategy to block blood . . ."** Ibid., p. 152.

**p. 173 "Frédéric Chopin . . ."** *New York Times*, June 6, 1995, p. C3.

**p. 173 "Genetic diseases take a particularly . . ."** Ibid., p. 30.

**p. 173 "As a result, the battle . . ."** Christopher Wills, *Exons, Introns, and Talking Genes*, Basic Books, New York, 1991, p. 310.

**p. 174 "The Talmud . . ."** Lyon and Gorner, *Altered Fates*, p. 38.

**p. 174 "It was, apparently . . ."** Steve Jones, *The Language of Genes*, Anchor Books, New York, 1993, p. 73; *New York Times*, June 6, 1995, p. C3.

**p. 174 " 'Our poor family . . .' "** Jones, *The Language of Genes*, p. 73. All told, perhaps as many as twenty of her royal female descendants were carriers and ten of her male descendants were hemophiliacs. (Steve Jones, Robert Martin, David Pilbeam, and Sarah Bunney, eds., *The Cambridge Encyclopedia of Human Evolution*, Cambridge University Press, Cambridge, Eng., 1992, p. 260.)

**p. 174 "Some historians claim Rasputin . . ."** *Science News*, Dec. 9, 1995, p. 394.

**p. 174 "As geneticist Steve Jones . . ."** Jones, *The Language of Genes*, p. 73.

**p. 174 "Some of them are truly . . ."** Lyon and Gorner, *Altered Fates*, p. 86.

**p. 174 "(The most famous victim . . ."** Merrick (who some claim might actually have suffered from a related disease, the Proteus syndrome) was immortalized in a Broadway play and a movie. Michael Jackson, intrigued by Merrick, even tried to purchase his bones. (Lyon and Gorner, *Altered Fates*, p. 337; Lee, *The Human Genome Project*, p. 195.)

**p. 175 "This is the most common genetic ailment . . ."** Chances are that among your friends, several are probably silent carriers of CF. Because each Caucasian has a 1/25 chance of having the CF gene, the chances that two carriers will mate is 1/25 × 1/25. Of their offspring, only 1/4 will carry both recessive genes for CF. So the rate of CF is roughly 1/25 × 1/25 × 1/4, or 1/2600, which is roughly the rate at which it is found among Caucasians.

**p. 175 "CF is a parent's nightmare . . ."** Wills, *Exons, Introns, and Talking Genes*, p. 195; Lyon and Gorner, *Altered Fates*, p. 384; Thomas F. Lee, *Gene Future*, Plenum Press, New York, 1993, p. 89.

**p. 175 " 'Woe to that child . . .' "** *Scientific American*, Dec. 1995, p. 52.

**p. 175 "Within this population, as many . . ."** Lee, *The Human Genome Project*, p. 263; Lee, *Gene Future*, p. 90; Lyon and Gorner, *Altered Fates*, pp. 3, 215.

**p. 175 "It is a chronic disease . . ."** Wills, *Exons, Introns, and Talking Genes*, p. 216; Lee, *The Human Genome Project*, p. 93; Lee, *Gene Future*, p. 96.

**p. 175 "At times, searching for these . . ."** *Time*, Jan. 17, 1994, p. 48.

**p. 176 "The mutation is caused by . . ."** Lyon and Gorner, *Altered Fates*, pp. 383, 398, 401.

**p. 176 "This, in turn, triggers . . ."** *Scientific American*, Dec. 1995, p. 52.

**p. 176 "This disease is caused . . ."** Lyon and Gorner, *Altered Fates*, p. 92. Wills, *Exons, Introns, and Talking Genes*, p. 63.

**p. 176 "In 1986, scientists finally isolated . . ."** Lyon and Gorner, *Altered Fates*, p. 355.

**p. 176 "It's one of the longest genes . . ."** Wills, *Exons, Introns, and Talking Genes*, p. 192.

**p. 177 "In fact, its extraordinary length . . ."** Lyon and Gorner, *Altered Fates*, p. 358.

**p. 177 "Nancy Wexler, who helped . . ."** *Time*, Special Issue, Fall 1996, p. 29.

**p. 177 "Before he died in 1984 . . ."** Lee, *Gene Future*, p. 182.

**p. 178 "David Rimoin of the Cedars-Sinai . . ."** *Time*, Special Issue, Fall 1996, p. 29.

**p. 178 "As Francis Collins says . . ."** Interview with Francis Collins, May 7, 1996.

**p. 178 "Although there were those who snickered . . ."** Interview with Michael Blaese, May 7, 1996.

**p. 179 "This disease, which strikes . . ."** *New York Times*, Oct. 31, 1995, p. C3. In the 1960s, in the mistaken belief that schizophrenics could be "cured" of their voices by antipsychotic or neuroleptic drugs like Thorazine, hundreds of thousands were discharged from state hospitals, eventually winding up homeless in the most decrepit urban areas, talking aimlessly to imaginary tormentors. (Lyon and Gorner, *Altered Fates*, p. 467.)

**p. 179 "However, the hope that this . . ."** Lyon and Gorner, *Altered Fates*, p. 468; Wills, *Exons, Introns, and Talking Genes*, p. 259.

**p. 179 "In 1995, another series . . ."** *New York Times*, Oct. 31, 1995, p. C3.

**p. 179 "Walter Gilbert believes that . . ."** Interview with Walter Gilbert.

## CHAPTER 9 Molecular Medicine and the Mind/Body Link

**p. 182 "Back in 1969, the Surgeon General . . ."** Randolph M. Nesse and George C. Williams, *Why We Get Sick*, Random House, New York, 1994, p. 52; Laurie Garrett, *The Coming Plague*, Penguin Books, New York, 1994, p. 33.

**p. 183 "One of the missions . . ."** *Viruses: The Greatest Threat to the Survival of*

*Our Species*, Pangea Digital Pictures, IVN Communications, in cooperation with *Discover* magazine.

**p. 183 "It has destroyed entire cultures . . ."** The mummy of the great Pharaoh Ramses, who is believed to have confronted Moses and banished the Jews from Egypt, shows signs of death by this disease. (Arnold J. Levine, *Viruses*, Scientific American Books, New York, 1992, p. 57.)

**p. 183 "By May 8, 1980 . . ."** *The World Paper*, May 1996, p. 1. Because the genome of smallpox, with about 190 genes, has been completely decoded for future study, most scientists feel confident in destroying this killer virus. As an added precaution, smallpox vaccines have been stored in case of an unexpected emergency.

**p. 183 "In June 1999 . . ."** The date has been changed several times and is still subject to change by a vote of the WHO's governing board. There are a minority of voices who believe that the smallpox samples should be kept alive for future study. Also, it is highly unlikely that other smallpox viruses exist on the planet, since no new cases have been reported to WHO in many years, much longer than the time needed for smallpox to infect new victims. The chain of infection has apparently been broken.

**p. 183 "Other diseases that may join . . ."** Executive Summary of the World Health Report 1996 of the WHO, 1996, p. 12, WHO Web Page: http://www.who.org.

**p. 184 "These viruses are so dangerous . . ."** Being studied in Building 15 is the Ebola virus, which is a type of hemorrhagic fever virus *(Scientific American*, Oct. 1995, pp. 56–57) causing a death by grotesque, uncontrolled bleeding. Other deadly hemorrhagic fever viruses studied at the CDC include the Lassa fever, hanta, dengue, yellow fever, and encephalitis viruses. Ebola is so mysterious that we do not even know what "vector" (e.g., mice, mosquito, lice) carries the disease before infecting humans.

**p. 184 "One doctor described the feeling . . ."** *New York Times Magazine*, Aug. 21, 1994, p. 37.

**p. 184 "This technique was first demonstrated . . ."** Ann Giudici Fettner, *The Science of Viruses*, William Morrow, New York, 1990, p. 125.

**p. 184 "By analyzing the three-dimensional . . ."** Unfortunately, knowing the structure of rhinovirus 14 is not decisive in the battle against the common cold. Because there are hundreds of different varieties of viruses that can cause a common cold, it would be prohibitively expensive at present to mount a frontal assault on the cold.

**p. 185 "We can also see why it is so easy to contract colds."** In addition, we can see why it's so easy to contract this type of cold virus. This soccer ball binds to lymphocytes that line our lung tissue by attaching on to receptor sites called intercellular adhesion molecules (ICAMs). This is a very clever mechanism. When our immune system reacts to the cold virus by increasing the number of ICAM sites, it simply makes more openings by which the virus can enter the cell, making the situation worse. (Nesse and Williams, *Why We Get Sick*, p. 42.)

**p. 185 "The influenza virus has . . ."** It is believed that the pandemic was probably caused by the social chaos of World War I, which displaced many refugee populations and weakened people's immune system.

**p. 185 "Fully half the people on the entire planet . . ."** Fettner, *The Science of Viruses*, p. 134.

**p. 185 "It was so virulent that . . ."** Enzo Russo and David Cove, *Genetic Engineering*, W. H. Freeman, New York, 1995, p. 62; *New York Times*, Jan. 27, 1996, p. 21.

**p. 185 "Pigs get infected by both duck and human . . ."** Fettner, *The Science of Viruses*, p. 132.

**p. 186 "Although it was reported in 1997 . . ."** *Washington Post*, Jan. 25, 1997, p. A1; *New York Times*, Feb. 28, 1997, p. A1.

**p. 187 "More significant is the number . . ."** *New York Times*, June 7, 1996, p. A3.

**p. 187 "This translates into a world tragedy . . ."** Even in the United States, the best estimate is that 630,000 to 900,000 people were infected with the HIV by 1993. In 1996, about 50,000 people died of AIDs-related diseases. In that year, about 40,000 to 80,000 people were infected with HIV. Into the next century, perhaps the main victims of this disease will be Africans and Asians. Unfortunately, governments and religious leaders are reluctant, even resistant to discussing this disease, and hence it will continue to spread almost unabated throughout large parts of the Third World.

In some areas, up to 40 percent of the adult population is infected with HIV. In Africa, sub-Saharan countries like Botswana, Uganda, and Zimbabwe could lose a quarter of their population to the disease.

Although Africa has gotten most of the bad news, the next and biggest frontier for AIDS is Asia, which holds two-thirds of the world's population. The World Health Organization, analyzing the alarming demographic trends in Asia, predicts only the worst. At present, WHO estimates that 3.5 million Asians are infected with the virus. By the year 2000, that number will rise to 12 million. The best estimate is that 10 million Asians will die of AIDS before 2015.

**p. 187 "Death usually follows within two years."** *Scientific American*, Aug. 1995, p. 60.

**p. 188 "In this way, one can use the HIV . . ."** *New York Newsday*, Dec. 21, 1993, p. 63.

**p. 188 "We know that each mutates . . ."** Meyers found the following subclasses of HIV-1:

| *Subclass of HIV-1* | *Area Affected* |
|---|---|
| Type A | Central and South Africa, India |
| Type B | North America, Peru, Europe, Brazil, southern Thailand, Africa |
| Type C | Malaysia, India, South America |
| Type D | Rwanda, Tanzania, Uganda |
| Type E | Africa, Thailand, India |
| Type F | Romania, Gabon, Zaire, and Brazil |

**p. 188 " 'And the various strains have evolved . . .' "** Ibid.

**p. 188 "Given the large number of GIs . . ."** *Discover*, June 1996, p. 69.

**p. 189 "With AZT substituting for thymidine . . ."** Ibid., p. 494.

**p. 190 "These individuals' immune cells lack . . ."** By examining the genes of these immune individuals, it was soon found that they possess a mutation in their immune system which prevents the HIV from grabbing on to their T cells. Normally, T cells possess a "docking site" on their protein surface which the HIV grab on to in order to enter the cell. However, mutant versions of these T cells lack the docking site and hence apparently give total immunity to the HIV.

From cloning the genes involved, scientists now know how this remarkable mechanism operates. Everyone possesses two copies of the *CKR5* gene, which manufactures the docking site, which is about 1,000 base pairs long. In mutated form, about 32 of these base pairs are actually missing, which interferes with the construction of the docking site.

If one inherits two copies of the mutated *CKR5* gene (one from each parent) then one is apparently immune from the HIV because they cannot grab on to the T cell's docking site. However, if one inherits only one mutated form, then the production of docking sites is lessened, but not enough to give immunity. People with only one version of the mutated *CKR5* gene live a bit longer, to about thirteen years, compared with the average of ten years.

**p. 190 "Viruses, which are tiny strands . . ."** By rights, bacteria should be sitting ducks. Bacteria, since they have their own metabolic machinery and require complex molecular processes in order to reproduce, can be attacked by disrupting any of these delicate processes.

**p. 190 "As James Hughes of the Centers . . ."** *Parade,* April 23, 1995, p. 10.

**p. 190 "For example, Legionnaires' disease . . ."** Legionnaires' disease is the result of the proliferation of air conditioning. It grows in an entirely artificial environment, the water within a hotel's air-conditioning system. Toxic shock syndrome is caused by a new superabsorbent type of tampon which created a novel environment with vastly increased surface area and plenty of oxygen. This created ideal conditions for the rapid growth of staphylococcal bacteria (called *Staphylococcus aureus,* which grows 10,000 times faster than normal and releases copious quantities of a deadly toxin called TSST-1). (Garrett, *The Coming Plague,* pp. 408, 553.) Lyme disease became a severe problem when suburbs began to spread into territory populated by deer and mice, which are carriers for ticks that spread Lyme disease (caused by the bacterium *Borrelia burgdorferi).*

**p. 191 "As one author puts it . . ."** G. Youmans, P. Paterson, and H. Sommers, *The Biological and Clinical Basis of Infectious Diseases,* W. B. Saunders, Philadelphia, 1980.

**p. 191 "Unfortunately, bacteria have the upper hand . . ."** Nesse and Williams, *Why We Get Sick,* p. 51.

**p. 191 "Today, more than 8,000 antibiotics . . ."** Bernard Dixon, *Power Unseen,* W. H. Freeman, New York, 1994, p. 21. *New York Times,* April 26, 1996, p. D1.

**p. 191 "About one in every 10 million . . ."** Russo and Cove, *Genetic Engineering,* p. 97.

**p. 192 "Today, there are mutant strains . . ."** Russo and Cove, *Genetic Engineering,* p. 96.

**p. 192 "As Robert E. Shope, professor of . . ."** *Washington Post,* June 27, 1995, p. A6.

**p. 192 " 'There are only so many ways . . .' "** *Discover,* Aug. 1994, p. 46.

**p. 192 "Because it often takes . . ."** *Washington Post,* April 14, 1996, p. H1. Abigail Salyers and Dixie Whitt are more blunt, laying the blame on the search for profits: "The pharmaceutical companies have been responding to market forces, in particular the current glut of antibiotics on the market, by shutting down or drastically cutting back their antibiotic discovery programs." (Abigail A. Salyers and Dixie D. Whitt, *Bacterial Pathogenesis,* ASM Press, Washington, D.C., 1994, p. 101.)

**p. 192 "That is a staggering 50 percent . . ."** *Washington Post,* June 27, 1995, p. A6.

**p. 193 "This can cause unintended effects."** For example, before genetic engineering was able to manufacture unlimited quantities of human insulin, scientists were forced to use pig insulin to control diabetes, which often had unpleasant side effects.

Similarly, antipsychotic drugs to control schizophrenia have to be carefully monitored because they sometimes cause bizarre movements of the tongue.

**p. 193 " 'It's like throwing a monkey wrench . . .' "** *Discover*, Aug. 1994, p. 49.

**p. 193 "Then they can design new antibiotics . . ."** Ibid.

**p. 193 " 'It's a straightforward counterattack by . . .' "** Ibid.

**p. 194 "Today, 90 percent of gonorrhea bacteria in . . ."** Nesse and Williams, *Why We Get Sick*, p. 54.

**p. 194 "New anesthetics, for example, have been found . . ."** *Scientific American*, June 1994, p. 84.

**p. 194 "The skin of an African frog . . ."** *New York Times*, April 26, 1996, p. D3.

**p. 194 "He sent the plant to . . ."** *Discover*, Nov. 1993, p. 60.

**p. 194 "In fact, 86 percent of the plants . . ."** *Scientific American*, June 1994, p. 84.

**p. 195 "New computerized methods can drive . . ."** *Washington Post*, April 14, 1996, p. H6.

**p. 195 "Edward Hurwitz, a biotechnology analyst . . ."** Ibid.

**p. 196 "Doctors followed 1,551 people for . . ."** *New York Times*, Dec. 17, 1996, p. C3.

**p. 196 "This indicated that social contact . . ."** Annika Rosengren et al., "Stressful Life Events, Social Support, and Mortality in Men Born in 1933," *British Medical Journal*, Oct. 19, 1993.

**p. 196 "By deliberately exposing students to . . ."** Sheldon Cohen et al., "Psychological Stress and Susceptibility to the Common Cold," *New England Journal of Medicine*, vol. 325 (1991).

**p. 197 "Stress even adversely affected the nervous system . . ."** Bruce McEwen and Eliot Stellar, "Stress and the Individual: Mechanisms Leading to Disease," *Archives of Internal Medicine*, vol. 153 (Sept. 27, 1993); M. Robertson and J. Ritz, "Biology and Clinical Relevance of Human Natural Killer Cells," *Blood*, vol. 76 (1990).

**p. 197 "flare-ups in herpes due to stress"** Ronald Glaser and Janice Kiecolt-Glaser, "Psychological Influences on Immunity," *American Psychologist*, vol. 43 (1988); H. E. Schmidt et al., "Stress as a Precipitating Factor in Subjects with Recurrent Herpes Labialis," *Journal of Family Practice*, vol. 20 (1985).

**p. 197 "incidence of colon cancer and stress"** Joseph C. Courtney et al., "Stressful Life Events and the Risk of Colorectal Cancer," *Epidemiology*, vol. 4 (5) (Sept. 1993).

**p. 197 "incidence of heart disease and hopelessness"** Robert Anda et al., "Depressed Affect, Hopelessness, and the Risk of Ischemic Heart Disease in a Cohort of U.S. Adults," *Epidemiology*, July 1993.

**p. 197 "surviving bypass heart surgery and optimism"** Chris Peterson et al., *Learned Helplessness: A Theory for the Age of Personal Control*, Oxford University Press, New York, 1993.

**p. 197 "surviving second heart attacks and anger"** Carl Thoreson, International Congress of Behavior Medicine, Uppsala, Sweden, July 1990.

**p. 197 "heart attack rate and depression"** Nancy Frasure-Smith et al., "Depression Following Myocardial Infarction," *Journal of the American Medical Association*, Oct. 20, 1993.

**p. 197 "survival rates from breast cancer . . ."** David Spiegel et al., "Effect of

Psychosocial Treatment on Survival of Patients with Metastatic Breast Cancer," *Lancet*, No. 8668, ii (1989).

**p. 197 "The list of experimental and epidemiological . . ."** We should also point out that some recent studies have contradicted some of these results. As with any science, only more studies with larger pools of human subjects will resolve the question. Science is always based on reproducible results.

**p. 197 "In 1996, scientists at the National Institute . . ."** *Washington Post*, Oct. 17, 1996, p. A8. *New England Journal of Medicine*, Oct. 1996.

**p. 199 "In the twenty-first century, however . . ."** M. Stehling, R. Turner, and P. Mansfield, "Echo-Planar Imaging: MRI in a Fraction of a Second," *Science*, 254 (1991): 2–11; Clement Bezold, Jerome A. Halperin, and Jacqueline L. Eng, *2020 Visions*, U.S. Pharmacopeial Convention Press, Rockville, Md., 1993, p. 121.

**p. 199 "Not only is this cheaper . . ."** *New York Times*, Nov. 19, 1996, p. C9.

## CHAPTER 10 To Live Forever?

**p. 201 "Leonard Hayflick of the University of . . ."** Leonard Hayflick, *How and Why We Age*, Ballantine Books, New York, 1994, p. 259. Interview, Nov. 15, 1996.

**p. 201 "It should be stressed that . . ."** That genes affect life span is not in question. What is controversial is whether a small handful of genes determine aging. Many scientists believe that evolution creates genes which have "trade-offs," so that a gene which is beneficial in one area may create problems in another. Thus, if evolution selects for a gene which gives us vitality and energy during our youth, the same gene may help to deteriorate our bodies during old age. Thus, even if we can modify certain genes which may regenerate our bodies, these genes may also accelerate the degradation of other organs of the body. Since evolution favors protecting the health of childbearing individuals, there may be no specific set of genes whose function is to prolong our life span. If this is true, then "fine-tuning" the genes of the body to ensure health and longevity will be a difficult process, perhaps involving the orchestration of hundreds or thousands of genes.

**p. 202 "Not only must science and medicine . . ."** Leonard Hayflick, for example, has expressed grave doubts about the wisdom of extending the human life span if it creates tremendous human suffering in the process. Living longer also means using up more resources which might be better used for the young. (Interview.)

**p. 202 "Perhaps the simplest clues . . ."** Animal studies show that there is a surprising diversity in life spans. Fruit flies live for only 3 weeks. Mice for 2–3 years. Elephants for 70 years. Bristlecone pines live for 5,000 years, according to radiocarbon dating. The world record is apparently held by the creosote bushes in the Mojave Desert in California, which have survived for 11,000 years, since the last Ice Age. (Jeff Lyon and Peter Gorner, *Altered Fates*, W. W. Norton, New York, 1995, p. 509.)

**p. 202 *"These animals are 'immortal'* . . ."** Hayflick, *How and Why We Age*, p. 21. Interview.

**p. 203 "However, the female flounder . . ."** Ibid.

**p. 204 "As Leonard Hayflick says . . ."** Hayflick, *How and Why We Age*, p. 86.

**p. 204 "Now, it is around seventy-six . . ."** Lyon and Gorman, *Altered Fates*, p. 516.

**p. 204 "As Joshua Lederberg notes . . ."** *New York Times*, Jan. 27, 1996, p. 21.

**p. 204 "And today, the fastest-growing . . ."** Randolph M. Nesse and George C. Williams, *Why We Get Sick*, Random House, New York, 1994, p. 108.

**p. 204 "(Recently, a curious new phenomenon . . ."** A study of 20,000 people aged sixty-five and up found that Americans are suffering fewer long-term disabilities than the previous generation. The more recent the generation of Americans, the fewer diseases at the same time in life, scientists found. This already has had an enormous impact on Medicare savings. Dr. Kenneth Manton of Duke University showed that if the elderly of 1995 became sick at the same rate as the elderly of 1982, then Medicare would have to shell out $200 billion more in health costs!

**p. 204 "The destiny of the universe . . ."** This assumes that our universe does not have enough matter to reverse the expansion of the Big Bang, which agrees with modern experimental data but is not conclusive, since we do not know the total amount of Dark Matter in the universe.

**p. 204 " 'molecular mischief' . . ."** Interview with Leonard Hayflick, Nov. 15, 1996.

**pp. 205–6 "The most common antioxidants include . . ."** Hayflick, *How and Why We Age*, p. 244.

**p. 206 "Studies have shown that antioxidants . . ."** Ibid., p. 246.

**p. 207 "From now to 2020 . . ."** Hayflick believes that these kinds of measures may help to protect us against disease and reduce the death rate per year, but eventually, as people approach their maximum life span, they will begin to die even with these therapies. (Interview.)

**p. 207 "The famous Nurses Health Study . . ."** Ibid.

**p. 207 "Other studies have shown . . ."** Ibid.

**p. 207 "For those taking estrogen . . ."** Ibid.

**p. 208 "As V. K. Cristofalo of the Center . . ."** Ibid., p. 514.

**p. 208 "Mice, for example, age thirty times . . ."** Ibid., p. 515.

**p. 208 "Isaac Schiff of the Massachusetts . . ."** Ibid.

**p. 208 "Already, men account for . . ."** *Newsweek*, Sept. 16, 1996, p. 75.

**p. 209 "Shrunken internal organs . . ."** Lyon and Gorner, *Altered Fates*, p. 508.

**p. 209 " 'We cannot recommend it' . . ."** *New York Times*, April 15, 1996, p. A13.

**p. 209 " 'We cannot recommend it' . . ."** What caused the discrepancy in the results? Papadakis believes it lies in the fact that the Rudman study was not blind. People in the Rudman study knew they were taking the growth hormone, and perhaps the power of suggestion or a placebo effect influenced the result.

So human growth hormone seems to have marginal effects on the body's composition, making it stronger, but does not improve performance. Furthermore, HGH has side effects, like swelling of tissue, aggravation of diabetes, and congestive heart failure. (*New York Times*, July 18, 1995, p. C1.)

**p. 210 "Skeptics point out that . . ."** Hayflick, *How and Why We Age*, p. 232.

**p. 210 "He says, 'In animals . . .' "** Lyon and Gorner, *Altered Fates*, p. 514.

**p. 210 "By raising their DHEA level . . ."** *Science News*, June 24, 1995.

**p. 210 "Richard J. Wurtman of MIT . . ."** *Washington Post*, Aug. 20, 1996, p. 7.

**p. 211 "However, as Hayflick stresses . . ."** Interview with Leonard Hayflick.

**p. 211 "As Johnson says: 'If something like . . .' "** Lyon and Gorner, *Altered Fates*, p. 521.

**p. 211 "She noted that the mutated worms . . ."** *New York Times*, Dec. 7, 1993, p. C13.

**p. 212 "But as Tom Johnson of the University . . ."** Ibid.

**p. 212 "In the body, the superoxide radical . . ."** Ibid.

**p. 212 "There may be a link . . ."** Lyon and Gorner, *Altered Fates*, p. 437.

**p. 212 "In support of this . . ."** Ibid., p. 521.

**p. 213 "Scientists have found that 87 percent . . ."** Hayflick, *How and Why We Age*, p. 64.

**p. 213 "Scientists have found that 87 percent . . ."** They found that having long-lived parents and grandparents did not guarantee a long life, but that people with long-lived ancestors had statistically better chances of inheriting that fortunate quality. They were even able to develop a mathematical formula predicting the chances of inheriting this quality.

**p. 213 "Research isolating these strange aging . . ."** The most important of these disorders is called progeria, which triggers a cascade of age-related illnesses. The disease is rare (only twelve or so cases are known in the United States) but the progression of the disease is remarkable. The growth of the infant is significantly retarded and the face begins to assume a birdlike appearance, with bulging eyes, hooked nose, shriveled skin, and balding scalp, although the person's mental development appears normal. Heart disease and atherosclerosis rapidly appear, cholesterol levels and blood pressure soar, and death occurs around the age of twelve or thirteen. Autopsies of these unfortunate children show that many of their internal organs appear to have aged at many times their normal rate. It is believed that progeria is caused by a defective dominant gene, although no firm cause for the disease has ever been found. (Hayflick, *How and Why We Age*, p. 107.)

The genetic origins of another aging disease among infants, however, has been firmly established. Werner's syndrome, which affects ten people per million in the world, is equally dramatic: the hair turns white and falls out, the skin wastes away, the bones become dangerously thin, and cataracts form. Death usually occurs from heart attacks and cancer in the person's forties. The disease has been shown to be recessively inherited from both parents.

**p. 213 "These genes have such a dramatic effect . . ."** Lyon and Gorner, *Altered Fates*, p. 517.

**p. 214 "One possibility is that having . . ."** *Scientific American*, Jan. 1995, p. 73.

**p. 214 "Wills believes that . . ."** Christopher Wills, *Exons, Introns, and Talking Genes*, Basic Books, New York, 1991, p. 5.

**p. 214 "A combination of studies . . ."** Evolutionary biologists, however, are more cautious than the molecular biologists in rushing to the marketplace with drugs that might influence these age genes. They point out that an age gene which accelerates the process of decay and aging may have an unintended purpose which is vital to our youth. Genes, they point out, often obey the pleiotropic theory—i.e., they often have more than one effect. Evolution, they stress, only selects for genes which keep us young and vigorous while we are still fertile and can have offspring. The genes which keep us healthy while we are young may, in turn, have a secondary effect, such as accelerating the aging process when we are past our fertile years.

For example, evolution may select for a gene that makes our bones heal faster. However, if this gene accomplishes this feat by accumulating calcium, it may have the unintended effect of depositing calcium in our arteries after many decades, long after we are fertile. Thus, a gene which keeps us healthy while we are fertile may accelerate the aging process when we are no longer fertile.

Another example is our immune system. While we are young, our genes may program the immune system to produce all sorts of potent chemicals to destroy invading germs. However, these same powerful chemicals may, over the years, create cumulative damage to our own cells and even cancer.

Thus, even if we find an age gene which accelerates the aging process when we are old, we should be careful about eliminating it—it may have the unintended effect of keeping us healthy while we are young.

As Randolph Nesse and George Williams state in *Why We Get Sick*, (Random House, New York, 1994): "Senescence is the price we pay for vigor in youth."

**p. 215 "This effect was first noticed . . ."** Lyon and Gorner, *Altered Fates*, p. 524.

**p. 215 "This effect was first noticed . . ."** One criticism of this work is that animals in the wild differ significantly from lab animals: wild animals do not eat the "regular diets" fed by MacKay and Walford, but instead are scavengers and hunters who eat a highly restricted, irregular diet. Perhaps, say the critics, these experiments prove the reverse, that the regular, rich diet fed to pampered, healthy animals in the lab actually shortens their normal life span. Perhaps scientists, instead of extending the life span of caloric-deprived animals, have actually been shortening the life spans of well-fed animals all along. However, it must be noted that even if this is true, a restricted diet, compared with a rich one, still correlates with a relatively longer life span. Unfortunately, controlled experiments on animals in the wild are difficult to perform and have not been done.

**p. 215 " 'We find that monkeys respond . . .' "** *New York Times*, April 30, 1996, p. C7.

**p. 216 "People on this kind of restricted diet . . ."** Nesse and Williams, *Why We Get Sick*, p. 118.

**p. 216 " 'Under calorie restriction, though . . .' "** Lyon and Gorner, *Altered Fates*, p. 528.

**p. 217 "This has made possible a promising . . ."** *Scientific American*, Sept. 1995, pp. 130–33; Robert Langer and Joseph P. Vacanti, "Tissue Engineering," *Science*, vol. 260 (May 14, 1993), pp. 920–26.

**p. 217 "This technology has already . . ."** *Scientific American*, June 1995, p. 46.

**p. 217 "The cells which seeded . . ."** See D. J. Mooney, G. Organ, J. Vacanti, and R. Langer, "Design and Fabrication of Biodegradable Polymer Devices to Engineer Tubular Tissues," *Cell Transplantation*, vol. 3, No. 2 (1994), pp. 203–10.

**p. 217 "As Marie Burk of Advanced Tissue . . ."** *New York Times Magazine*, Sept. 29, 1996, p. 152.

**p. 218 "Walter Gilbert predicts . . ."** Interview with Walter Gilbert, Dec. 30, 1996.

**p. 218 "Recently, a series of breakthroughs . . ."** *New York Times*, Oct. 22, 1996, p. C3.

**p. 218 "The step-by-step outline . . ."** *Scientific American*, Sept. 1995, p. 131.

## CHAPTER 11 Playing God: Designer Children and Clones

**p. 221 "He agonized that 'a race of devils . . .' "** Steve Jones, *The Language of Genes*, Anchor Books, New York, 1993, p. 235.

**p. 221 "Historically, the genetic manipulation . . ."** Thomas F. Lee, *Gene Future*, Plenum Press, New York, 1993, p. 288.

**pp. 221–22 "Selective breeding has split . . ."** Steve Jones, Robert Martin, David Pilbeam, and Sarah Bunney, eds. *The Cambridge Encyclopedia of Human Evolu-*

*tion*, Cambridge University Press, Cambridge, Eng., 1992, pp. 382–84; *Discover*, Oct. 1994, p. 94.

**p. 222 "The cat is the only domesticated . . ."** *The Cambridge Encyclopedia of Human Evolution*, p. 382.

**p. 222 "One lesson taught by the . . ."** When compared with their ancestors, domesticated animals (sheep, cattle, goats, ox, etc.) show the same dramatic effects of crossbreeding by humans: smaller body size, smaller brain, shortened face, and fatter body. The brain size for the dog, for example, decreased from about 400 cc to 250 cc after being domesticated. The crossbreeding is so dramatic that many domesticated animals probably cannot survive in the wild. *(Discover*, Oct. 1994, p. 98.)

**p. 222 "In the process, we created . . ."** *The Cambridge Encyclopedia of Human Evolution*, p. 376.

**p. 223 " 'We can put just about . . .' "** *New York Times*, Jan. 16, 1990, p. 241.

**p. 223 "Four million diabetics depend . . ."** Enzo Russo and David Cove, *Genetic Engineering*, W. H. Freeman, New York, 1995, pp. 74, 93.

**p. 223 "Since then, scores of other rare . . ."** Ibid., p. 95.

**p. 224 "The first breakthrough . . ."** Lee, *Gene Future*, p. 166.

**p. 224 "As expected, the mice were able . . ."** Ibid., p. 174.

**p. 224 "As expected, the mice were able . . ."** Palmiter clearly understood the vast implications of his discovery. In his paper in *Nature*, he stated that this technique could be used to "correct or mimic certain genetic diseases." He also said that it could "stimulate the growth of commercially valuable animals."

**p. 224 "Of these, a mere *nine* . . ."** Lee, *Gene Future*, p. 288.

**p. 224 "Because the population of the earth . . ."** Interview with Lester Brown, World Watch Institute.

**p. 225 " 'Sales for such products will be . . .' "** *New York Times*, March 3, 1996, p. 3–1.

**p. 225 "Pioneer Hi-Bred International . . ."** Ibid.

**p. 225 " 'It may be as important as . . .' "** Ibid.

**p. 225 "We can embed the Bt gene . . ."** Ibid.

**p. 225 "Human genes inserted into these plants . . ."** *New York Times*, Jan. 16, 1990, p. C1.

**p. 226 "The world was unprepared for his announcement . . ."** I. Wilmut et al., "Viable Offspring Derived from Fetal and Adult Mammalian Cells," *Nature*, Feb. 27, 1997, p. 810.

**p. 227 "But when asked how long it will be . . ."** *Time*, March 10, 1997, p. 65.

**p. 230 "Diabetes, heart disease, strokes . . ."** *Scientific American*, Aug. 1996, p. 88.

**p. 231 "The hope is that by studying . . ."** *New York Times*, July 30, 1996, p. C9.

**p. 231 "(Scientists believe that humans . . ."** *Science News*, June 3, 1995, p. 348.

**p. 231 " 'People are now finding these . . .' "** *New York Times*, Feb. 13, 1996, p. C7.

**p. 232 "One interesting development was the discovery . . ."** Lisa C. Ryner et. al., "Control of Male Sexual Behavior and Sexual Orientation in Drosophila by the *fruitless* Gene," *Cell*, Dec. 13, 1996, p. 1079.

**p. 232 "Scientists at MIT and Columbia University . . ."** Thomas J. McHugh et al., "Impaired Hippocampal Representation of Space in CA1-Specific NMDAR1 Knockout Mice," *Cell*, Dec. 27, 1996, p. 1339.

**p. 233 "But Gilbert also believes . . ."** Interview with Walter Gilbert.

**p. 233 "In 1996, a gene that contributes to 'anxiety' was . . ."** Klaus-Peter Lesch et al., "Association of Anxiety-Related Traits with a Polymorphism in the Serotonin Transporter Gene Regulatory Region," *Science*, Nov. 29, 1996, p. 1527.

**p. 233 "Although they did not locate . . ."** *Newsweek*, July 29, 1996, p. 78.

**p. 235 "One theory holds that the gills . . ."** Rudolf A. Raff, *The Shape of Life*, University of Chicago Press, Chicago, 1996, p. xv.

**p. 235 "In 1995, Walter Gehring and his colleagues . . ."** W. J. Dickinson et al., "Eye Evolution," *Science*, April 26, 1996, p. 5261.

**p. 236 " 'It's the paper of the year . . .' "** *New York Times*, March 24, 1995, p. A1.

**p. 236 " 'Everywhere we look, we're finding it . . .' "** *Discover*, July 1996, p. 114.

**p. 236 " 'All this points to a common . . .' "** Ibid.

## CHAPTER 12 Second Thoughts: The Genetics of a Brave New World?

**p. 242 " 'Today, it seems quite possible . . .' "** Aldous Huxley, *Brave New World*, Harper, New York, 1946, p. xvii.

**p. 243 "For example, the flow of enriched uranium . . ."** The avowed nuclear powers are the United States, Great Britain, France, Russia, and China, although the nuclear status of the various parts of the former Soviet Union are still being negotiated. The South African government admitted to creating seven atomic bombs, which it has since dismantled. Israel is reputed to have about two hundred atomic bombs. India exploded an atomic bomb in the 1970s. Pakistan is believed to have nuclear weapons. The status of North Korea is not clear.

**p. 244 "Subsequent testing by the company . . ."** Interview with Rebecca Goldburg.

**p. 244 " 'I think they've taken a good idea . . .' "** *New York Times*, Aug. 27, 1995, p. 30.

**p. 244 "From 1987 to 1995 . . ."** Interview with Rebecca Goldburg.

**p. 245 "Many critics point out . . ."** Ibid.

**p. 245 "A case in point . . ."** Ibid.

**p. 246 "Daniel Cohen, of the Center . . ."** Thomas F. Lee, *Gene Future*, Plenum Press, New York, 1993, p. 301.

**p. 246 "It also argued that . . ."** *Discover*, Jan. 1997, p. 78.

**p. 246 " 'When the Human Genome Project . . .' "** Interview with Francis Collins.

**p. 246 " 'Should you, if you're running for President . . .' "** Loic Wingerson, *Mapping Our Genes*, Penguin Books, New York, 1990, p. 297.

**p. 247 "David Sidransky of Johns Hopkins . . ."** *Newsweek*, Dec. 23, 1996, p. 47.

**p. 247 "Arthur Caplan of the Center for Bioethics . . ."** Interview with Arthur Caplan.

**p. 248 "The OTA report stated . . ."** Jeff Lyon and Peter Gorner, *Altered Fates*, W. W. Norton, New York, 1995, p. 484.

**p. 248 " 'Those concerns are somewhat lessened . . .' "** Interview with Francis Collins.

**p. 248 "While he believes that math ability . . ."** Interview with Arthur Caplan, July 21, 1996.

**p. 249 "Christopher Wills of the University . . ."** Christopher Wills, *Exons, Introns, and Talking Genes*, Basic Books, New York, 1991, p. 10.

**p. 249 "But an exhaustive study of 331 people . . ."** *Washington Post*, Nov. 4, 1996, p. A2.

**p. 249 "The initial controversy was sparked . . ."** Usually, males have XY, and females have XX chromosomes.

**p. 249 "The press dubbed the Y chromosome . . ."** Wingerson, *Mapping Our Genes*, p. 95.

**p. 250 "Said psychiatrist Peter Breggin . . ."** *Washington Post*, Jan. 29, 1995, p. C4.

**p. 250 "He concludes: '*The Bell Curve* was . . .'"** Interview with Arthur Caplan.

**p. 251 "One congressman said, 'The primary reason . . .'"** Enzo Russo and David Cove, *Genetic Engineering*, W. H. Freeman, New York, 1995, p. 170.

**p. 251 "President Calvin Coolidge . . ."** Thomas F. Lee, *The Human Genome Project*, Plenum Press, New York, 1991, p. 276.

**p. 251 "In 1988, the European Research . . ."** Lee, *Gene Future*, p. 160.

**p. 251 "There are, however, some disagreements . . ."** At the Council for International Organizations of Medical Sciences, meeting in Japan in 1990, the participants took a different position: "Although germ-cell gene therapy is not contemplated at present, continued discussion of germ-cell gene therapy is nonetheless important. The option of germ-cell therapy must not be prematurely foreclosed. It may someday offer clinical benefits attainable in no other way." (Ibid., p. 161.)

**p. 251 "'Are you kidding? Yes!'"** Interview with Arthur Caplan.

**pp. 251–52 "'I think there is no doubt . . .'"** Ibid.

**p. 252 "Caplan predicts that . . ."** Ibid.

**p. 252 "'The use of this technology . . .'"** Interview with Francis Collins.

**p. 252 "In other words, 3.6 percent . . ."** *Washington Post*, May 11, 1996, p. A1.

**p. 252 "On the southern coast of China . . ."** *New York Times*, June 7, 1996, p. A11.

**p. 253 "Apparently, anxious parents . . ."** *Science News*, Sept. 7, 1996, p. 154.

**p. 255 "'I don't see how you can stop . . .'"** *Time*, March 10, 1997, p. 72.

**p. 256 "In 1883, he coined the word 'eugenics' . . ."** Lee, *The Human Genome Project*, p. 275.

**p. 256 "According to geneticist Steve Jones . . ."** Steve Jones, *The Language of Genes*, Anchor Books, New York, 1993, p. 224.

**p. 256 "'Society must protect itself' . . ."** Ibid., p. 224.

**p. 256 "Justice Oliver Wendell Holmes . . ."** Ibid., p. 150.

**p. 257 "Gregory Kavka, a philosopher . . ."** Cranor, *Are Genes Us?*, p. 170.

**p. 257 "The deep fracture lines of society . . ."** The irony in all of this is that the Morlocks have their ultimate revenge. They eat the Eloi.

**p. 258 "And in the eighteenth century . . ."** Suzuki, *Genetics*, p. 197.

**p. 258 "During World War I, 100,000 tons . . ."** Ibid.

**p. 258 "During World War II, the Nazis . . ."** Ibid.

**p. 258 "Some scientists' greatest fears were voiced . . ."** Ibid.

**p. 258 "D. A. Henderson, who helped . . ."** Ibid.

**p. 259 "'It can easily be done . . .'"** Ibid.

**p. 259 "Barbara Rosenberg of the Federation . . ."** Ibid.

**p. 259 "Within ten days, it reaches . . ."** Ibid., pp. 93–94.

**p. 259 "Recalling that chilling exercise . . ."** Ibid.

**p. 259 "Ethnic weapons were first proposed . . ."** Charles Piller and Keith R. Yamamoto, *Gene Wars*, William Morrow, New York, 1988, p. 99.

**p. 259 "Recently declassified documents . . ."** These secret tests were conducted at the military's Mechanicsburg, Pennsylvania, supply depot. The document says, "Within this system there are employed large numbers of laborers, including many Negroes, whose incapacitation would seriously affect the operation of the supply system. Since Negroes are more susceptible to *Coccidioides* than are whites, this fungus was simulated by using *Asperfillus fumigatus.*" (Ibid.)

**p. 259 "Military planners once considered . . ."** Interview with Charles Piller; Piller and Yamamoto, *Gene Wars*, p. 100.

**p. 259 "Francis Collins says . . ."** Interview with Francis Collins.

**p. 261 "In 1995, in a report . . ."** *Scientific American*, Dec. 1996, p. 62.

## CHAPTER 13 The Quantum Future

**p. 267 "Some of the ways . . ."** *Scientific American*, April 1996, p. 94.

**p. 268 "He challenged the physics community . . ."** Richard Feynman, "There's Plenty of Room at the Bottom," *Engineering and Science*, Feb. 1960. The precise challenge was for someone to take a page of a book and shrink it by a factor of 25,000 times, such that it can be read by an electron microscope. A second $1,000 prize would go to the person who could make an operating electric motor the size of a 1/64-inch cube.

**p. 269 "Istvan Csicsery-Ronay, Jr., an editor . . ."** *Scientific American*, April 1996, p. 99.

**p. 269 "Although they are still a far cry . . ."** *New York Times*, Jan. 27, 1997; *High Technology Careers*, Feb.–March, 1997, p. 1.

**p. 270 " 'MEMS is a really intriguing . . .' "** *New York Times*, Jan. 27, 1997, p. D12.

**p. 271 "The needle of the scanning . . ."** *New York Times*, Nov. 19, 1996, p. C1.

**p. 276 "Nobel Laureate Philip Anderson . . ."** *Science News*, March 9, 1996, p. 157.

**p. 277 "They hope to have . . ."** *Scientific American*, Sept. 1995, p. 100B.

**p. 278 " 'By the middle of the next century . . .' "** *Scientific American*, Sept. 1995, p. 174.

**p. 278 "Barring new significant discoveries . . ."** The date is subject to many uncertainties, including the discovery of new oil deposits and the rate of future oil consumption, neither of which can be precisely calculated.

**p. 278 "Since there is virtually unlimited . . ."** Deuterium is an isotope of hydrogen. Ordinary hydrogen contains a single proton in its nucleus while deuterium contains a proton and a neutron in its nucleus. The simplest fusion process being studied in fusion machines involves fusing deuterium and tritium, another isotope, which contains a proton and two neutrons.

**p. 282 "About 40 to 50 percent of its uranium . . ."** T. J. Thompson and J. G. Beckerley, *The Technology of Nuclear Reactor Safety*, MIT Press, Cambridge, Mass., 1964, vol. 1, p. 631.

**p. 282 "It created a major emergency . . ."** Because the core consisted of 25 percent enriched uranium, much greater than the 3 percent enriched uranium used today, there was concern that the melted uranium could re-form and reach critical mass, capable of releasing energy and worsening the accident. As a consequence, scientists very carefully probed the core to prevent the melted uranium from re-forming. Eventually, they found that the accident was due to the loss of sodium coolant, caused by a piece of zirconium that had broken off and jammed the coolant pipes. Operators of the plant successfully prevented news of this accident from leaking to the public.

**p. 284 "Each year, the energy the earth . . ."** *Scientific American*, Sept. 1995, p. 170.

**p. 284 "At a cost of 10 cents . . ."** Ibid., p. 172.

**p. 284 "They claimed that the $150 million . . ."** Ibid., p. 173.

**p. 285 "Conversely, batteries which . . ."** *The Economist*, June 22, 1996, p. 8.

**p. 285 "The average gasoline-powered car . . ."** The World Watch Institute estimates that air pollution is so bad in the United States that only one in five city dwellers enjoys healthy air. Interview with Lester Brown.

**p. 285 "Cars and other motor vehicles . . ."** *Scientific American*, Nov. 1996, p. 54.

**p. 287 "The quotas for zero-emission vehicles . . ."** Ibid., p. 58.

**p. 287 "Daniel Sperling, director of the . . ."** Ibid.

**p. 288 "In 1989, the first dramatic demonstration . . ."** *The Computer in the 21st Century*, Scientific American Books, New York, 1995, p. 62.

**p. 289 "The X-ray pulse from a hydrogen bomb . . ."** Recalibration of the energy of a nuclear X-ray laser showed that it generated less power than expected, too small to be used successfully in a Star Wars program. More important, a Star Wars system can be neutralized by cheap countermeasures, such as releasing millions of metallic balloons or decoys to confuse ground radar, so that the X-ray laser will not know where to point its beam.

**p. 290 "The most intriguing of these . . ."** The charge of antimatter is the opposite of that of ordinary matter. Thus, antielectrons (positrons) have positive charge. In an antiatom, positive antielectrons circle around negative antiprotons, creating neutral antiatoms. In principle, we can have antimolecules, anti-DNA, and even antipeople. Antimatter annihilates on contact with ordinary matter, creating a burst of gamma rays and other particles.

**p. 291 "The collisions between the antiprotons . . ."** *Science News*, Jan. 13, 1996, p. 20.

## CHAPTER 14 To Reach for the Stars

**p. 296 "Ten NASA space probes . . ."** Because "windows" of opportunity open up for sending spacecraft to Mars depending on its closest approach, NASA will send two spacecraft every other year to Mars.

**p. 297 "This astonishing claim . . ."** Since then, there have been a number of independent studies, both supporting and criticizing the original study. This debate will continue for years, until one can determine if these structures have cell walls or until genuine Mars rocks are retrieved from that planet.

**p. 297 "This would significantly reduce . . ."** *Space News*, Sept. 2–8, 1996, p. 3.

**p. 298 " 'The surface of Mars . . .' "** *Scientific American*, Oct. 1994, p. 97.

**p. 298 " 'Who is to say . . .' "** *Time*, Aug. 19, 1996, p. 62.

**p. 298 " 'Who is to say . . .' "** Unfortunately, because Mars is a small planet, its weak gravitational field was not sufficient to keep the atmosphere from leaking into space. As the atmospheric pressure gradually decreased, water could no longer stay in liquid form. As the atmosphere began to leak out, the lakes and seas probably evaporated into space, or seeped underground into the permafrost, or went to the polar ice caps. Any form of life that developed 3 billion years ago would also have followed the path taken by the water: into space, into the permafrost, or to the poles.

**p. 298 " 'I'm sure there's life there . . .' "** *N.Y. Daily News*, April 10, 1997, p. 5.

**p. 300 " 'It's a job program . . .' "** *Newsweek*, April 11, 1994, p. 30.

**p. 300 "The National Research Council . . ."** Ibid.

**p. 300 "From a purely scientific point . . ."** *New York Times*, June 29, 1995, p. A7; *Washington Post*, June 24, 1995, p. A8.

**p. 301 " 'They are bigger, slower . . .' "** *Discover*, July 1994, p. 74.

**p. 301 "Also, the Russians . . ."** *New York Times*, Jan. 27, 1997, p. B9.

**p. 301 " 'From the day we started . . .' "** Ibid.

**p. 301 "The Space Shuttle has been called . . ."** *Time*, July 15, 1996, p. 58.

**p. 301 "The Space Shuttle has been called . . ."** Interview with John Lewis, Dec. 11, 1996.

**p. 303 "By 2008, says Gene Austin . . ."** *Time*, July 15, 1996, p. 58.

**p. 303 "By 2012, it should . . ."** *Space News*, July 15–21, 1996, p. 4.

**p. 303 "Ultimately, the goal . . ."** Ibid.

**p. 305 "Several competing designs . . ."** Although nuclear rockets have specific impulses on the order of 1,000 to 2,000 seconds, they are perhaps the most unstable and dangerous. For several decades, the U.S. government has tinkered with the nuclear rocket, often in total secrecy. A nuclear rocket, using a nuclear reactor to heat up gases for thrust, would generate large amounts of power, but it would also cause enormous contamination of the earth if it ever blew up like the *Challenger*. Falling nuclear debris from a nuclear rocket would make large portions of the earth uninhabitable.

The latest nuclear rocket was the Timberwind, which was secretly being developed by the U.S. military for the Star Wars program, until its existence was finally revealed by the Federation of American Scientists. This embarrassing revelation caused President Clinton to cancel it.

**p. 305 "Several competing designs . . ."** The rail gun blasts payloads to thousands of miles per hour via electromagnetic induction.

The rail gun evokes memories of Jules Verne's prophetic novel *From the Earth to the Moon*. Unfortunately, his design for the gun also violated several laws of physics.

First, the velocity generated by a chemical explosion is not great enough to reach 25,000 miles per hour, the velocity needed to escape the earth's gravitational pull. The shock wave generated by a chemical explosion is roughly on the order of magnitude of the speed of sound, which is far too small to boost a payload into space.

Second, the near-instantaneous acceleration generated by being shot out of a gun would have crushed the passengers with horrible g forces.

**p. 305 "Physicist Freeman Dyson . . ."** *Scientific American*, Sept. 1995, p. 116.

**p. 305 "(By convention, the unit . . ."** Specific impulse is the thrust multiplied by the time divided by the mass of the propellant and the gravitational constant. The unit of the resulting number is in seconds. Eugene Mallove and Gregory Matloff, *The Starflight Handbook*, John Wiley, New York, 1989, p. 44.

**p. 306 " 'A diversified system . . .' "** *Scientific American*, Sept. 1995, p. 116A.

**p. 306 "By 1997, scientists were ecstatic . . ."** Recently, one of the extra-solar planets has been challenged by other scientists. Although it remains to be seen how many of these planets will survive other challenges, thousands of these planets will soon be discovered.

**p. 307 "A Jupiter-sized planet . . ."** *New York Times*, June 12, 1996, p. A24.

**p. 307 "Its close proximity . . ."** *Washington Post*, June 12, 1996, p. A3.

**p. 309 "Since about 200 million comets . . ."** *Discover*, Nov. 1995, p. 83.

**p. 310 "Astronomer John Lewis . . ."** Interview with John Lewis.

**p. 311 "Second, according to Einstein . . ."** More precisely, no information can travel faster than the speed of light in a local frame. There are several things which go faster than light, but they cannot be used to send information, or they rely on global effects deriving from Einstein's more powerful general theory of relativity. For example, the phase velocity of a wave and measurements taken in the Einstein-Rosen-Podolsky experiment can exceed the speed of light, but they cannot be used to send messages. Likewise, the Big Bang expanded faster than light, but general relativity involves a global, not local, frame. Similarly, wormholes may allow one to violate special relativity because they act globally, not locally, on the space. In the latter case, we caution that wormholes may be possible but they are not practical at present: the energy necessary to open up a wormhole exceeds the entire energy output of the earth.

**p. 312 "After twenty-five shipboard . . ."** Observers on the earth, looking at the spaceships soaring to the nearby planets, will see them frozen in time, due to the dilation effect of special relativity.

**p. 313 "But because ramjets do . . ."** Mallove and Matloff, *The Starflight Handbook*, p. 112.

**p. 314 "(Theodore Taylor, the nuclear bomb . . ."** Interview with Theodore Taylor.

**p. 314 "Many of these problems . . ."** Laser beams do not dissipate as fast as sunlight, which decreases as the inverse square of the distance. But laser beams do dissipate. A typical laser beam flashed onto the moon, for example, spreads to roughly five miles.

**p. 315 "Fish, for example, can swim . . ."** *Discover*, Aug. 1994, p. 39.

**p. 316 "More important, frogs have . . ."** Ibid.

**p. 316 "The upper limit for the existence . . ."** Five billion years from now, our sun, a typical main-sequence hydrogen-burning yellow star, will exhaust its hydrogen fuel and will make the transition to a helium-burning star, or a red giant. The atmosphere of a red giant can extend out to the orbit of Mars, so the earth will probably be vaporized as the sun's atmosphere expands.

**p. 317 "The astronomers at the University . . ."** Interview with John Lewis and Neal Tyson. John S. Lewis, *Mining the Sky*, Addison-Wesley, Reading, Mass., 1996, p. 83.

**p. 317 "(Fortunately, these asteroids . . ."** Ibid.

**p. 317 "(Two students at the University . . ."** *Time*, June 3, 1996, p. 61.

**p. 317 "If it 'hit on the West Coast' . . ."** *New York Times Magazine*, July 28, 1996, p. 17.

**p. 317 "About 15,000 years ago . . ."** Ibid., pp. 114–15.

**p. 318 "Most of the work locating . . ."** Some people advocate using hydrogen bombs to blow NEOs to pieces. This is not a wise idea. The resulting fragments could pose an even greater threat to the earth than the original NEO. If we have enough warning, it might be better to simply deflect the NEO while it is still far enough away from the earth.

**p. 319 "However, the SETI . . ."** Interview with Paul Shuch, executive director, SETI League.

**p. 319 "Astronomers have scanned . . ."** In 1978, astronomer Paul Horowitz scanned all sun-like systems, 185 in all, within 80 light-years of the earth. In 1979, 600 star systems were scanned. None showed any conclusive evidence of intelligent life.

## CHAPTER 15 Toward a Planetary Civilization

**p. 323 "Since the universe . . ."** The precise number is subject to change, since the Hubble constant, which measures the expansion of the universe, is not known very well. In 1977, an estimate made by a European satellite indicates the universe may be as young as 10 to 12 billion years.

**p. 323 "Russian astronomer Nikolai Kardashev . . ."** Nikolai Kardashev, "Transmission of Information by Extraterrestrial Civilizations," *Soviet Astronomy AJ*, vol. 8 (1964), pp. 217–21.

**p. 323 "At such a growth rate . . ."** Freeman Dyson, *Disturbing the Universe*, Harper & Row, New York, 1979, p. 212.

**p. 325 "Consider the modification . . ."** Chaos theory tells us that the weather cannot be precisely predicted by even the largest computer, so the best that a Type I civilization can attain is the modification of its weather.

**p. 326 "Because a Type I civilization . . ."** There is also the possibility that a Type I civilization can decide to alter its own genetic makeup.

**p. 327 "Finding a Type II civilization . . ."** For example, it may choose to use vast networks of fiber optics and cables rather than satellite emissions for communication.

**p. 327 "However, it cannot violate . . ."** The Second Law of Thermodynamics here says that any machine operating between two different temperatures necessarily creates waste heat. Even if the Type II civilization seals off its sun with a Dyson sphere, eventually the sphere starts to heat up and emit waste heat.

**p. 331 "The threat of global warming . . ."** UN Web Page. *Science News*, Nov. 4, 1995, p. 293.

**p. 331 "The report presented a grim tale . . ."** Malaria alone infects 300 to 500 million people, or almost 10 percent of all humanity. Warming the earth by 3 to 5 degrees C could spread malaria to 60 percent of the earth's surface. Other dreaded diseases, such as cholera, can spread through global warming by encouraging the growth of algae and bacteria in contaminated water supplies.

**p. 331 "The report presented a grim tale . . ."** At present, the carbon dioxide level in the atmosphere is the highest it has been in the past 150,000 years, on the basis of studies done on ice buried deep under the polar ice cap. In fact, charts which analyze the changing levels of carbon dioxide and the temperature of the earth show a direct correlation. Behind global warming lies the fact that every year the burning of fossil fuels adds 6 billion tons of carbon to the atmosphere. (170 billion tons of carbon have accumulated since the industrial revolution.)

**p. 331 "The report presented a grim tale . . ."** Interview with Lester Brown, World Watch Institute.

**p. 331 "The report presented a grim tale . . ."** Interview with Michael Oppenheimer, Environmental Defense Fund.

**p. 332 "Similarly, world grain production . . ."** Interview with Lester Brown.

**p. 332 "Meanwhile, the total area . . ."** *Scientific American*, Oct. 1994, p. 116.

**p. 332 "The UN estimates . . ."** *New York Times*, Nov. 17, 1996, p. 3.

**p. 332 "At present, thirty nations . . ."** World Watch Institute, *State of the World, 1996*, Washington, D.C., 1996, p. 12.

**p. 332 " 'Finally, it improves . . .' "** *Scientific American*, Oct. 1994, p. 120.

**p. 334 "Alvin Toffler writes . . ."** Alvin Toffler, *The Third Wave*, Bantam Books, New York, 1980, p. 80.

**p. 334 "Before the industrial revolution . . ."** Ibid.

**p. 335 "Kenichi Ohmae, author of . . ."** Kenichi Ohmae, *The End of the Nation State*, Simon & Schuster, New York, 1995, p. 5.

**p. 335 " 'Over time, arteries harden . . .' "** Ibid., p. 142.

**p. 335 " 'Not one of our 28 . . .' "** Toffler, *The Third Wave*, p. 230.

**p. 335 "Toffler adds: 'We are . . .' "** Ibid., p. 327.

**p. 336 "He feels that when . . ."** Ohmae, *The End of the Nation State*, p. 44.

**p. 336 " 'Equally important, they will start . . .' "** Ibid., p. 45.

**p. 336 "He believes only 250 . . ."** *Science News*, Feb. 25, 1995, p. 117.

**p. 336 "As columnist William Safire . . ."** *New York Times Magazine*, Sept. 29, 1996, p. 61.

**p. 336 "As columnist William Safire . . ."** *The Computer in the 21st Century*, Scientific American Books, New York, 1995, p. 4.

**p. 337 "Bill Gates claims . . ."** Bill Gates, *The Road Ahead*, Viking, New York, 1995, p. 263.

## CHAPTER 16 Masters of Space and Time

**p. 339 "In such a grand vision . . ."** The temperature of the core of the sun is not high enough to forge the elements of our body. Even in a white dwarf, temperatures are not high enough to forge elements beyond iron. Creating the higher elements that make up the atoms of our body requires trillions of degrees, which can be found only in a supernova explosion. Thus, our sun is actually a recycled star. Our sun was made of the ashes of a supernova which exploded before the creation of the solar system.

**p. 340 "But the net effect . . ."** According to the special theory of relativity, an object cannot go faster than light. But the theory is just a special case of the general theory of relativity. According to the general theory, one might be able to pass through a wormhole and wind up on the other side of the universe, thereby, in effect, going faster than light. The only restriction coming from the special theory is that we cannot go faster than light as we enter the wormhole.

**p. 342 "Putting your hand through . . ."** One problem with entering the Kerr black hole is the question of stability. It is possible that an object falling through the Kerr metric will cause the wormhole to close up.

**p. 344 "This strange form of matter . . ."** By charge conjugation, physicists believe, antimatter will fall down, not up.

**p. 348 "Similarly, Weinberg concludes . . ."** Steven Weinberg, *Dreams of a Final Theory*, Pantheon Books, New York, 1992, p. 231.

**p. 348 "Gell-Mann has said . . ."** Interview with Murray Gell-Mann. See also John Brockman, *The Third Culture*, Simon & Schuster, New York, 1995, p. 256.

**p. 351 "Although the Big Bang theory . . ."** Some journalists have recently cast doubt on the Big Bang theory on the basis of the "clumpiness" of the universe, but they misunderstand the physics. The COBE data shows that the Big Bang was a remarkably smooth explosion. However, the universe today looks quite irregular, with galaxies clumped into superclusters, leaving large voids in the heavens. This clumpiness apparently was created about a billion years after the Big Bang, which is a very small time scale for this to occur, which has led some to criticize the Big Bang theory. However, a careful analysis of the COBE data shows that there were tiny fluctuations in the original Big Bang, consistent with quantum fluctuations, which are sufficient to explain the current clumpiness of the universe. In other words, the present distribution of galaxies, including our own Milky Way galaxy, is probably a direct by-product of the quantum theory of the Big Bang.

**p. 352 "In fact, the sum of the two is zero . . ."** For example, when we say that the gravitational energy of the earth going around the sun is negative, by this we mean that zero energy is measured at a distance away from the sun. Since we must add energy in order to remove the earth from the sun, the gravitational energy of the earth is negative.

More precisely, in a closed universe, the total energy is zero, while in an open universe the total energy is infinite. Since energy is just one component of the second-rank tensor (the energy-momentum tensor), it is not an invariant quantity by itself, and hence depends on the local frame in which we measure it.

Thus, it takes no net energy to create a closed universe out of Nothing.

The idea of the universe originating as a quantum fluctuation was first proposed by Edward Tryon of Hunter College.

# Recommended Reading


PART ONE **The Computer Revolution**

Allman, William F., *Apprentices of Wonder: Inside the Neural Network Revolution.* Bantam Books, New York, 1989.

Asimov, Isaac, *Robot Dreams.* Ace Books, New York, 1986.

Caudill, Maureen, *In Our Own Image: Building an Artificial Person.* Oxford University Press, Oxford, 1992.

Coveney, Peter, and Roger Highfield, *Frontiers of Complexity: The Search for Order in a Chaotic World.* Ballantine Books, New York, 1995.

Crevier, Daniel, *AI: The Tumultuous History of the Search for Artificial Intelligence.* Basic Books, New York, 1993.

Dertouzos, Michael, *What Will Be: How the New World of Information Will Change Our Lives.* HarperCollins, San Francisco, 1997.

Freeman, David H., *Brainmakers: How Scientists Are Moving Beyond Computers to Create a Rival to the Human Brain.* Simon & Schuster, New York, 1994.

Gates, Bill, *The Road Ahead.* Viking, New York, 1995.

Gelertner, David, *Mirror Worlds.* Oxford University Press, Oxford, 1991.

Gibilisco, Stan, ed., *The McGraw-Hill Illustrated Encyclopedia of Robotics and Artificial Intelligence.* McGraw-Hill, New York, 1994.

Hafner, Katie, and Matthew Lyon, *Where Wizards Stay Up Late: The Origins of the Internet.* Simon & Schuster, New York, 1996.

Halberstam, David, *The Next Century.* William Morrow, New York, 1991.

Harrar, George, *Radical Robots.* Simon & Schuster, New York, 1990.

Horgan, John, *The End of Science.* Addison-Wesley, Reading, Mass., 1996.

Johnson, R. Colin, and Chappell Brown, *Cognizers: Neural Networks and Machines That Think.* John Wiley, New York, 1988.

Kaku, Michio, and Daniel Axelrod, *To Win a Nuclear War.* South End Press, Boston, 1987.

Kauffman, William J., and Larry L. Smarr, *Supercomputing and the Transformation of Science.* Scientific American Books, New York, 1993.

Kelly, Kevin, *Out of Control: The New Biology of Machines, Social Systems, and the Economic World.* Addison-Wesley, Reading, Mass., 1994.

Lebow, Irwin, *The Digital Connection: A Layman's Guide to the Information Age.* W. H. Freeman, New York, 1991.

Lubar, Steven, *Info Culture: The Smithsonian Book of Information Age Inventions.* Houghton Mifflin, Boston, 1993.
Minsky, Marvin, *The Society of Mind.* Simon & Schuster, New York, 1985.
Moravec, Hans, *Mind Children.* Harvard University Press, Cambridge, Mass., 1988.
Negroponte, Nicholas, *Being Digital.* Alfred A. Knopf, New York, 1995.
Pagels, Heinz R., *The Dreams of Reason: The Computer and the Rise of the Sciences of Complexity.* Bantam Books, New York, 1988.
Reichardt, Jasia, *Robots: Fact, Fiction, and Prediction.* Penguin Books, New York, 1978.
Scientific American, *The Computer in the 21st Century.* Scientific American Books, New York, 1995.
Shasha, Dennis, and Cathy Lazere, *Out of Their Minds: The Lives and Discoveries of 15 Great Computer Scientists.* Springer-Verlag, New York, 1995.
Simons, Geoff, *Robots: The Quest for Living Machines.* Cassell, London, 1992.
Stoll, Clifford, *Silicon Snake Oil: Second Thoughts on the Information Highway.* Doubleday, New York, 1995.
Thurow, Lester C., *The Future of Capitalism: How Today's Economic Forces Shape Tomorrow's World.* William Morrow, New York, 1996.

## PART TWO The Biomolecular Revolution

Bezold, Clement, Jerome A. Halperin, and Jacqueline L. Eng, eds., *2020 Visions: Health Care Information Standards and Technologies.* U.S. Pharmacopeial Convention Press, Rockville, Md., 1993.
Cavalli-Sforza, Luigi Luca, and Francesco Cavalli-Sforza, *The Great Human Diasporas: The History of Diversity and Evolution.*Addison-Wesley, Reading, Mass., 1995.
Cook-Deegan, Robert, *The Gene Wars: Science, Politics, and the Human Genome.* W. W. Norton, New York, 1994.
Cranor, Carl F., *Are Genes Us? The Social Consequences of the New Genetics.* Rutgers University Press, New Brunswick, N.J., 1994.
Dixon, Bernard, *Power Unseen: How Microbes Rule the World.* W. H. Freeman, New York, 1994.
Drlica, Karl A., *Double-Edged Sword: The Promises and Risks of the Genetic Revolution.* Addison-Wesley, Reading, Mass., 1994.
———, *Understanding DNA and Gene Cloning: A Guide for the Curious.* John Wiley, New York, 1992.
Fettner, Ann Giudici, *The Science of Viruses: What They Are, Why They Make Us Sick, and How They Will Change the Future.* William Morrow, New York, 1990.
Frank-Kamenetskii, Maxim D., *Unraveling DNA.* VCH Publishers, New York, 1993.
Garrett, Laurie, *The Coming Plague: Newly Emerging Diseases in a World Out of Balance.* Penguin Books, New York, 1994.
Goleman, Daniel, *Emotional Intelligence: The Groundbreaking Book That Redefines What It Means to be Smart.* Bantam Books, New York, 1995.
Goodsell, David S., *Our Molecular Nature: The Body's Motors, Machines, and Messages.* Springer-Verlag, New York, 1996.
Gould, Stephen Jay, *Ontogeny and Phylogeny.* Harvard University Press, Cambridge, Mass., 1977.
———, *The Mismeasure of Man.* W. W. Norton, New York, 1996.
Hayflick, Leonard, *How and Why We Age.* Ballantine Books, New York, 1994.
Huxley, Aldous, *Brave New World.* New York: Harper, 1946.

Jones, Steve, *The Language of Genes: Solving the Mysteries of Our Genetic Past, Present, and Future.* Anchor Books, New York, 1993.

———, Robert Martin, David Pilbeam, and Sarah Bunney, eds., *The Cambridge Encyclopedia of Human Evolution.* Cambridge University Press, Cambridge, Eng., 1992.

Kendrew, Sir John, ed., *The Encyclopedia of Molecular Biology.* Blackwell Science, Cambridge, Eng., 1994.

Kevles, Daniel J., *In the Name of Eugenics: Genetics and the Uses of Human Heredity.* Harvard University Press, Cambridge, Mass., 1995.

———, and Leroy Hood, eds., *The Code of Codes.* Harvard University Press, Cambridge, Mass., 1992.

Kimbrell, Andrew, *The Human Body Shop: The Engineering and Marketing of Life.* HarperCollins, San Francisco, 1993.

Kitcher, Philip, *The Lives to Come: The Genetic Revolution and Human Possibilities.* Simon & Schuster, New York, 1996.

Kleinsmith, Lewis J., and Valerie M. Kish, *Principles of Cell and Molecular Biology.* HarperCollins, New York, 1995.

Lee, Thomas F., *Gene Future: The Promise and Perils of the New Biology.* Plenum Press, New York, 1993.

———, *The Human Genome Project: Cracking the Code of Life.* Plenum Press, New York, 1991.

Levin, Arnold J., *Virus.* Scientific American Books, New York, 1992.

Lyon, Jeff, and Peter Gorner, *Altered Fates: Gene Therapy and the Retooling of Human Life.* W. W. Norton, New York, 1995.

Meyers, Robert A., *Molecular Biology and Biotechnology: A Comprehensive Desk Reference.* VCH Publishers, New York, 1995.

Moore, Thomas J., *Lifespan: New Perspectives on Extending Human Longevity.* Simon & Schuster, New York, 1993.

Morse, Stephen S., ed., *Emerging Viruses.* Oxford University Press, Oxford, 1997.

Murphy, Michael P., and Luke A. J. O'Neill, *What Is Life? The Next Fifty Years.* Cambridge University Press, Cambridge, Eng., 1995.

Neel, James V., *Physician to the Gene Pool: Genetic Lessons and Other Stories.* John Wiley, New York, 1994.

Nesse, Randolph M., and George C. Williams, *Why We Get Sick: The New Science of Darwinian Medicine.* Random House, New York, 1994.

Nicholl, Desmond S., *An Introduction to Genetic Engineering.* Cambridge University Press, Cambridge, Eng., 1994.

Old, R. W., and Primrose, S. B., *Principles of Gene Manipulation: An Introduction to Genetic Engineering.* Blackwell Science, Cambridge, Eng., 1994.

Pillar, Charles, and Keith R. Yamamoto, *Gene Wars.* William Morrow, New York, 1988.

Raff, Rudolph A., *The Shape of Life: Genes, Development, and the Evolution of Animal Form.* University of Chicago Press, Chicago, 1996.

Russo, Enzo, and David Cove, *Genetic Engineering: Dreams and Nightmares.* W. H. Freeman, New York, 1995.

Sagan, Carl, *The Dragons of Eden: Speculations on the Evolution of Human Intelligence.* Ballantine Books, New York, 1977.

Salyers, Abigail A., and Dixie D. Whitt, *Bacterial Pathogenesis.* ASM Press, Washington, D.C., 1994.

Steen, R. Grant, *DNA and Destiny: Nature and Nurture in Human Behavior.* Plenum Press, New York, 1996.

Suzuki, David, and Peter Knudtson, *Genetics: The Clash Between the New Genetics and Human Values.* Harvard University Press, Cambridge, Mass., 1990.

Varmus, Harold, and Robert A. Weinberg, *Genes and the Biology of Cancer.* Scientific American Books, New York, 1993.

Weinberg, Robert A., *Racing to the Beginning of the Road. The Search for the Origin of Cancer.* Random House, New York, 1996.

Wills, Christopher, *Exons, Introns, and Talking Genes: The Science Behind the Human Genome Project.* Basic Books, New York, 1991.

Wingerson, Lois, *Mapping Our Genes: The Genome Project and the Future of Medicine.* Penguin Books, New York, 1990.

## PART THREE The Quantum Revolution

Barrow, John, *Theories of Everything.* Oxford University Press, Oxford, 1991.

Brockman, John, *The Third Culture: Beyond the Scientific Revolution.* Simon & Schuster, New York, 1995.

Brown, Lester R., *State of the World.* W. W. Norton, New York, 1996.

Crease, R., and C. Mann, *The Second Creation.* Macmillan, New York, 1986.

Davies, Paul, *Superforce: The Search for a Grand Unified Theory of Nature.* Simon & Schuster, New York, 1984.

Dyson, Freeman, *Infinite in All Directions.* New York: Harper & Row, 1988.

———, *Disturbing the Universe.* New York: Harper & Row, 1979.

Gribben, J. *In Search of Schrödinger's Cat.* Bantam Books, New York, 1984.

Hawking, S. W., *A Brief History of Time.* Bantam Books, New York, 1988.

Kaku, Michio, *Hyperspace: A Scientific Odyssey Through Parallel Universes, Time Warps, and the 10th Dimension.* Anchor Books, New York, 1995.

———, *Introduction to Superstrings.* Springer-Verlag, New York, 1988.

———, and Jennifer Thompson, *Beyond Einstein: The Cosmic Quest for the Theory of the Universe.* Anchor Books, New York, 1996.

Lewis, John S., *Mining the Sky: Untold Riches from the Asteroids, Comets, and Planets.* Addison-Wesley, Reading, Mass., 1996.

———, *Rain of Iron and Ice.* Addison-Wesley, Reading, Mass., 1996.

Mallove, Eugene, and Gregory Matloff, *The Starflight Handbook: A Pioneer's Guide to Interstellar Travel.* John Wiley, New York, 1989.

McRae, Hamish, *The World in 2020: Power, Culture, and Prosperity.* Harvard Business School Press, Cambridge, Mass., 1994.

Moore, Walter, *Schrödinger: Life and Thought.* Cambridge University Press, Cambridge, Eng., 1989.

Ohmae, Kenichi, *The End of the Nation State: The Rise of Regional Economies.* Simon & Schuster, New York, 1995.

Pagels, Heinz R., *Perfect Symmetry: The Search for the Beginning of Time.* Bantam Books, New York, 1986.

———, *The Cosmic Code: Quantum Physics as the Language of Nature.* Bantam Books, New York, 1983.

Pais, Abraham, *Inward Bound: Of Matter and Forces in the Physical World.* Oxford University Press, Oxford, 1986.

———, *Subtle Is the Lord: The Science and the Life of Albert Einstein.* Oxford University Press, Oxford, 1982.

Petersen, John L., *The Road to 2015: Profiles of the Future.* Waite Group Press, Corte Madera, Calif., 1994.

Regis, Ed, *Nano: The Emerging Science of Nanotechnology.* Little, Brown, Boston, 1995.

Sagan, Carl, *Pale Blue Dot: A Vision of the Human Future in Space.* Random House, New York, 1994.

Sheffield, Charles, Marcelo Alonso, and Morton A. Kaplan, *The World of 2044: Technological Development and the Future of Society.* Paragon House, St. Paul, 1994.

Toffler, Alvin, *The Third Wave: The Classic Study of Tomorrow.* Bantam Books, New York, 1980.

Waldrop, M. Mitchell, *Complexity: The Emerging Science at the Edge of Order and Chaos.* Simon & Schuster, New York, 1992.

Weinberg, Steven, *Dreams of a Final Theory.* Pantheon Books, New York, 1992.

———, *The First Three Minutes: A Modern View of the Origin of the Universe.* Bantam Books, New York, 1977.

# Index

Acquired immune deficiency syndrome, 181, 186–90
drugs for, 189
and gene therapy, 163
and molecular medicine, 182
screening, 159
Adelman, Leonard, 104, 106
Agent, intelligent, 58–60
Aging, 163, 200–219
caloric theory, 215–16
and cancer, 207–9
characteristics of, 201
and evolution, 203
gene theory, 216
and hormones, 206–7
and menopause, 207
mitochondria theory, 216
and molecule damage, 211
new organ growth, 202, 217–19
oxidation theory, 205–6, 216
research, 214
reversing, 212
role of genes, 201, 203
unified theory of, 212
AIDS. *See* Acquired immune deficiency syndrome
Aliens, 319–21
Alzheimer's disease, 144, 163, 173, 214, 233
Anderson, W. French, 164–65
Angiogenesis, 172
Antibiotics, 182, 190, 191
careless use of, 191–92
for livestock, 192
natural sources of, 194
Antimatter, 290–92, 314–15, 344
Antioxidants, 171, 205–6, 211, 216
and aging, 201
and free radicals, 211
production of, 212
superoxide dismutase, 211
ARPANET, 48, 49
Asimov, Isaac, 133, 299, 339
Asteroids, 316–18
Atomic
bomb, 6
clocks, 39
distances, 108
Atoms, 4, 7, 42, 53
decaying, 291
nuclei, 7
quantum theory of, 146
radioactive, 291
smashing, 102
and transistors, 108
Attila robot, 71, 72, 72*il*, 73, 87
Automatons, 76–77, 98

Bacteria, 144, 181, 182, 190–91, 223
mutations to, 182
necrotizing fasciitis, 181, 190
resistant, 190–91, 193–94
streptococcus, 190, 192
Behavior, 231–34, 248–51
Bell Laboratories, 8, 103, 106, 110
Berg, Paul, 164, 181
Beta-carotene, 206
Big Bang theory, 9, 350–51
Binary code, 110
Biocompatibility, 113
Bioengineering, 217, 241, 244, 245
Biogenetics, 16–17
Biogerontology, 201, 206
Biological warfare, 257–59
Biological Weapons Convention (1972), 260–61
Biology
computational, 157–58
law of evolution in, 82
molecular, 74, 153, 158, 160, 218
Biomolecular revolution, 9, 11*il*, 74, 115–16, 165, 176, 221
Bionics, 112–14
Biotechnology, 4, 13, 16, 222, 227, 228, 241, 242, 244, 328
growth curve of, 15
wartime misuse of, 257–59
Birth
options, 242
parthenogenesis, 254
by selective breeding, 242, 256
surrogate parent, 242
test tube, 242, 253
Black holes, 40, 41, 204, 341–42, 354
Blaese, Michael, 178
Bohr, Niels, 265
Boneh, Dan, 106
Boss, Alan, 306
Botstein, David, 156
Brain
aging, 213
calculational power, 79–80
damage, 115
functioning, 74, 84
interface with, 112
layers in, 78–79
limbic system, 93
mapping, 115
memory function in, 232
neurons, 17, 112, 113
and quantum physics, 82
replacements, 116
research, 74
size, 250
structure of, 79
surgery, 77
Breeder reactors, 282–83

Breggin, Peter, 250
Brigham, Carl, 251
Brinkley, John "Doc," 206
Broder, Samuel, 170
Bromley, Allan, 301
Brooks, Rodney, 59, 71, 72, 73, 75, 87
Brown, Karen, 102
Brzezinski, Zbigniew, 45–46
Burk, Marie, 217
Bussard, Robert, 311

Callahan, Daniel, 255
Cancer, 15, 145, 162–80
  and aging, 207–9
  angiogenesis, 172
  and antioxidants, 206
  causes of, 166
  curing, 143
  detection, 171
  development rate, 168
  and hormones, 207–9
  immune system enhancement for, 171–72
  natural fighters, 171
  pervasiveness of, 166
  proliferation of, 166
  risk reduction, 208
  targeting genes in, 172
  telomeres in, 169–70
  therapy, 171
  unified theory of, 166–67
  vaccines, 172
Caplan, Arthur, 227, 241, 247, 248, 250, 251, 252, 260
Carter, Jimmy, 45
Casimir effect, 344
CAT scans, 6, 198, 266
Cells
  aging, 170
  apoptosis, 168
  death of, 168, 170
  decoding nuclei, 7
  destruction, 212
  development, 15
  forgetfulness in, 226–27
  fuel, 287
  interactions, 144
  nerve, 218, 233
  photovoltaic, 283, 284
  place, 232
  sex, 251
  solar, 283–84
  universal donor, 217
  viral attacks on, 188–90
Centers for Disease Control, 183, 190, 192, 258
Cerf, Vinton, 49
Chernobyl reactor, 281, 283
Christian, Jeffrey, 122
Chromosomes, 140, 147, 170, 236
  criminal, 249
Chu, Paul, 274
Civilizations
  extraterrestrial, 17, 323
  galactic, 18
  planetary, 17–19, 322–37
  and population growth, 332
  Type I–III, 17–19, 323–29
Clark, Jim, 45
Clarke, Arthur, 130, 135, 295, 299–301
Clinton, Bill, 155, 248, 255, 295, 301
Cloning, 16, 225–27, 242, 253–56
  banning, 255
  gene, 143
  legal rights in, 255
Codons, 149
Cog robot, 59, 87–90, 89*il*
Cohen, Daniel, 246
Cohen, Mitchell, 192
Colbert, Daniel, 272
Collins, Francis, 139, 140, 141, 143, 146, 149, 162, 175–76, 176, 178, 246, 248, 252, 259
Common sense, 62–66
Communication
  dominant modes of, 93
  electronic, 48
Computer revolution, 4, 8, 11*il*, 115, 118–35, 337
Computers and computing, 6, 35, 80, 99, 100, 107
  architecture, 24, 27, 98
  boards, 32–33
  and common sense, 62–66
  conscious, 28
  control of environment by, 131
  data encryption standard code, 106
  digital, 74, 105
  DNA, 15, 100, 104–7, 156–57
  economic necessity for, 118
  electronic, 4, 7
  growth of, 15
  identification, 36–37
  influence of, 119
  interactions, 24
  invisible, 23–42
  mechanical, 8, 109
  in medicine, 33, 36, 62
  memory, 103–4
  microchips in, 14, 15, 16, 24, 25, 29, 31, 37, 44, 99, 101
  molecular, 15, 100, 110
  Moore's law, 14
  NETalk, 81–82
  operating systems, 23
  optical, 15, 100, 102–3
  organic, 107
  pads, 32
  phases of, 26–28, 58–59, 76, 98
  power of, 4, 14, 29
  processing speed, 80
  protein, 107
  quantum, 16, 100, 109–10
  in search for new medications, 194–95
  secrecy, 120
  self-aware, 28
  self-diagnostic, 96–97
  sensors, 31
  simulations, 40–41, 144

software programming, 123–24
speech recognition by, 58–59, 90
strong job prospects, 123–24
tabs, 32
threats to jobs, 121–23
ubiquitous, 24, 25, 27, 28, 30, 31, 32, 40
ultimate, 99–117
wearable, 35–36
Consciousness
degrees of, 96–98
and goals, 97, 98
plant, 97
in robots, 93–96
"seat" of, 95
transfer of, 116
Constructivism, 96
Cook-Deegan, Robert, 151, 157, 158
Cooper, Leon, 86, 276
Cosmology, 9
quantum, 351
Counter, Christopher, 170
Cox, Paul, 194
Crevier, Daniel, 130, 133
Crick, Francis, 9, 146, 147, 156
Cristofalo, V.K., 208
Crocker, Steve, 48
Crooke, Elliot, 200
Csicsery-Ronay, Istvan Jr., 269
Cyber
babble, 59
jobs, 118
libraries, 44
marketplace, 51
medicine, 42
science, 40–41
serfs, 127
yuppies, 127
Cyborgs, 100, 115–17
Cyc, 64, 65, 88

Darwin, Charles, 9, 221, 235, 256
Davenport, Charles, 256
Davis, Randall, 66
Dean, Thomas, 75
Delany, John, 298
Delbrück, Max, 146, 156
Dennet, Daniel, 94
Deutch, David, 109, 111
Diseases
age-related, 213
Alzheimer's disease, 144, 163, 173, 214, 233
amyotrophic lateral sclerosis, 173
autoimmune, 16, 144
cancer, 15, 162–80
cardiac, 144, 173
cholera, 191
chronic, 16, 206
curing, 143, 164
cystic fibrosis, 140, 143, 159, 173, 175, 176
diabetes, 173
Ehlers-Danlos syndrome, 173
emergent, 184
of eyes, 113–14
genetic, 158, 166, 247, 251, 260
genetic codes for, 143
hemophilia, 143, 173, 174
hereditary, 143, 172–75
Huntington's chorea, 143, 173, 174–75, 176, 177
infectious, 15, 182
influenza, 152, 185–86
inherited, 212, 213
Legionnaires', 190
leprosy, 183
Lesch-Nyhan syndrome, 174, 176
Lyme disease, 190
malaria, 183, 191
measles, 152
meningitis, 191
muscular dystrophy, 143, 175, 176
neurofibromatosis, 174
pneumonia, 191, 192
polio, 183
polygenic, 16, 144, 178–79
porphyria, 173, 174
pycnodysostosis, 173
rabies, 152
resistant, 193
SCIDS, 177–78
sickle-cell anemia, 143, 173, 175
smallpox, 152, 183
syphilis, 191
Tay-Sachs, 143, 173, 175
tuberculosis, 183, 191
DNA
in Acquired immune deficiency syndrome, 188–90
atomic structure, 9
base pairs, 148, 149, 151, 176
chips, 158–59
computers, 15, 100, 104–7
data banks, 155
decay, 106
folded, 226
functions of, 16
information in, 104
molecules, 9, 104
patterns in, 157
personalized codes, 143, 145, 164, 201
and quantum physics, 147–49
repair mechanisms, 213, 216
research, 142, 156–57
revolution, 4
sequencing, 14, 15, 141, 144, 145, 149–51, 156, 158, 161, 170–79, 201, 214, 223, 227, 228, 249
testing, 154–56, 246–48
Doherty, Rick, 56
Doolittle, Russell, 157
"Doomsday Virus," 183
Drake, Frank, 319
Dreams, 85–87
Dubos, René, 241
Dyson, Freeman, 17, 305, 306, 328

Ehrenreich, Barbara, 127

Einstein, Albert, 9, 290, 294, 311, 312, 338, 340, 341, 342, 345, 346, 348, 349, 351
Einstein-Rosen bridge, 342, 343
Electric cars, 285–87
Electricity, 25, 43
  conductors, 270, 271
  solar, 284
  speed of, 100
Electrons, 8, 29, 42, 53, 101, 102, 103, 107, 108, 290, 347, 354
E-mail, 47, 59
Embryos
  frozen, 242
  research, 255
  test tube, 242
Emotions
  "focusers," 91
  genetic basis for, 233
  in robots, 90–93
Energy
  consumption, 18, 278, 325, 330
  in dreams, 85
  efficiency, 105
  fluctuating, 85
  laws, 17
  minimum state, 160
  negative, 344
  noncontinuous nature of, 7
  nuclear, 243–44
  Planck, 329
  production, 17
  in quanta, 7–8
  scale, 345
  solar, 283–84, 327
  sources, 18
  stellar, 18
  supply, 18
  terrestrial, 18
  usable, 34
Engines
  antimatter, 290–92, 314–15
  gasoline, 285
  internal combustion, 286
  ionic, 16, 305
  photonic, 314–15
  ramjet fusion, 311, 312*il*, 313
  steam, 4
  warp, 315
Environment
  collapse of, 19
  computer control of, 47, 131
  contaminants in, 173
  interactions, 144
  pollution of, 324, 331
  triggers for disease, 144
Enzymes
  helicase, 213
  HIV integrase, 188
  HIV protease, 188
  molecular structure, 188
  restriction, 150, 223
  reverse transcriptase, 188
  telomerase, 170, 172
Ethical issues, 17, 240, 241–61
  cloning, 225–27, 253–56
  eugenics, 256–57
  gene patenting, 246
  germ-line manipulation, 251–53
  privacy, 246–48
Eugenics, 256–57
Evolution, 9
  and adaptation, 116
  and aging, 203
  and cyborgs, 100
  eyes in, 235
  gene manipulation in, 221–22
  and homologous genes, 152–54
  laws of, 82, 252, 326
  and unnatural selection, 116
Eyes
  artificial, 114
  genes for, 235–37
  macular degeneration, 114
  retinitis pigmentosa, 114

Fancher, Carol, 37
Fermi-1 reactor, 282
Fernald, Russel, 236
Feynman, Richard, 109, 268, 272
Finer, S.E., 334
Fleming, James, 212
Force fields, 293
Foundation for Economic Trends, 245
Franklin, Rosalind, 147
Franson, James, 120
Frazier, Gary, 108–9
Free radicals, 205, 211, 212
  hydroxl, 216
Freud, Sigmund, 85
Fryxell, Bruce, 41
Fuels
  costs of, 278
  fossil, 278, 284, 331
  reserves, 278
Furth, Harold P., 278
Fusion, 16, 277–82
  and fission, 281–82
  and Lawson's criterion, 280
  machines, 311
  research, 280

Galbraith, John Kenneth, 335
Galileo probe, 298
Galton, Francis, 256
Gamow, George, 146, 156, 350
Gates, Bill, 27, 28, 45, 52, 127, 337
Gehrels, Tom, 317
Gehring, Walter, 235, 236
Gel electrophoresis, 150
Gell-Mann, Murray, 348
Genes
  age, 16, 201, 203, 211, 213, 214, 217
  base pairs, 156
  and behavior, 231–34, 248–51
  for body shape, 229–30
  cancer, 166–69, 246
  "clock," 212
  cloning, 143

damage to, 227
defective, 176
for eyes, 235–37
for face and scalp, 230–31
functional genomics, 228
functions of, 16
for hair growth, 230–31
homeobox, 237–38
homologous, 152–54, 233
and inbreeding, 221, 222
and intelligence, 250–51
interactions, 16, 228
interrelations of, 144
manipulation of, 221–22, 251
mapping, 140, 141, 144
master, 235, 236
memory, 233
mortality, 213
mutations to, 15, 160, 163, 167, 168, 170, 173, 190
oncogenes, 166, 167
patenting, 246
and polygenic traits, 227–29, 234, 235
and privacy, 246–48
splicing, 241
system numbers, 142*il*
therapy, 143, 145, 162–64, 168, 172, 178, 190, 241, 251
transfers, 17, 223, 238
transgenic research, 222–25
tumor suppressors, 166, 167, 246
for violence, 249
Genetic
codes, 9, 146, 147, 149–51
defects, 176, 222, 253
discrimination, 247, 248, 260
distances, 153, 186
engineering, 223, 224, 226–27, 253
overlaps, 153
research, 160
revolution, 243–44
screening, 158–59
"wallflowers," 248
Genetocracy, 257
George, Thomas, 29
Germ-lines, 179–80, 251–53
therapy, 260
Gerschman, R., 205
Gershenfeld, Neil, 33
Gilbert, Walter, 143, 144, 146, 149, 151, 156, 158, 159, 179, 218, 233
Glashow, Sheldon, 10
Gleick, James, 37
Global Positioning System, 36, 39, 45, 47, 67
Gödel, Kurt, 343
Goldburg, Rebecca, 244
Goldin, Daniel, 299, 301
Gore, Al, 50
Gould, Stephen Jay, 146
Graham, Ronald, 106
Gravitational confinement, 279
Greenhouse effect, 41, 309, 310
Grove, Andrew, 29, 43, 122

Halberstam, David, 129
Hammond, Charles, 207
Harman, Denham, 205
Haseltine, William, 145
Hawking, Steven, 172–73, 345, 352
Hayflick, Leonard, 201, 204, 211
Hebb's rule, 81
Heisenberg, Werner, 7, 346
Heisenberg Uncertainty Principle, 107, 351
Hekimi, Siegfried, 211–12
HelpMate robot, 77
Hemophilia, 143
Henderson, D.A., 258
Hernstein, Richard, 250
Heuristics, 60, 61, 94, 133
expert systems, 61–62, 65, 66–69, 133
Hiatt, Andrew, 223
HIV. *See* Acquired immune deficiency syndrome
Hoagland, William, 284
Hoffman, Randy, 39
Hofstadter, Douglas, 61
Holmes, Oliver Wendell, 256–57
Holograms, 103–4
Homologues, 152–54
Hood, Leroy, 156, 158, 164
Hopfield, John, 82, 83, 84, 85, 86, 87
Horgan, John, 9
Hormones, 206–7
and aging, 201, 209
and cancer, 207–9
DHEA, 210
estrogen, 207
fads, 210
growth, 207, 209, 223, 253
leptin, 230
lifesaving, 223
melatonin, 210
sex, 166, 207, 210
side effects, 201, 208, 209, 210
testosterone, 208, 249
therapy, 206, 207, 208
Horn, Berthold, 75
Hughes, James, 190
Human Genome Project, 115, 139, 140, 141, 143, 151, 157, 246
Ethical, Legal, and Social Implications Branch, 243, 247, 260
Hurwitz, Edward, 195
Huxley, Aldous, 241–42, 253, 337
Hyperspace, 349

Ice ages, 318–19
Imagery
robotic, 90
three-dimensional, 103
Immortality, 100, 116, 202–3
Immune system, 171–72, 178, 187, 191
and antioxidants, 206
effects of stress, 196–97
mind/body link, 196–98
in organ transplant, 217
Industrial revolution, 4, 126, 332
second, 16, 274

Third World, 126
Inertial confinement, 279
Information
access to, 119
control of, 119
digital, 102–3
in DNA, 104
economy, 42
genetic, 59, 247
ghettos, 126
haves/have nots, 119
Internet transmission of, 52, 53
laser transmission, 54
processing speed, 80
pure, 37
revolution, 19
services employment, 124
storage, 104
stratification, 127
Information highway. *See* Internet
Information revolution, 19
Intel Corp., 14, 29, 43, 122
Intelligence
artificial, 7, 28, 44, 58, 59, 60, 64, 71, 74, 75, 76, 83, 93, 130, 131, 132, 328
bottom-up school, 73–76, 89, 90
codifying, 60
expert systems in, 61–62, 65, 66–69, 133
genetic basis for, 250–51
heuristics, 60, 61
manipulation of, 8
and race, 250–51
top-down approach, 74, 75, 89, 90
Internet, 8, 14–15, 24, 27, 31, 33, 35, 38, 42, 43, 49, 110, 337
artificial intelligence on, 60
bottlenecks on, 52–54
business on, 121–27
communicating with, 44
criticism of, 50
directories, 45
eavesdropping on, 119–21
effects of, 51–52
filters, 59
"graphic jams" on, 49
historical significance of, 49–51
influence of, 119
killerapps, 52–53
limits of, 54
merger with television, 49, 55–56
multimedia access to, 48
news provision, 126
on-line services, 51–52
range of, 51
size, 4
Invisibility, 294

Jacoby, George, 193
James, Ron, 227
Jazwinski, S. Michael, 212
Jeremiah, David, 268
Johnson, Karl, 258, 259
Johnson, Lyndon, 299
Johnson, Thomas, 211, 212
Jones, David, 269
Jones, Steve, 174, 256
Jordan, Pascual, 146, 156
Jupiter, 295, 298, 306, 310

Kakadon, Harvey, 189
Kaluza, Theordr, 349
Kardashev, Nikolai, 17, 323
Karp, Richard, 157
Kates, Robert, 332
Kavka, Gregory, 257
Kemeny, John, 26
Kennedy, John F., 298
Kenyon, Cynthia, 211
Kerr, Roy, 342
Kimble, Jeff, 112
Klausner, Richard, 171
Krauss, Michael, 336
Krauthammer, Charles, 125
Krugman, Paul, 123
Kuiper Belt, 306, 309

Lander, Eric, 141
Langer, Robert, 217
Lap-Chee Tsui, 176
Lasers, 6, 8, 31, 102, 103, 280, 287–90, 292
carbon dioxide, 291
fiber optics, 53–54, 287
giant, 288
and the Internet, 54
microscopic, 104, 287
printers, 23
total internal reflection, 54
X-ray, 288, 289, 293
Law of Unintended Side Effects, 253
Laws
economic, 28
energy, 17
of evolution, 82, 252, 326
of Increasing Returns, 51
of logic, 64
of physics, 63, 99, 268, 292–94
of robotics, 133–35
of rocketry, 296
of thermodynamics, 17
of unintended side effects, 253
universal, 84
Lawson's criterion, 280
Lenat, Douglas, 62, 64, 65, 88, 90
Lesch-Nyhan syndrome, 174, 176
Lewis, John, 303, 310
Life expectancy, 3–4, 203, 204
Life span, 16, 144, 201, 202, 203, 206, 210, 212, 214, 215
Light
infrared, 121
polarization of, 120
and quantum cryptography, 120
speed of, 103, 328
Light emitting diodes, 35
Lipshutz, Robert, 159
Lipton, Richard, 105, 106
Lloyd, Seth, 111
Longevity, 212–14

Lozano-Perez, Tomas, 76

Machines
  conscious, 96
  fusion, 311
  molecular, 266–73, *267il*
  self-replicating, 266–69
  thinking, 70–98
  time, 344–45
  Turing, 74, 80, 81, 105, 109, 110, 111
  universal computing, 84
MacKay, Clive, 215
MacLean, Paul, 79
Maes, Pattie, 59, 60
Maglev, 273
  trains, 276–77
Magnetism, 273
Mars, 295, 296, 297, 298–99, 309, 310
  Global Surveyor missions, 297
  Observer, 299
  Pathfinder, 73
  Rover, 73, 297
Massachusetts Institute of Technology, 111, 129, 132
  Artificial Intelligence Laboratory, 70–73, 76
  Media Laboratory, 33–37, 42, 59, 289–90
  Whitehead Institute, 141
Matter, 7
  defining, 7
  quantum theory of, 9
Measles, 152
Medicine
  computers in, 33, 36, 62, 145
  cyber, 42
  effects of stress, 196–97
  and microchips, 114–15
  mind/body approach in, 196–98
  molecular, 144–45, 165, 173, 181–99
  prevention-based, 145
  robots in, 77
  stages in, 165
  telemedicine, 124
  theoretical, 144
  treatment-based, 145
Memories, 84
  and sleep, 86
  spurious, 85, 86
MEMS. *See* Microelectromechanical systems
Mercury, 310
Merkle, Ralph, 115, 116
Meteors, 296, 318
Microchips, 8, 15, 24, 29, 44
  cost of, 14, 25, 29, 31, 37
  DNA, 158–59
  heat problem, 101, 103, 105
  and medicine, 114–15
  miniaturization, 30
  modification of, 101
  and neurons, 112
  point one barrier, 41
  power of, 16, 79, 99
  size, 15
  viability of, 102
Microelectromechanical systems, 269–70
Microelectronics, 13, 268
Microgravity, 301
Microinjection, 224
Microlasers, 288
Micromachines, 269–70
Minsky, Marvin, 65, 75, 91, 94, 95, 100, 116
Molecules, 4
  designer, 192–94
  DNA, 9, 104
  genetic code in, 9, 146, 147
  repair of, 211
  size of, 268
  smart, 143
Monju reactor, 283
Moore, Gordon, 14, 270
Moore's law, 14, 28–30, 42, 99, 100, 115, 118, 141–42, 158, 270
Moravec, Hans, 76, 77, 90, 116, 118, 131
MRI scans, 6, 163, 198, 199, 266
Multiverse, 352–53

Nanotechnology, 16, 266–73
Nanotubes, 271–72
NASA. *See* National Aeronautics and Space Administration
National Academy of Sciences, 151, 157
National Aeronautics and Space Administration, 73, 296, 297, 298, 303, 308
National Cancer Institute, 170, 171, 194
National Institute of Standards and Technology, 112
National Institutes of Health, 139, 164, 181, 210, 215
  National Institute on Aging, 207
National Research and Education Network, 50
Near Earth Objects, 316, 317
Negroponte, Nicholas, 33, 49
Neptune, 306
Netscape, 45
Neural networks, 83–87, 115–16, 141
Neurofibromatosis, 174
Neurons, 80
  decoding, 112
  electronic, 81–82, 116
  manipulation of, 116
  measuring, 113
  and microchips, 112
Neutrinos, 347, 350, 354
Newton, Isaac, 3, 7, 19, 340, 342
NREN. *See* National Research and Education Network
Nuclear
  annihilation, 324
  breeder reactors, 282–83
  fission, 281–82
  fusion, 16, 277–82
  rockets, 305, 313–14
  terrorism, 282–83
  war, 19, 45, 46, 47, 49
  waste, 281
Nuland, Sherwin, 165

Ohmae, Kenichi, 335
Old, Lloyd, 172
Onnes, Heike Kamerlingh, 273
Oppenheimer, J. Robert, 347
Organs
  control of development, 237
  laboratory growth of, 15, 17, 217–19
  replacement, 227
Orwell, George, 119
Owen, Frank, 127
Ozone, 330–31

Pagels, Heinz, 82, 83, 96
Parthenogenesis, 254
Particle accelerators, 291
Pauling, Linus, 146, 156
Pesticides, 244, 245
PET scans, 6, 95, 115, 198, 199, 266
Photolithography, 30, 159, 287
Photons, 8, 111, 283
Physics
  laws of, 268, 292–94
  solid-state, 83
  theoretical, 74
  *See also* Quantum physics
Plank's law, 102
Pluto, 306
"Point one" barrier, 41–42, 100, 101
Primates, 92–93
Progress and Freedom Foundation, 127
Proteins
  and aging, 207
  apolipoprotein, 214
  for bone growth, 218
  and cancer cells, 171
  folding, 41, 160, 161, 228
  interactions, 144

Quantum
  chemistry, 146
  cryptography, 120
  electronics, 108
  future, 265–94
  mechanics, 160, 268
  revolution, 4, 7–8, 74, 115
  transistors, 107–9
Quantum physics, 74, 266, 278, 283
  breakthroughs by, 6
  and DNA, 147–49
  laws of, 15, 99
  and robotics, 82–84
Quantum theory, 11–12, 17, 29, 39, 42, 83, 107, 120, 290, 294, 346, 347
  of atoms, 146
  in gene research, 160
  practical applications, 121
Quarks, 8, 347, 348, 350
Qubits, 110, 111

Radar, 38
Radio, 6, 31
  CB, 50–51
Reagan, Ronald, 289, 292, 303
Reductionism, 10, 11–12
Replicators, 293–94
Research
  on aging, 213, 214
  atomic energy, 243
  biomolecular, 15
  brain, 74
  DNA, 142
  fusion, 280
  genetic, 160
  human embryo, 255
  models for, 74
  transgenic, 222–25
Resonance, 108
Retroviruses, 188
Rifkin, Jeremy, 127, 245, 246
Rissler, Jane, 244, 245
RNA, 147, 188, 189
Robo-Surgeon robot, 77
Robotics, 78–81
  in drug laboratories, 195
  laws of, 133–35
  and quantum physics, 82–84
Robots, 13, 16, 59, 72*il*
  Cog, 59, 87–90, 89*il*
  conscious, 76, 93–96
  control of, 119
  dangers from, 130–35
  in DNA sequencing, 15
  and emotion, 90–93
  and imagery, 90
  industrial, 76, 77
  as killing machines, 131–32
  malfunctions in, 132–33
  in medicine, 77
  mesa effect, 133
  and pattern recognition, 78, 86, 87
  preprogrammed, 76–78
  protection from, 133–35
  with reasoning capabilities, 90
  remote-controlled, 76, 77
  self-aware, 16, 76, 98, 119, 130–31
  self-teaching, 71–73, 88–90
  submarines, 77
  talking, 81–82

Saffo, Paul, 30, 66
Safire, William, 303, 336
Sagan, Carl, 298, 309, 316
Sakharov, Andrei, 279
Salmon, Dennis, 166
Sanger, Frederick, 149, 151
Satellites, 53
  communications, 337
  Global Positioning System, 36, 39, 45, 47, 67
  low-earth-orbit, 303–4
  Navstar, 39
Saturn, 306, 310
Scanning tunneling microscope, 270, 271
Schiff, Isaac, 208
Schilling, Andreas, 275
Schrieffer, J. Robert, 276
Schrödinger, Erwin, 7, 9, 146, 156
Schrödinger wave equation, 8, 107, 146, 276

Schwartz, Arthur, 210
SCIDS, 177–78
Science
acceleration of, 12
advanced degrees in, 129
budgets, 129, 130
competition in, 246
cyber, 40–41
development of, 12
education, 129
employment prospects, 124
experimental, 40, 41
fiction, 130
and increased wealth, 128
information, 157
investment in, 129
material, 13
pillars of, 7
and religion, 95–96
theoretical, 40
Searle, John, 96
Second Law of Thermodynamics, 204, 205, 211, 327, 354
Sejnowski, Terry, 81, 82, 87
Semiconductors, 99
Sensors, 31, 38, 39
miniature, 269
Sentry robot, 77
Shakey robot, 78, 132
Shapiro, Gary, 55
Shockley, William, 257
Shope, Robert, 192
Shor, Peter, 110
Silicon-germanium, 102
Simon, Herbert, 75, 94
Sky hooks, 272
Smart
cards, 37–38
cars, 31, 38–40
homes, 32–33
molecules, 143
offices, 32–33
rooms, 36–37
tables, 34
wearables, 34–36
Smith, Adam, 129
Space
aliens, 319–21
colonization, 17, 296, 310
curved, 340, 341
earth like planets in, 307–8
exploration, 4
extra-solar planets, 306–7
manipulation of, 346
multiply connected, 340
probes, 296
programs, 73
time, 338, 339–41
transport, 304–6
travel, 272, 291
warps, 294, 339
wormholes, 339, 340, 341–42
Space Interferometry Mission, 308
Space Shuttle, 295, 299–301
Space Station Alpha, 295, 299–301, 300*il*
Space-time theory, 9
Specialization, 11
Sperling, Daniel, 287
Spreitzer, Bill, 38
Standard Model, 8, 347
Stars
dead, 204
neutron, 41
supernova, 8, 41, 318–19, 326
Starships, 311–15
Star Wars program, 289, 292–93
Stewart, William, 182
Stoll, Clifford, 50, 51
Supercomputers, 100, 107, 161, 304
Superconductors, 16, 273–76
Supernovas, 8, 41, 318–19, 326
Superoxides, 216
Superphoenix reactor, 283
Superstrings, 9, 348–49
Surveillance dust, 270
Suspended animation, 315–16
Symmetric self-electro-optic effect, 103, 104
Synchrotron, 102
Synergy, 11–12, 156–57

Tamm, Igor, 279
Taylor, Theodore, 314
Technology
acceleration of, 12
development, 130
employment prospects, 124
etching, 104
funding, 130
investment in, 129
known, 15
secret, 45, 46
solar, 284
and wealth, 13, 128
X-ray/electron beam, 102
Telecommunications, 4, 13, 44
global, 337
Teleconferencing, 33, 46
Telematics, 39–40
Telemedicine, 124
Telescopes, 296, 307–8, 341
Television, 6
active-matrix screen, 57
analog, 55
cathode ray tube, 55, 56–57
digital, 54, 55
holographic, 16, 209, 290
interactive, 32
LCD screen, 56–57
merger with Internet, 49, 55–56
plasma screen, 56, 57
three-dimensional, 287, 289
wall screen, 56–57
Teller, Edward, 293
Telomeres, 169–70
Tenover, Fred, 192
Tenover, Joyce, 208
Tesler, Larry, 54, 59
Testing

DNA, 154–56, 246–48
field, 244
food, 244
and health costs, 247
Test tube babies, 242
Theories
aging, 204
Big Bang, 9, 350–51
caloric restriction, 215–16
of dreams, 85
of everything, 5, 345–46
evolution, 9
genetic, 212
germ, 204
gravity, 343, 349
information, 204
of intelligence, 251
mitochondria, 216
neural network, 81, 83, 84, 86
oxidation, 205–6
quantum, 7, 8, 11–12, 17, 29, 39, 83
relativity, 283, 290, 312, 340, 346
space-time, 9
space warps, 294
superstring, 9, 348–49
unified field, 9–10
vitalism, 9
Therapy
cancer, 171
gene, 143, 145, 162–64, 168, 172, 178, 190, 241, 251
germ-line, 179–80, 251–53
hormone, 206, 207, 208
Things That Think project, 33–35
Three Mile Island reactor, 77, 281, 283
Thurow, Lester, 13, 129
Time
concept of, 63
machines, 344–45
manipulation of, 346
travel, 342–43
Toffler, Alvin, 334, 335
Townsend, Paul, 350
Transgenic research, 222–25
Transistors, 6, 8, 15, 29
micron-sized, 287
miniaturization of, 29–30, 57
molecular, 270, 271–72
optical, 103
quantum, 107–9
semiconductor, 29
Transporters, 293–94
Traveling Salesman problem, 106
Turing, Alan, 84, 96
Turing machine, 74, 80, 81, 105, 109, 110, 111

Ulam, Stanislaw, 156, 313
Unified field theory, 9–10
Union of Concerned Scientists, 244

Vacanti, Joseph, 217
Vaccines, 165, 172, 223
Van Allen, James, 301
Varmus, Harold, 3
VentureStar, 295, 301–2, 302*il*
Venus, 309
Vernam cipher, 119–20
Virasoro, Miguel, 79, 80
Virtual reality, 15, 34, 40–41, 45, 46–47, 124, 144, 182, 193
Viruses, 144, 178
antiviral agents, 194
cytomegalovirus, 152
decoding, 184, 185
DNA sequencing of, 151
Ebola, 181, 183, 184, 258
HIV, 9, 183, 186–90
mutations to, 17, 182, 184, 189
neutralization of, 177–78
origins of, 185–86
rhinovirus, 184
Vitalism, 9, 146
Vlahos, Michael, 127
Vogelstein, Bert, 167
Von Neumann, John, 103

Walford, Roy, 215
Walter, William, 299
Waterson, Robert, 157
Watson, James, 9, 139, 146, 147, 220, 241, 246
Wealth, 12–14
creation of, 128–30
increasing, 128
Weinberg, Robert, 162
Weinberg, Steven, 348, 352, 353
Weiser, Mark, 23, 24, 25
Weizenbaum, Joseph, 132
Wells, H.G., 257, 292, 294, 295, 296
West, Michael, 213
Wexler, Nancy, 177, 247
Wheelon, Albert, 300
Williams, Robert, 284
Wills, Christopher, 214, 249
Wilmut, Ian, 226, 227, 253, 255
Wineland, David, 112
Winston, Patrick, 76
Winter, Robin, 231
Witten, Edward, 350
World Health Assembly, 183
World Health Organization, 183
World Wide Web, 6, 32, 35, 48, 56
Wormholes, 294, 339, 340, 341–42, 352
Wurtman, Richard, 210

Xerox PARC, 23–24, 25, 26, 27, 115
X-ray, 6, 42, 102
crystallography, 9, 41, 147, 160, 228
lasers, 288, 289, 293

Yoon, Barbara, 86

Zare, Richard, 298
Zero-emission vehicles, 287
Zuker, Charles, 236

ABOUT THE AUTHOR

Michio Kaku is the Henry Semat Professor of Theoretical Physics at the City College of New York. An internationally acclaimed physicist, he is the cofounder of string field theory. He graduated from Harvard and received his Ph.D. from Berkeley. He is the author of the critically acclaimed and bestselling *Hyperspace*, as well as *Beyond Einstein* (with Jennifer Thompson), *Quantum Field Theory: A Modern Introduction*, and *Introduction to Superstrings*. He hosts a weekly hour-long radio science program that is nationally syndicated.